Prometheus Wired

Prometheus Wired

The Hope for Democracy
in the Age of Network Technology

Darin Barney

UBCPress · Vancouver · Toronto

Printed in Canada on acid-free paper ∞

ISBN 0-7748-0796-2

Canadian Cataloguing in Publication Data

Barney, Darin David, 1966-

Prometheus wired

Includes bibliographical references and index.
ISBN 0-7748-0796-2

1. Democracy. 2. Computer networks – Political aspects. 3. Political participation – Computer networks. 4. Computers and civilization. I. Title.
QA76.9.C66B37   2000   321.8'0285'4678   C00-910219-1

This book has been published with the help of a grant from the Humanities and Social Sciences Federation of Canada, using funds provided by the Social Sciences and Humanities Research Council of Canada.

UBC Press acknowledges the financial support of the Government of Canada through the Book Publishing Industry Development Program (BPIDP) for our publishing activities.

Canadä

We also gratefully acknowledge support of the Canada Council for the Arts for our publishing program, as well as the support of the British Columbia Arts Council.

UBC Press
University of British Columbia
2029 West Mall, Vancouver, BC V6T 1Z2
(604) 822-5959
Fax: (604) 822-6083
E-mail: info@ubcpress.ubc.ca
www.ubcpress.ubc.ca

In memory of Evangeline Pauline Barney (1941-93)

The kindest of all God's dispensations is that individuals cannot predict the future in detail.

GEORGE GRANT, 1965

contents

Acknowledgments / ix

1 Prometheus Wired / 3

2 On Technology / 27

3 Networks / 58

4 The Political Economy of Network Technology 1:
   The Mode of Production / 104

5 The Political Economy of Network Technology 2:
   Work, Consumption, and Exchange / 132

6 A Standing-Reserve of Bits / 192

7 Government, Politics, and Democracy:
   Network Technology as Stand-in / 236

Notes / 269

Bibliography / 307

Index / 325

acknowledgments

I am indebted to Edward Andrew for his wisdom, generosity, courage, and friendship; Mary Stone for her love and intelligence; Ronald Beiner and Richard Simeon for their encouragement, advice, and criticism; Langdon Winner, Hugh Thorburn, Reg Whitaker, David Cook, and Ron Deibert for their thoughtful reading and commentary; Emily Andrew for her craft and care; David Barney for his strength; Beverley Endersby for her precision; Camilla Jenkins for her patience and attention; Peter Milroy for his faith and determination; Richard Bingham for flexing his media; and Matthew C. Lella for drawing beauty from the standing-reserve of bits. Thanks also to the family of 115 Major Street and 284 Euclid Avenue, for setting a table more hospitable to thought than to fashion.

Prometheus Wired

prometheus wired

. . . . .

THERE IS A MANTRA among those invigorated by the emergence of network technology. John Perry Barlow, formerly a songwriter for the Grateful Dead and co-founder of the Electronic Frontier Foundation, expresses it this way: "Everything we know is wrong."[1] Recently, I appeared as a guest on a television show to discuss the question "Are we becoming cyborgs?" and, after referring to *Frankenstein* as a potential source of instruction about the perils of dabbling in human creation, I was upbraided by a learned colleague and co-panellist for being mired in "old narratives" that were "useless" in the present context. Similarly, after putting forward my considered criticisms of the "teledemocracy" program developed by one of Canada's major political parties, a party MP informed me that "most of what you have been taught about traditional politics will be of little value in the years ahead ... The old ways don't work any more."[2] What follows is based on nearly the opposite assumption to these – namely, that a great deal of what we already know is *not* wrong, and is therefore still useful. Even if the advance of network technology fundamentally alters social, economic, and political structures, and even if it radically affects the way we communicate and perceive ourselves or our world, this does not necessarily mean that our amassed knowledge – in particular, what we already know about technology and politics – is an unsound basis for understanding or forming judgments about these changes. In short, we know quite a bit, and it can't all be wrong. In the chapters that follow, I will attempt to bring some of what we already know about technology and politics to bear on a number of the questions facing us as we head into the age of networks.

The movement of digitized information over computer networks is, according to Barlow, "the most profound technological shift *since the capture of fire*."[3] Judging by the many volumes heralding the onset of a new "information society," the rush of governments to dispense public resources in developing digital infrastructure, the reconfiguration of education systems in observance of perceived technological imperatives, and the sustained buzz emanating from mainstream media, Barlow is not alone in thinking so. Predictions such as this capture our attention because of their audacity, but the comparison of computer networks to fire is interesting for another reason. Fire, of course, is at the very heart of the modern technological mythology.

The myth of Prometheus the fire-giver is an ancient one, but the drama it depicts illuminates much about the modern technological spirit.[4] Basically, the story is as follows: After being insulted by Prometheus, Zeus exacted revenge by punishing his rival's human children. Zeus "hid the livelihood of men ... hid the bread of life ... and hid fire." Seeing the toil this deprivation caused, Prometheus concealed a flame in a fennel stalk and "stole again for men" the instrument that had been taken from them.[5] The theft did not concern Zeus enough for him to punish Prometheus directly, and he worried so little about humans possessing fire – after all, humans had used fire instrumentally well before the gods starting playing games with them – that he did not bother to retrieve it. Instead, out of spite, he visited evil upon men in the form of Pandora, the "all-gifted" female who released among the Titans all the grievous gifts of her pestilence jar, save one: "Only hope abode within her unbreakable chamber under the lips of the jar, and flew not forth."[6] Deprived of hope, human beings could make little use of the fire that had been restored to them. It was at this point that Prometheus – whose name translates literally as "forethought" – was moved to commit the crime that ultimately brought the wrath of Zeus upon him:

> *Prometheus:* I caused mortals no longer to foresee their own doom.
> *Chorus:* Of what sort was the cure thou didst find for this affliction?
> *Prometheus:* I caused blind hopes to dwell in their breasts.[7]

For this, Prometheus was chained to a rock, his ever-regenerating liver to be devoured in perpetuity by an insatiable eagle.

Why was this such a heinous crime – indeed, more heinous than the theft of fire itself – and what does it have to do with the modern technological spirit? Fire illuminates the physical world, but it is hope that relieves people of their spiritual limits and entices them to impose themselves, blindly, on the future. When beings who are mortal by nature no longer foresee their own death, they begin to regard themselves as *im*mortal: as having no natural limits, like gods, which they are not. Hope thus seduces human beings into overestimating and overreaching themselves, with tragic consequences. Beings who recognize their limits can use instruments such as fire (or computer networks) in a healthy and responsible way; but instrumental, hopeful beings who believe themselves to be free of limits are dangerous to themselves and, ultimately, to their gods. Fire was a significant instrument, but without the added fuel of hope its flames could be contained. With hope in their breasts, and brandishing a fiery torch, human beings thought themselves free to light the way to their own destiny, and would act accordingly. The dominion of Zeus was doomed.

Hope enlightens, but it also blinds. It lights the way to the future but, unmoderated by reason, it renders progress toward our self-made destiny reckless, delusional, and dangerous. Just as hope causes us to regard ourselves as *more* than we are, it also thrusts us into the future as irrational, that is, as *less* deliberative and reasonable than we are capable of being. Blindness is an extreme condition: blind hope is an immoderate, feverish, and desperate substitute for prudent, thoughtful, responsible deliberation. We *hope* for the best when we are unable or unwilling to *think* about what is best. Despair – the absence of hope – has its own pathological consequences for human agency in the world, but this does not mean that blind hope is the best, or even a good, disposition for beings in a world that gives them access to very powerful technical instruments. Far better would be a modest appreciation of the abiding human appetite for a good life, and prudent deliberation about appropriate means for achieving that end.

Nevertheless, it is hope that has consistently animated humanity's collective and public approach to the development of technology. It is not without reason that the Prometheus myth has been so resonant for those who have thought about the technological spirit of the modern age. Francis Bacon, the father of modern science, felt the need to recast Prometheus as a hero rather than a warning; Karl Marx, the great "progressive," invoked the Promethean creed in his earliest work; Mary Shelley, the romantic, subtitled her cautionary tale "The Modern Prometheus"; Friedrich Nietzsche, who saw clearly into the heart of modernity, found Prometheus waiting there.[8] It comes as no surprise that one of the most influential studies of the Industrial Revolution – the cradle of technological development in the West, referred to by the author as a "new age of promise" – bears the title *The Unbound Prometheus*.[9] In the modern era, Prometheus has been released from his chains, his spirit set free. The story of modern technology is the story of Prometheus's people writ large: the story of humanity blindly wielding instruments to command and transcend that which is given, in the hope of creating its own future. It is my contention that network technology is part of, rather than a departure from, this trajectory. In the age of digital networks, Prometheus is certainly *unbound*, but he is also *wired*. It is, I would suggest, imperative that we subject our hopes for this technology to the sort of thoughtful consideration that, in moderating hope, befits our nature as rational beings.

Technologies of Hope and Fire

To begin, I would like briefly to situate networks historically, in relation to the technologies that have preceded them. If, for heuristic purposes, endowment is divided into that which is given in space, time, matter, biological life, and the capacity for consciousness, it becomes possible to identify certain prominent modern technologies as emblematic of the human desire for transcendence, command, or creativity in relation to these categories. It is telling that this spirit has been equally present both in the technologies that inaugurated modernity, and in those that attend its culmination. The transcendence of sensory spatial perceptions was initially a function of the development of glass technologies: spectacles in the thirteenth century, mirrors and microscopes in the sixteenth, mass-produced clear glass windows and the telescope in the seventeenth – all extended humanity's view beyond what it could see with its own eyes alone. Fantastic dreams about expanses imagined and unimaginable were replaced by a conscious desire to command space by travelling over distances seen, which drove the continued development of transportation technologies such as the steamboat in the eighteenth century, the railway in the nineteenth, and the airplane and rocketship in the twentieth. Even the modern city itself can be understood in these terms – the aggregation of human labour required for early, large-scale, industrial production necessitated an overcoming of the distances separating those labourers, and so the urban city was born as a technology of spatial concentration. And as urban space threatened to grow too large for efficient enterprise, a civic reorientation around technologies of transit such as the automobile, superhighway, bus, and subway provided late modernity with its own basis for an obsession with the command of space.

Prior to its conquest by technology, time was more or less conceived of as a boundless eternity, punctuated only by organic rhythms beyond the control of human beings. External rhythms such as the falling of night, the rising of the sun, and the turning of the seasons were clearly beyond human competence, and even those cadences that were internal – the beating of hearts, the welling of hunger, the ageing of bodies – defied human command. The introduction of regularized time in the fourteenth century via the mechanical clock represented an attempt to transcend the organic necessities of time by applying a technology that rendered it subject to human regulation.[10] As David Landes describes, mechanical time emerged in Benedictine monasteries to regulate the ringing of bells marking canonical hours. Though such clocks established a liturgy independent of natural cycles, they were not impious: time belonged to God, and the clock ensured it would not be wasted. It was not until time was secularized, and the clock was enlisted to habituate newly urbanized labourers to the cycles of industry and commerce, that this instrument became emblematic of modernity's conquest of Nature itself.[11]

In a sense, modern humanity transcended time by creating it in a form that could be commanded; the mysteries of Eternity were evaded with help from the mathematics of infinity. Few modern inventions have achieved the near-universal generality of mechanical time, which perhaps explains the privileged place enjoyed by considerations of speed in the design of most technologies developed since its adoption. It also perhaps explains the ease with which the World Congress standardized chronometric measurement across a revolving and rotating planet in 1885, by simply drawing lines on a map to create time zones that allowed for the coordination, in time, of activities separated by vast spaces and political priorities.[12] The unpredictability and sluggishness of pre-modern transportation meant that coordination had been previously unnecessary – things simply arrived when they arrived. The increasing speed of modern transport, rooted as it was in time's mechanization, necessitated its standardization as well. Waiting was simply no longer an option.

This spirit of human ingenuity is also in evidence with regard to modernity's relationship to matter. The difficulties of creating and destroying matter have not deterred modern humanity from setting out to transform it in ways deemed productive and profitable, through the use of technology. The generalization of the clock as a definitive attribute of modernity is matched only by the proliferation of industrial technologies, particularly from the mid-eighteenth century onward, designed and engineered to transcend the limitations imposed on the transformation of matter by human labour. Three principles guided the Industrial Revolution – mechanization of production, increased power generation, and enhanced exploitation of a greater variety of raw materials – and in combination they set the course for the period's technological development.[13] Mechanization came to the textile industry in the late eighteenth century with the invention of the spinning jenny and cotton gin; proceeded through the nineteenth century with the development of a variety of milling, reaping, drilling, lifting, sewing, and pressing machines; and culminated in the twentieth with the arrival of automated robotics. Reliance on animate sources to drive the great machines defeated their very purpose, and so industrialists turned instead to a series of technologies of motive power, the succession of which from the eighteenth to the twentieth century can be followed as if it were the bouncing ball of modernity: from the steam engine to the dynamo, to the internal combustion engine, to the turbine, to the nuclear reactor.

Much like humans and horses, these machines required fuel, and the pursuit of energy has been one of the dominant themes of the modern desire to command raw matter by turning it into something else. For the most part, this pursuit has entailed an extension of the dominion that God granted men over the earth to include dominion *under* it as well. The art of mining pre-dates the Industrial Revolution, but it was the escalating demand for coal that accompanied steam-driven production and transport, and the smelting of ore that established the mine as a key seam in the fabric of modern industrialism. The ascendancy of the internal combustion engine, augmented by the mass production and consumption of automobiles, brought with it a return to the ground – only with great drills and pumps this time, instead of shovels and hammers – to suck petroleum from its hidden natural stores. Electric power called for similar interventions. Unable to defy gravity, humans proved that they could, at least, harness it for their own ends by enlisting it to cause a vast volume of water to cascade over turbines and create hydro-electricity. And when the efficiency or supply of coal, gas, and water came into doubt, the realm below was once again scoured for unleashable energy. It finally yielded plutonium and uranium – substances whose atoms could be split with powerful results, including some that would ultimately be returned to the ground, whence they came.

The exploitation of raw materials has thus been intimately linked with the search for power throughout modernity. However, it would be wrong to suggest that the modern transformation of matter was concerned solely with the fuelling of machines, and did not also involve the use of raw materials to create objects or things. The shift from organic animal and vegetable matter to inorganic minerals as the primary material of production is a key marker of the modern industrial age.[14] The replacement of wood, in particular, by metals smelted from ores – primarily iron and copper – that had been dug from the earth characterized the material preoccupation of the early modern industrial era. By the beginning of the twentieth century, these metals were to be replaced, to a degree but not entirely, by metal alloys and lightweight, strong, and plentiful aluminum. But the penultimate modern century also brought with it perhaps the most modern of materials: inorganic synthetics, also known as plastics. Plastics had been around since the invention of celluloid in 1868, but it was not until the synthetic resin known as "Bakelite" was patented in 1909 that they become the modern material of choice and began to find their way into everything from tableware to clothing. Not only was plastic about as close to being truly artificial as anything could be, it was also seemingly impervious to other "natural" forces, due to its impermeability, electrical resistance, and flexibility. Plastic was a material sign of human creativity and durability achieved by technological means: "Here in unexpected form was a surrogate for the long-sought secret of transmuting and creating matter."[15]

The matter of transcending, commanding, and creating the biology of human life has proved more difficult for modern humanity. Human beings have always been able to reproduce by copulating, but this method has become a distraction to the modern scientific spirit, which seeks to transcend the role Nature provides in the ongoing generation of living beings and instead investigates the means of creating life itself. It is this obsession that drove Dr. Frankenstein, who could easily have produced a child by natural means with his admiring and presumably fertile Elizabeth, but instead pursued an obsessive desire to become "capable of bestowing animation upon lifeless matter ... the creation of a human being," through technology.[16] Dr. Frankenstein succeeded where real modern scientists have thus far failed, although the sophisticated new genetic and reproductive technologies appearing at the close of the millennium suggest the possibility of an impending meeting between fact and fiction. This is still not the case at the other end of the biological cycle of human life. The ancient healing arts have given way in modernity to a medicalization of the human body that has had as its express purpose the extension of human life – indeed, acceptable rates of mortality and life expectancy have become a required attribute of a fully "modernized" society – but we have yet to isolate the elixir that will do so indefinitely. The most concerted efforts to secure the conditions of everlasting life have been exerted in the area of cryogenics. Ironically, it was an early investigation in this field that cut short the life of Francis Bacon, whose Promethean fire was apparently not quite so hot that it could melt away the fatal escalation of a common cold.

However, as Dr. Frankenstein observed, "to examine the causes of life, we must first have recourse to death."[17] Human beings have never been at a loss to devise creative ways of killing themselves and their enemies, and the modern era has certainly featured its share of technological leaps in this regard. In fact, it is probably fair to say that, next to profit, the more effective waging of warfare has been the chief stimulant of modern technological development. The impact of militarism was felt far and wide in modernity, with technological spin-offs in sectors including manufacturing, engineering, and even industrial organization itself.[18] However, it was in the development and deployment of weaponry that the technological hand of modernity turned toward the snuffing out of human lives. Firearms in the form of powder-fired guns and cannons began to appear in the fourteenth century, and were followed in the fifteenth by numerous embellishments from the hand of Leonardo da Vinci. The sixteenth century brought with it the first mobile tanks, which led to advances in fortification technology in the seventeenth. Guns continued to merit technological attention, with the bayonet appearing in the seventeenth century, the mass-produced musket in the eighteenth, and Gatling's machine-gun in the nineteenth. The twentieth century contributed its own share of killing technologies. Some, such as the submarine boat, the warship, the fighter plane, various types of missiles, and the mechanized tank, were simply improvements on old designs, while others, including a variety of poisonous gases and chemical defoliants, were entirely new. However, the discovery of a quintessentially modern killing technology was to come very late in the game, with the development in the mid-twentieth century of the atomic bomb. Up to this point, weapons technology had developed along a trajectory of increasing efficiency in terminating larger numbers of individual lives. Through efforts to direct the energy produced by splitting atoms toward destructive ends, the ability to eliminate life itself – all life – was finally realized. In lieu of God-like powers of creation, modern humanity had discovered, and settled for, those of ultimate wrath.

Fortunately, transcendence achieved through the exercise of such wrath would leave nothing of Nature left to command, and so that exercise has been generally avoided. In the modern era, the command of human minds has proved to be a far less catastrophic option than the obliteration of human bodies. This is not to suggest that modernity has suffered from a lack of slaughter. It is merely to point out that, when the limitations of annihilation and brutality have been reached, for one reason or another the technological seizure of consciousness has presented itself as a quite sustainable and fruitful alternative. The success of the modern technologies of consciousness – often referred to as "communications" technologies – is attributable, in part, to the ways in which they have complemented efforts to transcend time, command space, and transform matter through industry. For example, the arrival of the printing press in the late fifteenth century is widely credited with smashing the monopoly on knowledge previously held by the clergy, and in so doing facilitating the birth of both modern individual consciousness and the nation-state.[19] However, it also instigated a reorientation of people's relationship to time and space, in that the printed word was more permanent than speech, and the book easier to transport than previous storage media.[20] It is for this reason that Rousseau described printing as "the art of immortalizing the errors and extravagances of the human mind," whereby the "pernicious doctrines" of would-be philosophers could be made to "last forever."[21] The technologies of the nineteenth century would far surpass printing in their ability to transcend space and time: telegraphy and telephony liberated communication from questions of transportation, allowing instantaneous conversation across vast distances; audio recording and photography facilitated the timeless registration and collection of images and sounds that previously would have been condemned to the uncertainty of memory. In concert, these technologies contributed to the modern perception of the unique position occupied by the individual self in both time and place.

However, it was not until the twentieth century that technologies of consciousness would appear that were able to meet the idiosyncratic requirements of the maturing modern era. These were the broadcast technologies and, with the possible exception of the Holy Bible, their effectiveness as instruments for shaping human minds was unprecedented. By this time, the influence of the clergy had been eclipsed in the West by that of various secular authorities and, though the army, prisons, and asylums provided disciplinary support for the state, the church had yet to be replaced with a satisfactory institution of primary and deep socialization. Furthermore, while the Industrial Revolution enabled the production of vast quantities of an increasing variety of goods, the question remained as to how their consumption could be incited on an equivalent scale. Having recognized the integrity of the free individual, and having developed means to manufacture a surfeit of commodities, the powerful interests of modernity had reached the point of requiring technologies to facilitate the manufacture of consent, and the manufacture of needs. Able to instantly transmit complex aural, and eventually visual, messages from a central source to a multitude of distant receivers simultaneously, broadcasting emerged as the perfect technological solution to both these distinctly modern problems. As the lives of modern individuals became increasingly isolated and private, radio and television provided one-way information conduits directly into their homes. Here were powerful technological means by which large numbers of individual citizens could be assembled into a great mass, ready to receive political instruction in the form of "news," and basic socialization in the form of "entertainment."[22] Additionally, these technologies were perfect media for the stimulation of consumption. Not only were they platforms for advertising particular products, they also enabled the promotion of consumptive behaviour itself, and regularized a climate of need by routinely presenting images of lives made happy, normal, and fulfilling through the possession of consumer goods.[23]

Ultimately, the "product" of broadcast technologies was neither programming nor advertising, but, rather, the audiences that could be gathered and delivered to advertisers – be they political or commercial – in the form of a saleable commodity like any other.[24] It is at this point that the socialization and commercial roles of broadcast technology dovetail to evoke one of modernity's great ironies: the paradox of mass pluralism, in which millions of consumers are convinced they can assert their essential individuality by purchasing the same running shoes in the same shopping malls as millions of other people. The discourse of radical individualism was necessary to shake loose the grip of premodern organic collectivism, but, if manifested socially in the form of *genuine* pluralism, it would be simply unable to meet the socialization requirements of powerful, modern political and economic elites. The broadcast technologies were configured to assist in this regard, but only if the plurality of modern individuals could be collected into relatively undifferentiated masses, via the cultivation of a sameness that was antithetical to the modern spirit of individuality. It is precisely this interest, common to political elites seeking to manufacture consent, industrialists seeking to manufacture consumers, and broadcasters seeking to manufacture audiences, that accounts for the clock-rivalling success of television among modern technologies.

Networks as Postmodern Technology?

As noted above, the prophets of the network revolution believe we are on the cusp of a new world in which the spark is being replaced bit by *bit*.[25] If the modern technological world was driven by the alchemy of fire *and* hope, claims about the revolutionary nature of network technology force us to ask whether we might also be in the midst of a parallel shift in mythology that will sustain an entirely new way of being, predicated on the use of these instruments. That is, if bits come after fire, then what comes after hope?

One set of answers to this question has been offered by what is known loosely as "postmodernism." Postmodernists, especially those who write explicitly about information and communications technology, tend to agree with network gurus when they proclaim that nothing we already know can be of much use to us: "the political metanarratives of emancipation from the eighteenth and nineteenth centuries that have served as frames and reference points for the disciplines of history, literature, philosophy, sociology, anthropology and so forth now appear to be losing their powers of coherence, their ability to provide a groundwork of assumptions that make it appear natural to ask certain questions, and to think that the answers to those questions define the limit and extent of the problem of truth."[26] Postmodern writers such as Mark Poster believe that the "mode of information" characteristic of the "second media age" requires that we "propose new questions that the old ones subordinate."[27] Poster, for example, looks to Michel Foucault for new questions about the panoptic tendencies of database surveillance, to Jacques Derrida for new questions about electronic writing, to Jean-François Lyotard for new questions about computer science, and to Jean Baudrillard for new questions about electronic advertising. In a similar spirit, George Landow informs us that "we must abandon conceptual systems founded upon ideas of center, margin, hierarchy and linearity and replace them with ones of multilinearity, nodes, links and networks."[28] This exhortation represents only the very thinnest edge of the postmodern wedge when it comes to the terms in which contemporary technology is often discussed. Along with those terms singled out by Landow, one could list the following as common to the discourse animating both postmodernism and network technology: discontinuity; simulation/ virtuality/hyperreality; decentred, unstable, multiple and dispersed identities; pastiche, play, and gaming; the demise of authority/authorship; sovereignty as an anachronism; contingency, uncertainty, and irreferentiality; decentralization; intersubjectivity/intertextuality; irony; and radical democracy. Thus, there is much to suggest that postmodernism and network technology go together. As Ronald Deibert has observed, there is a certain "fitness" between postmodern social epistemology and the hypermedia communications environment.[29]

However, because this is an examination of the politics of network technology, I will refrain from engaging in a comprehensive critique of postmodernism *as theory*. In the first place, the world does not suffer from a lack of commentaries on this subject. Postmodernism is everywhere, and there are enough people rushing ahead either to embrace or to vilify it that I think it is probably safe for some of us to stay back and pick over what is being left behind. Second, I am more interested in the politics that network technologies – sometimes in conjunction with certain aspects of postmodern discourse – inspire, and in coming to grips with these, using the resources of political philosophy, than I am in critiquing postmodern theory on its own. This means that I will also refrain from using postmodernist theory to understand network technology and its politics. Instead, I opt for a few ancient and modern tortoises over the postmodern hare. It is true that postmodernism has something to tell us about many aspects of emerging computer and network technologies, in so far as it provides a lexicon of "therapeutic redescription" for naming new phenomena and for renaming old, but changed ones.[30] The utility of this lexicon is enhanced when it is used in conjunction with other resources drawn from the tradition of critical theory.[31] It does not necessarily follow, however, that we have nothing left to learn from that which was thought and written in the long period before Baudrillard's "ecstasy of communication" allegedly supplanted meditation on the substance of a good life.[32] Therapy becomes necessary when the possibility of understanding through the use of established resources has been exhausted. I am not convinced we have reached that point vis-à-vis the politics of technology.

At the very least, we should sort carefully through the piles of theoretical rubbish before we consign them to the postmodern recycling bin.

Political judgment places a premium on caution and is wary of enthusiasm. My choice to employ the resources of traditional political philosophy to understand the politics of network technology and the discourse accompanying it, rather than to use postmodernism to simply describe their appearance, reflects the caution proper to reasoned judgment. A reluctance to embrace the "newness of the new" championed by Poster and others is appropriate in this case precisely because so much of what passes for postmodern "theory" sounds so much like the technology and politics it is meant to be theorizing.[33] Take, for example, this statement about hypertext: "[postmodern] critical theory promises to theorize hypertext and hypertext promises to embody and thereby test aspects of theory."[34] I am not sure what the words "test" and "promise" are intended to mean in this sentence, but it is clear that the fitness between theory and the theorized presented here – that of a convergence – is perhaps too snug for honest comfort. It is not a sin to write enthusiastically about the manner in which a technology vindicates one's perspective, but it may not be theory either.

"Theorist" is from the Greek *theōros*, meaning "spectator." Sometimes, postmodern interventions appear to have too much invested in what they are supposed only to observe. According to Baudrillard, "it is not enough for theory to describe and analyze, it must itself be an event in the universe it describes. In order to do this *theory must partake of and become the acceleration of this logic*."[35] This is why it is easier to say "postmodern*ism*" than it is to say "postmodern *theory*." From vantages closer than any spectator's, postmodernism can provide descriptions (therapeutic or otherwise), and even participate, but it cannot gain the distance that is crucial for judgment. The issue of the conditions and practice of judgment is a complicated one. However, as Ronald Beiner writes in his effort to articulate a philosophy of political judgment that respects both the Kantian and the Aristotelian traditions, "political judgment must embrace the standpoint of both the spectator and the actor: it calls for both distance and experience."[36] Judgment requires theory to "clarify what is at stake, and disclose the conditions that render efforts towards a satisfactory resolution possible"; it also requires prudential reflection upon direct experience, because questions of judgment "can only be resolved in the concrete, when confronted with particulars."[37] In so far as it is immersed in the very technology it describes, postmodernism can and does provide close-up accounts of the various particularities of digital networks. However, what postmodernism gains in proximity, it lacks in critical, theoretical distance. Indeed, for the most part, postmodernists reject the idea that distance of this sort is even possible. Fredric Jameson, in one of the definitive statements of postmodernism, concluded that "distance in general (including 'critical distance' in particular) has been very precisely abolished in the new space of postmodernism ... our now postmodern bodies are bereft of spatial coordinates and practically (let alone theoretically) incapable of distantiation."[38] Network technology and the postmodern are, as Chris Gray describes, thoroughly *embedded:* "As a weapon, as a myth, as a metaphor, as a force multiplier, as an edge, as a trope, as a factor, and as an asset, information (and its hand-maidens – computers to process it, multimedia to spread it, systems to represent it) has become the central sign of postmodernity."[39]

If information technology is so central to postmodern*ity* – if the latter cannot exist without the former – then postmodern*ism*, on its own, cannot be expected to provide the tools for a disinterested understanding and judgment of this technology. For this, we require theories that, from a distance, help us "clarify what is at stake" in committing ourselves to this particular technology. To their credit, postmodernists such as Poster admit that postmodernism is a "fledgling position" capable only of "registering changes" in society.[40] The registration of particularity is an indispensable element of judgment, and while postmodern writers certainly contribute to this process, they do not enjoy a monopoly over it. Part of this book involves a discussion of the particularities of network technologies, and in some instances I draw on the work of postmodernist writers who have paid attention to these. For the second, crucial aspect of the foundation for judgment – the distance that accrues to theory – I rely on spectators of past technologies in the hope they can help us clarify what is at stake in our own.

Network Technology and the Discourses of Change and Democracy

Curiously, network technology taps into the most stubbornly modern aspects of the postmodern narrative – namely, the popular discourses of irresistible change and ineluctable democracy. The valorization of change over endurance – in particular, change that is deemed progressive – is a hallmark of modern politics. Change is the ground upon which modern political actors bearing a variety of ideologies meet: liberals believe change expresses freedom; socialists believe change is necessary before freedom can be won; and even so-called neoconservatives, who resist changes in public morality and domestic life, believe external limits on market freedom – the freedom to fluctuate or change – should be avoided. Modern political subjects not only desire change but also are certain that change is within their grasp and theirs to make.

Network technology has escalated, but not altered fundamentally, the fetishization of change that characterizes modernity. According to two recent heralds of the new age, "there is no disagreement on the essentially revolutionary nature of the forces unleashed by the new technology. And there can be no doubt that the Digital Revolution is going to change the way knowledge is gained and the way wealth is created."[41] Celebrated futurists Alvin and Heidi Toffler agree: "what is happening now is nothing less than a global revolution, a quantum leap ... we are the final generation of an old civilization and the first generation of a new one."[42] From former banker Walter Wriston we hear that "the rules have been changed forever" by network technology;[43] and George Gilder is certain that the possibilities for change are "bounded only by the reach of the mind and by the span of the global ganglion of computers and cables, the new world wide web of glass and light."[44] The government of the United States has determined that "a seamless web of communications networks, computers, databases and consumer electronics ... will change forever the way people live, work, and interact with each other."[45] The chair of Canada's Information Highway Advisory Council concurs: "As Adam and Eve left the Garden of Eden, one said to the other, 'We are in a period of profound change.' So are we today ... Today's information revolution will be as deep and momentous as any other scientific movement in history."[46] If there is one thing that network technology has left intact, it is the abiding faith that change is immanent, and that things will never be the same.

There also appears to be what one might call a *marginal* consensus among the technology's proponents about the character of the changes that will be wrought by digitization and computer networks. It is only a marginal consensus because, while the vast majority of people who have written about this technology agree that it necessarily portends change, most of them are less willing to make a case for what kind of change this will be, preferring instead to trot out ambivalent platitudes about the commingling of peril and promise, danger and delight, or benefits and detriments. There are certainly exceptions, but the account of change that seems to have captured the public discourse about net-works is one that suggests the change this technology instigates will be revolutionary in the truest sense: it is believed that networks will fun-damentally alter relationships of power in society.[47] So we are assured that "the information revolution is *profoundly threatening* to the power structures of the world."[48] More specifically: "The force of micro-electronics will blow apart all the monopolies, hierarchies, pyramids, and power grids of established industrial society."[49] And this view is not simply an indication that the users of networks harbour revolu-tionary intentions. Instead, rebellion is said to inhere in the essence of the technology itself: "There appears, in fact, to be a *core conflict* be-tween the *basic nature* of the Internet and the demands of organized, large-scale commerce," due to "digital technology's natural tendency to promote decentralized, non-hierarchical social relations and organi-zational forms."[50] If this is true – if network technology is *inherently* revolutionary – it leads one to wonder why existing governmental, bu-reaucratic, corporate, and financial elites are so enthusiastic about, and so heavily invested in, the success of this technology.

Perhaps the key article of faith concerning the essentially revolu-tionary series of social, economic, and political changes promised by digital networks is the conviction that these are democratic media par excellence, a faith augmented by an anticipation that the democracy of networks will be contagious and impossible to quarantine. The my-thologies of democracy constitute the dominant elements in the narra-tive accompanying the proliferation of network technology. I use the plural "mythologies" because democracy – the great empty vessel of contemporary political discourse – means different things to different people. To some it means consumer capitalism; to others it means anarchy. To some it means liberalism; to others it requires socialism. To some it means voting; to others it means deliberating. For some it is based on rights; for others it evokes the duties of citizenship. The point is that digital networks appear amenable to presentation in ways that capture the imagination of nearly every kind of democrat and, as a result, democracy has figured prominently in the discourse that names not only the essential characteristics of this technology, but also the wider societal changes it promises to precipitate.

This is hardly surprising. In the modern era at least, developments in communications technology, including the telegraph, telephone, and television, have characteristically inspired renewed democratic aspirations.[51] Examples of how the narrative of network technology has been colonized by the mythologies of democracy are harder to miss than they are to find, but nowhere have I found a more expressive articulation of this dynamic than in a book titled *The Electronic Republic,* by Lawrence Grossman, formerly a leading citizen in the world of television.[52] According to Grossman, telecommunications networks "make it possible for our political system to return to the roots of Western democracy as it was first practiced in the city-states of ancient Greece," and will also facilitate "a modern-day extension of Jeffersonian participatory democracy."[53] Network technology has created a situation whereby "members of the public are gaining a seat of their own at the table of political power," and in which they are becoming "increasingly involved in day-to-day decision-making alongside the President and Congress."[54] Crucially, network technology promises to overcome the obstacles of scale that have traditionally thwarted vigorous democratic participation by providing for "keypad democracy": "Time and distance will be no factor. Using a combination telephone-video screen computer, citizens will be capable of participating in audio- and videophone calls, teleconferences, tele-debates, tele-discussions, tele-forums and electronic town meetings."[55] Also, networks are credited with the potential to obliterate the correspondence between economic means and political participation because, as Walter Wriston puts it, "information has always been society's great equalizer."[56]

This perception of the likely impact of network technology has been bolstered by political actors at the governmental level who – hoping to catch the wave and trying to avoid being stigmatized as anti-democratic – have joined the rush to digitality by placing themselves and at least part of their work "on-line." Typically, this has taken the form of making government documents and services available via computer networks; maintaining party, ministerial, and departmental Web sites; and assigning electronic-mail addresses to elected representatives.[57] In some cases, it includes the facilitation of on-line discussion groups and electronic plebiscites, and the provision of a variety of information resources.[58] Our political leaders have been no less enthusiastic about the potential of network technology than the captains of industry, the futurists, and the pundits have been. Preston Manning, the leader of the Reform Party of Canada, can barely contain his enthusiasm in announcing, "We're building the Athens of the twenty-first century."[59] Newt Gingrich, as Republican leader of the House of Representatives, christened the US Congress's new on-line document system with the name "Thomas," after Thomas Jefferson. Even those who are wary of the corporate bogeyman lurking among the wires nevertheless maintain an enduring faith that this is primarily a democratic technology. Howard Rheingold, a proponent of "virtual communities" and a high-profile member of the Whole Earth 'Lectronic Link (WELL), expresses this faithful hope in more measured terms than those favoured by the partisans noted above when he says: "The political significance of computer mediated communication lies in its capacity to challenge the existing political hierarchy's monopoly on powerful communications media, and perhaps thus revitalize citizen-based democracy."[60] Rheingold, and others like him, recognize the possibility that digital networks could be colonized by the same commercial interests that dominate other communications media. Nevertheless, their implication is that this would represent a corruption of what is *originally* and *essentially* a democratic medium. Therefore, the consensus is clear: the new technology is a democratic technology as surely as we are democrats. Like any other consensus, this one cries out for scrutiny.

A Standard of Democracy

Notwithstanding the aforementioned variety and contention concerning the meaning of democracy, conducting an investigation such as this requires that one adopt a particular understanding of what the word means. The popular currency and bastardization of the term have reached the point where attempts to establish any one standard of democracy as definitive are basically futile. Nor is this the place to undertake a review of the considerable breadth of definitions, descriptions, and categorizations available in the tradition of democratic political theory.[61] Nevertheless, it is important when discussing democracy that one at least specify what one means. Accordingly, in the present investigation, "democracy" refers to a form of government in which citizens enjoy an equal ability to participate meaningfully in the decisions that closely affect their common lives as individuals in communities.

I adopt this definition because it captures three elements that are essential in *any* serious definition of democracy: equality, participation, and a public sphere from which sovereignty emanates. Recognizing that most conflicts over the nature of democracy stem from differences regarding the specific content of these three elements, I have also tried to indicate in this definition something of what I consider their content to be. For example, this definition suggests that the equality that is an essential attribute of democracy refers to an equality of *ability* to participate, rather than simply to an equal *opportunity* to do so. Equality of opportunity is an attribute of liberalism that denotes an absence of formal or legal constraints preventing participation, and it is often substituted for democracy's more demanding standard of equal ability. However, the absence of formal or legal barriers is only a necessary, and not a sufficient, condition for the equal ability to participate: citizens who are not constrained from doing so by law may nevertheless be unable to participate equally with their fellows for other reasons. In a situation where resources such as wealth or expertise provide access to crucial sites and modes of civic participation, those who are deprived of these resources (i.e., the poor and uneducated) cannot participate equally with those who possess them. The absence of practical, as distinct from legal, barriers to equal participation is a condition of full democracy. It is for this reason that all liberal states are not necessarily democratic. It is also why the word "equality" in the definition of democracy must refer to ability, rather than merely opportunity. This being said, it should be noted that this definition of democracy does not require that people *in fact* participate equally; it simply requires that they have an equal ability (which includes, but is not exhausted by, equal opportunity) to do so should they so desire.

The definition adopted here also stipulates that citizen participation must be *meaningful* in order for it to qualify as democratic. Admittedly, the adjective "meaningful" is somewhat indefinite. However, it is meant to suggest that a political arrangement cannot be called democratic if the participation it allows is frivolous, or merely symbolic. Democratic participation must be clearly and decisively connected to the political decisions that direct the activity of the participants' community. By this definition, polities in which citizens' participation is limited to legitimizing deliberations and decisions made without their participation is not a democracy. Thus, democracy requires that citizen participation be specifically linked to policy outcomes, rather than relegated to the general role of system legitimation.

Finally, this definition insists that democracy denotes a form of governing the public and common affairs of individuals in communities. Again, this stipulation is meant to distinguish the requirements of democracy from those of liberalism. The latter is not a system of government, it is an ideology whose chief concern is to assert those areas of individual human endeavour from which public government should be forever absent. Democracy is defined by the constitutionalized practice of gathering together the private individuals who make up a particular community to decide publicly on courses of action and inaction regarding their common affairs. This means that democracy is not constituted wholly by freedom of consumer choice in a market or the freedom to do privately whatever one lists. Instead, democracy is about the public taking of collective decisions that are to govern the common and public practices of the members of a community. This is not to say that every decision taken by a democracy must be unambiguously in the general interest. Aristotle defined "democracy" as rule by the many in their self-interest – a definition that is not hostile to the one I am advancing here. Democracy does not require that the interests brought before it as a system of governing are public; what is essential is that these interests contribute to decisions that are binding on the public and common life of individuals in the community, via a process in which each of them has an equal ability to participate in a meaningful way.

This definition is not intended as a shorthand theory of democracy, but simply to specify what I mean when using the word. The definition I have articulated is almost certainly not the one employed by all those who have placed their hopes in the democratic potential of network technology. In most cases, what I mean by democracy is not what they mean. To conduct an investigation of their claims about the technology against this measure is somewhat unfair, in so far as it demands they live up to a standard to which they do not profess to ascribe. Nevertheless, the popular discourse surrounding this technology is long on inflammatory rhetorical claims about its political potential (i.e., that it is the instrument of a democratic revolution), but short on definitional substance (i.e., what constitutes a democracy? what qualifies as a revolutionary change?). Thus, some common definition must be brought to bear in assessing the political claims of network proponents. The claims being made by the technology's prophets suffice to raise the questions that will direct this investigation, but finding the answers requires the technology be assessed in light of more demanding standards. Ultimately, what is at issue here are the politics of network technology as manifest in current tendencies and practices, not the political rhetoric of those amazed by it. The latter is merely a convenient starting point for the journey toward discovering the former.

Admittedly, this definition of democracy entails criteria of qualification that are quite exacting. Indeed, most contemporary governments that call themselves democratic would fail to meet them. However, it is a standard that I think honours, rather than ignores, the long history of this name in the tradition of political philosophy. To the charge that the requirements of this definition of democracy are *exceedingly* stringent, and that network technology – or any other technology, for that matter – simply has no hope of meeting them, there is only one defence: network technology and its various manifestations are either democratic or they are not, and, if they are not, they tend either toward democracy or away from it. A clear, if demanding, definition can only help us in the very important task of deciding which is the case.

Finally, I would like to point out that, while this definition posits democracy as something that is very difficult to achieve, it does not necessarily imply that it is the best, or even a good, form of human government. It may be the case that democracy is the best way we can govern ourselves, and it just as well may be that it is the worst; it is more likely that it is better than some ways and worse than others. Whatever the case may be, this is not the question to which the present investigation is addressed – this is an examination of the politics of network technology, not a work of democratic theory. Despite their consistent inattention to its substance, most of those who see a political revolution among the wires believe democracy to be an unambiguous and unquestionable good. Indeed, it often appears that their primary rhetorical strategy is to throw the considerable discursive weight of democracy's near-universal popular appeal upon the scales that will measure the desirability of a continued proliferation of network technology. When one commands such an unalloyed good as democracy, the progression is simple: democracy is undeniably good; network technology is democratic; therefore, network technology is also good. To avoid falling into this sort of dubious argumentation, the definition adopted here acknowledges that democracy is a technical, rather than a normative, designation. The question is whether its technicalities complement, or conflict with, those of networked computers.

In the chapters that follow, I examine the political implications of network technology with a view to determining whether this technology and the world it makes are likely to live up to the hopes for change and democracy suggested by the discourse supporting it. This discussion is predicated on the understanding that technology and democracy share a relationship that is essentially ambiguous in character. In some respects, there is a strong affinity between technology and democracy: the technological urge arises from the human appetite for mastery and control of the future; genuine democracy does not specify any content to what is considered good, beyond that which people decide for themselves, as sovereign masters of their own future. Thus, in hope, technology and democracy seem to share common ground. However, there is also a crucial antagonism between democracy and technology: democracy does not require substantial expertise as a qualification for participation in decision making, and so it allows for government by mass ignorance; technology, as it becomes increasingly complex, requires for its control and deployment levels of expertise that exceed the capacity of most citizens and, thus, it defies democratic governance. As Ronald Beiner has put it, "the possibility looms that technological society makes a nonsense of democratic theory. We are mocked by our own technical powers, while the very idea of democracy lingers on only as an embarrassing recollection."[62] Apparently, though both democracy and technology spring from the hope for mastery, somewhere along the way their respective hopes cause them to collide. In what follows, I attempt to sort out where computer networks are situated in terms of the complementary and contradictory aspirations of democracy and technology, and to determine whether the present situation represents a significant change from previous technologies.

I begin by exploring the writing of five political philosophers – Plato, Aristotle, Marx, Heidegger, and Grant – all of whom thought deeply about the relationship between technology and politics. Blind hope attached to technology is essentially an opinion about what the outcome of our encounter with that technology might be; since its origins in ancient Greece, political philosophy has always presented itself as a means for proceeding from belief to understanding by asking questions of opinion. If understanding is our goal, the questions asked by these philosophers about politics and technology are the questions we should ask of the opinion that the politics of network technology are, and will continue to be, democratic and revolutionary.

on technology

. . . . .

IN HIS *LETTER ON HUMANISM*, first published in 1947, Martin Heidegger remarked that the "essence of technology" was something "about which much has been written but little has been thought."[1] With due respect for the care with which Heidegger used words like "essence," and for the precise distinction he drew between writing and thinking, I think it is safe to say things have both changed and remained the same since this particular observation was made. Certainly, the subject of technology has not suffered from a lack of written attention in the latter half of the twentieth century, as various scribes have stepped up to document, describe, extol, and explain the spectacular parade of gadgetry that has marked this period of history. Not all of this reflection has been specifically devoted to the essence of technology, and even less of it has been particularly thoughtful. However, some of it – including Heidegger's own contribution – has been, and it is to this body of thoughtful writing I wish to turn to discover the questions we should ask about the democratic potential of networked information and communication technologies. This approach takes for granted that it is impossible to speak thoughtfully about telecommunications networks *as technologies* without first establishing what a technology is, and what a technology does.

The best place to begin is with the words themselves. "Technology" combines the ancient Greek words *technē* and *logos*. *Technē* refers to the useful arts – sometimes called "crafts" or "sciences" – that are involved in making or fashioning a thing, and thus pertains to the realm of artifice. The complex meaning of *logos* has been well expressed by Edward Andrew, as follows: "*Logos*, the Greek for speech, word, or reason,

. . .

derives from *legein*, which means to gather, collect or pick up as well as to utter or say. The *lect* in the English words collect and select as well as lecture and dialect, comes from *legein*. Thus speech for the Greeks was a gathering, a collection; the Logos was the one unifying the many: the word collected the phenomena observed, the things named."[2] *Logos*, then, is a manner of reasoned speaking, a form of discourse that gathers or unites objects that are given. As Andrew points out, *Logos* was "the Word" that gathered and unified the people of God, "in the beginning." For the ancient Greek philosophers, the gathering that distinguished *logos*, or discourse, was that of knowledge of the human condition, and, in particular, of the essential nature of a good and just life in a human community. Philosophers such as Plato and Aristotle did not combine *technē* and *logos* into a single compound, because, to their minds, these words had specific, distinct meanings that should not be casually collapsed into one: *technē* makes things that do not already exist and that are, therefore, artificial; *logos* attempts to gather that which always-already exists in Nature and is wholly true. As I discuss below, while Plato and Aristotle thought that certain *technai* could complement *logos*, they declined to think of the two things as identical.

Despite the counsel of the ancients, we moderns combine *technē* and *logos*, and this alone should signal that technology represents more than just the mechanized or electronic devices to which this word commonly refers in contemporary parlance. Etymology itself suggests that technology is a useful art that not only produces some sort of material object but also entails a kind of speaking about or gathering of what we consider to be important to the human condition. Along with making things, technology stands for something about what we are, or wish to be, and about the manner in which we live together. Of course, this view is complicated by technology's tendency to produce other technologies, and by its predisposition to gather us as technological beings. But this is jumping ahead. For now, I wish to point out simply that technology is clearly more than the sum or operation of its parts.

Consideration of the etymology of the other two words that name important aspects of the so-called network revolution – information and communication – yields complementary results. Those who are concerned to defend network technologies against charges of precipitating information overload, or "glut," are quick to distinguish between information and raw "data." Information, by this account, is data made useful by organization and classification. While this may be a helpful distinction, it is not an adequate definition. The root of "information" is *form*, from the Latin *formare*. In English, we tend to emphasize the qualities of "form" as a noun, rather than as a verb. Thinking of it as the elaboration of *a* form, we customarily define "information" as an item or items of knowledge. However, the English word "information" is derived from the Old French verb *enformer*, which meant *to form* or, more specifically, to give shape to, or to fashion. There are English usages of "information" that suggest its capacity to inspire, impart a quality, or permeate that to which it is applied, but they are not commonplace. I would like to suggest that these are meanings of "information" that should be emphasized, particularly in light of the intimate connection between information and technology. Information, in the current context, is more than organized data, more than an item of knowledge, and more than a thing that is told. It is, crucially, a telling that shapes or forms. Consequently, it becomes important to investigate not only that which is told, but also the impact, manner, and medium of its telling.

Telling is one-half of the act of "communication," which happens to be another word customarily linked with contemporary network technologies. As a noun, *communication* denotes an object – usually information (also as a noun) – of exchange. Telling is the half of communication involved in transmitting information. However, communication requires reception and understanding for its completion: *communicate* is also a verb, from the Latin *communis*, for "common," and the Old French verb *comuner*, which means "to share." Communication, in this sense, shares an etymology with words like "communion," "community," and "commitment," which connote somewhat more expansive and enduring relationships than those entailed in simple market exchange. It is the character of these relationships that is of interest in the present context, and so it becomes important to pay attention to the ways in which they are influenced by particular communicative practices or forms.

The words "technology," "information," and "communication" constitute the holy trinity of the age of computerized networks. If, as I propose, "technology" is defined as a productive practice that simultaneously tells us something significant about our collective selves, "information" as a practice that shapes or gives form, and "communication" as the locus of relationships in which we share that which is common, then I think it is safe to say there is something of social, economic, and political import to be gleaned from their coincidence at present. What will the shape of a society enmeshed in a network of computerized information and communications technologies be? What relationships will it encourage or discourage, and what will it tell us of the important things we gather to and about ourselves? Specifically, are the society and relationships encouraged by these technologies likely to achieve the democratic potential hoped for them?

To answer these questions, one must establish clear ideas about the nature of technology. This stipulation is based on the premise that of the three aspects of the network trinity – information, communication, and technology – it is the impact of the last of the three terms that is, at once, most hidden and most determinate. Validation of this premise constitutes the subtext of much of the present investigation, but, I submit, the intuition that information and communication are influenced decisively by the technologies that mediate them is enough to warrant paying special attention to what technology is, and what it is capable of. For this attention, I now turn to the tradition of thinking about technology alluded to earlier. In particular, I wish to review aspects of this tradition that admit the richness of the very word.

Plato, Aristotle, and *Techné*

As mentioned above, the ancient Greeks did not render "technology" as a single word, choosing instead to separate the useful arts and practical sciences (*techné*) from reasoned discourse about the true nature of goodness (*logos*). This distinction was meant not only to specify the exact practices designated by these words, but also to indicate the right relationship between them. This is clear in Plato's refusal to attach the label *techné* to a range of practices that today would be called "techniques." When the sophist Gorgias attempts to define oratory as the "art of speech" about "right and wrong," "the greatest and best of all human concerns" that "embraces and controls all other spheres of human activity" – the *techné* of *logos* – Socrates responds that oratory is not a *techné* at all.[3] Instead, it is an *empeiria*: a knack, habit, or routine based on empirical results gleaned from experience.[4] To "have a knack" for something is to be handy without really understanding what you are doing, and to act routinely is to act without thinking. Thought is not required when the end of an activity is simple gratification, and *empeiria* "makes pleasure its aim instead of the good."[5] Plato lists things such as cookery and fashion among those *empeiriai* that, positing pleasure as an end in itself, and deriving from habituation undirected by reason, are liable to neglect what is good in favour of what is simply pleasing.[6]

True *technai*, on the other hand, have as their object the "highest welfare of body and soul."[7] They seek not what is agreeable, but what is best, and place knowledge gained through rational thought ahead of habits formed in response to repeated experiences of physical pleasure. Medicine and physical exercise are true *technai* because they seek to understand what is good for the body (not merely what pleases it at any given moment) and proceed rationally toward the end of delivering this goodness; they are mirror opposites of cookery and fashion. Nevertheless, *empeiriai* often stand in for *technai*, in particular, when the spirit's moderating capacities have atrophied, and the appetite is left vulnerable to flattery and deception with regard to certain essential goods. However, devoid of any real knowledge of their professed object, *empeiriai* can only ever be counterfeits of *technai*; people with a knack for making others feel better with chicken soup do not really know what doctors know, and so can only imitate them.[8] With respect to health, cookery is to medicine an understudy who does not bother to study. With respect to beauty, fashion is a poor, but pretty, surrogate for exercise. And, in terms of justice, oratory is "to the soul what cookery is to the body": an unreflective knack that panders to the appetites of the *dēmos,* however irrational and unhealthy they might be.[9] The *technē* that contests oratory for the direction and care of souls is the art of government: the practice of legislation, according to reason, directed by and toward what is best for the political community.

Socrates is clear on this point: "I refuse to give the title of art [*technē*] to anything irrational."[10] *Empeiriai* are irrational because reason is not involved in their execution: the person exercising a knack "has no rational account to give of the various things which it offers."[11] However, it is not simply irrationality that disqualifies oratory and the others from being labelled *technai*. In pandering to pleasure exclusive of any consideration of goodness, these and other *empeiriai* are also "dishonourable."[12] For Plato, there is a distinctly *philosophical* aspect to the designation *technē*, in so far as it refers only to those human arts that have as their end the encouragement or realization of goodness. In this respect, practices that pander to human appetites beyond the requirements of necessity, and that do so simply for the pleasures of indulgence, do not qualify as *technai*. Efficient gratification is not a reliable indicator of the virtue that defines a *technē*; it may simply indicate a knack, or *empeiria*, an unintelligent routine.

By way of contrast, we can note that, in modern usage, the word "technique" bears little resemblance to its Platonic forebear. In modern usage, "technique" refers to the repeated application of a skill in the production of a predetermined end, and the word applies to such practices regardless of what their ends might be. Techniques are typically judged as "good" simply on the basis of whether or not they yield desired results consistently and efficiently. Efficiency and consistency are the fruits of experience, and – whether we are talking about the pressing of rivets on an assembly line, the fabrication of desire through advertising, or the honing of a slapshot on the backyard pond – technique is optimized through repetition and habituation. Modern technique bears many of the characteristics Plato identified as peculiar to *empeiriai,* and corrosive of *technē*. This is not to say that *none* of our modern techniques would have qualified as a *technē* in Plato's mind (indeed, some – many surgical techniques, for example – would have). Rather, it is simply to assert that many, maybe even most, of them would have been deprived of this distinction, and that attention to this difference might assist our understanding of contemporary technologies and the politics surrounding them.

In the *Nichomachean Ethics,* Aristotle specifies the character and role of *technē* in the context of his broader discussion of the contents of a good ethical and political life. According to Aristotle, *technē* was involved in the transformation of objects into a state that differs from their natural one. Of these "things which admit of being other than they are," there are two kinds: "things made and things done."[13] Aristotle distinguishes between making and doing, producing and acting, and suggests that *technai* are pursuant to the category of production, making, or fabrication. The useful arts are, thus, *art*ificial in the sense that they help us transform Nature to produce that which Nature does not produce, in that form, on its own. As Aristotle says, "all art [*technē*] is concerned with the realm of coming-to-be, i.e., with contriving and studying how something which is capable both of being and of not being may come into existence, a thing whose starting point or source is in the producer and not in the thing produced. For art is concerned neither with things which exist or come into being by necessity, nor with things produced by nature: these have their source of motion within themselves."[14] The products of *technē*, then, do not exist necessarily, and whether or not they "come into being" depends upon the volition of "the producer." For many, this observation by Aristotle suggests that he views *technai* as strictly instrumental: the technical arts are essentially neutral means to ends determined by the fabricator or artificer.[15] This is the supposed Aristotelian source of the view that there is nothing of a political or ethical nature to be discussed with regards to technologies themselves.

It is true that Aristotle regarded *technai* as means or instruments to ends generated by human beings, but it is an embellishment of this position to suggest that it indicates he believed the useful arts and practical sciences to be ethically neutral. Aristotle opens *Nichomachean Ethics* with the stipulation that, while various *technai* are instrumental to a variety of corresponding goods or ends, they are all subordinate to "the most sovereign and most comprehensive master science," the end of which is "the highest good," which "we desire for its own sake."[16] This master science is *politikē*, the science of politics, which, according to Aristotle, contains all the others. In light of this point, the claim that Aristotle thought *technai* to be neutral is questionable, primarily because, as subordinate elements of the science that pursues the highest good for its own sake (i.e., not as instrumental to some other good), the productive arts can be judged according to their contribution to that pursuit. That which admits of being judged as good or bad is not, by definition, neutral. Further, Aristotle lists *technē*, along with *phronēsis* (practical wisdom), *epistēmē* (pure science), *nous* (intelligence), and *sophia* (wisdom), as "faculties by which the soul *expresses* truth by way of affirmation or denial."[17] Affirmation or denial precludes neutrality. Finally, Aristotle states that *technē* can be considered true *technē* only if

· · ·

it "is identical with the characteristic of producing under the guidance of true reason."[18] Thus, the definitive attribute of *technē* is that it is truly rational in its productive capacity and, if it is judged to be otherwise, it should be designated as something else. The assumption that such judgment could be brought to bear against a particular art or practical science appears to indicate that Aristotle was far from positing the essential neutrality of *technē*. Indeed, along with Plato, he seems to be saying the opposite: the arts and sciences either accord with true reason or do not, and only those determined to be contributory to the attainment of the highest good can properly be labelled *technē*.

The master science that makes these determinations is *politikē*. It is *politikē* that "determines which sciences ought to exist in states; what kind of sciences each group of citizens must learn, and what degree of proficiency each must attain."[19] For Aristotle, decisions about *technē* are political decisions. To conflate Aristotle's designation of the practical arts and sciences as instruments with an assumption of their essential neutrality is to run the risk of neglecting his fundamental teaching regarding *technē*: political judgments about *technai* are implied in judgments of the ends they serve as instruments. Technical practices are not neutral, because the ends to which they are connected are not neutral, and, therefore, *technē* is not exempt from political judgment. Contemporary critics of technology generally take for granted the argument that because technology has obvious kinds of social impact, its advance should be subject to political and ethical deliberation. They typically contrast themselves to those who believe technological development should be fettered only by what the market can bear, and that the progress of disinterested scientific invention should continue unabatedly. My point here is not to insist that these critics acknowledge their debt to Aristotle's philosophy for a conviction they hold to be self-evidently true. Instead, I wish simply to point out that Aristotle's instrumentalism is not reducible to a belief in the neutrality of technology. Indeed, a theory that understands technology to be, simultaneously, always instrumental and never neutral provides a formidable resource to those who wish to think critically about technology.

Karl Marx and the Technologies of Capitalism

Of the numerous mythologies that have arisen out of the political and economic thought of Karl Marx, the charge that he was an adherent of or, even worse, that he originated the doctrine of technological determinism stands out as particularly unfortunate.[20] To label someone a technological determinist is to condemn him or her as thoughtless and uncritical, adjectives that hardly seem descriptive of Marx and his work. In the past few decades, however, attempts have been made to defend Marx against this charge, and they are largely convincing.[21] Nevertheless, the potential this mislabelling has to obscure the real contribution Marx has made to understanding technology is such that I think it is important to review the ways in which it is inaccurate.

Much of the misunderstanding about Marx's alleged determinism is attributable to his famous maxim "The hand-mill gives you society with the feudal lord; the steam mill, society with the industrial capitalist."[22] The source of the confusion appears to be the French verb *donner* in the original, which, though correctly translated as "gives," for some reason is equated by Marx's detractors with "makes," "produces," or "creates." The act of giving is far less deterministic than the act of creation, and is more akin to indicating or signifying than to making or producing. When we *give* someone a gift, we *indicate* our love for them, but we do not *make* love to, with, or for them. What Marx appears to be saying in this aphorism is that certain technologies are indicative of, or significant to, particular productive relations. He may be going so far as to posit that these technologies *facilitate* particular relations, but, unlike the determinist reading, this is well within what is suggested by "giving."

That Marx was not a crude technological determinist is made clear both in the remarks immediately preceding this notorious aphorism and in the broader context of his theory of historical materialism. Two sentences prior, Marx observes that "social relations are closely bound up with productive forces."[23] It is not by accident that he chooses "bound up with" rather than "determined by." He then proceeds to explain exactly what this means: "In acquiring new productive forces men change their mode of production; and in changing their mode of production, in changing the way of earning their living, they change all their social relations."[24] This is the indispensable context that should serve to dispel any charges of technological determinism brought against Marx. For Marx, the social existence of human beings is conditioned by the

organization of their material life or, as he termed it, the mode of production.[25] The mode of production consists of an ensemble of the relations of production (the distribution of the ownership of the means of
production) and the forces of production. Technology is one very important – but not the only – element of the forces of production.[26] Labour
power, natural resources, expertise, knowledge, and skill are also included among these productive forces. Thus, Marx appears to be suggesting simply that technology, as one among many forces of
production, contributes, along with the relations of production, to the
determination of social life by the mode of production. Phrased differently, Marx's message is that technology *gives* something to the mode
of production that enables *it* to determine the shape and character of
social and political life – a far cry from arguing that technology bears
this capacity *on its own*.

This is not to say that Marx minimized the impact or importance of
technology. On the contrary, he understood that the role technology
plays in productive life implicates it in human nature as well. According to Marx, "as individuals express their life, so they are. What they
are, therefore, coincides with their production, both with *what* they
produce and with *how* they produce."[27] The German word translated as
"express" here is *äussern*, from the root *äusser*, meaning "outer" or "ex-
terior." This implies that, for Marx, what we are is represented in how
we interact materially with the world that is outside our selves. For the
most part, this is accomplished when a person actively applies productive labour to his external surroundings, "in order to appropriate
Nature's productions in a form adapted to his own wants."[28] And, in so
doing, it is not simply external Nature that is altered: "By thus acting
on the external world and changing it, he at the same time changes his
own nature."[29] It would seem, then, that productive activity is important to Marx for two reasons: it is the medium through which human
beings express what they essentially are; simultaneously, its form has
a decisive impact on the very essence it expresses. Technology, as one
of a number of forces of production, plays a crucial role in this process.
As Marx puts it, "technology discloses man's mode of dealing with
Nature, the process of production by which he sustains his life, and
thereby also lays bare the mode of formation of his social relations,
and the mental conceptions that flow from them."[30]

To illustrate this, Marx discussed the development and impact of a specific manifestation of technology, in the context of a particular mode of production – the machine in modern industrial capitalism. The movement from simple manufacture to the modern industrial mode of production occurs, Marx says, when "the instruments of labour are converted from tools into machines."[31] The mechanization of production was an extension of the division of labour prevalent in small-scale manufacture. In the latter, productive activity was divided into constituent handicrafts in which individual workers applied tools to the transformation of matter. However, this division of labour "gradually transforms the workers' operations into more and more mechanical ones, so that at a certain point a mechanism can step into their places."[32] In the industrial mode of production, the handicrafts themselves are broken down into constituent elements, and tools are collected and passed from the workers' hands to the fasteners, levers, and gears of a machine. As Marx describes it, "the machine proper is therefore a mechanism that, after being set in motion, performs with its tools the same operations that were formerly done by the workman with similar tools ... The machine, which is the starting-point of the industrial evolution, supersedes the workman, who handles a single tool, by a mechanism operating with a number of similar tools, and set in motion by a single motive power, whatever the form of that power may be."[33] Thus, as a technology, the machine is not equivalent to the tool as an instrument of the worker's labour, because the labourer does not apply the machine to the transformation of matter using his or her skill. The machine replaces the labourer's skill with repetition and mechanization. Marx's description of this confiscation is rich:

> In no way does the machine appear as the individual worker's means of labour ... Not as with the instrument, which the worker animates and makes into his organ with his skill and strength, and whose handling therefore depends on his virtuosity. Rather, it is the machine which possesses skill and strength in place of the worker, is itself the virtuoso, with a soul of its own in the mechanical laws acting through it ... The worker's activity, reduced to a mere abstraction of activity, is determined and regulated on all

· · ·

37

sides by the movement of the machinery, and not the opposite ... What was the living worker's activity becomes the
activity of the machine.[34]

It is not the technology itself that necessitates this transformation. It
is, instead, the result of a conscious decision by those who enjoy economic power to reshape the "traditional, inherited means of labour
into a form adequate to capital."[35] The form of labour most adequate to
capital is, of course, that form which most efficiently produces the greatest quantity of exchangeable goods, and which adds as much surplus
value to those products as possible. It was these goals that mechanization met so adequately in the transition from manufacture to industrial capitalism.[36] In Marx's view, the technology of the machine resulted
from a design to increase profits, not some mysterious, disembodied
force. Indeed, it was this design that constituted the "soul of the machine" to which Marx referred. The culmination of this design was
reached in the development of the factory, in which machinery was
"organized as a system," and modern industry "stood on its own feet"
as these factories began to "construct machines by machines."[37]

Marx expressed the ontological impact of this technology in a variety of ways, most of which centred on the difference between freely
self-determined labour and labour that is an expendable part of a process over which the labourer has little control. In the mechanized industrial factory, human labour itself becomes an instrument, to a greater
extent even than it had been in the manufacturing system. In the automated factory, the workman is "dismembered,"[38] converted "into a living appendage of the machine,"[39] and, ultimately, "enslaved by the
machine."[40] Thus, according to Marx, technology is implicated in human nature to the extent that it makes a particular contribution to the
general alienation experienced by the majority of people labouring in
the context of industrial capitalism.[41] In this respect, machine technology completes capitalism's separation of working persons from their
ability to determine freely the manner, conditions, and outcomes of
their own productive endeavours – endeavours that remain, nevertheless, expressive of what those working persons are. As Marx puts it,
"the special skill of each insignificant factory operative vanishes as an
infinitesimal quantity before the science, the gigantic physical forces,

and the mass of labour that are embodied in the factory mechanism and, together with that mechanism, constitute the power of the 'master.'"[42] Technology, then, is an instrument that not only facilitates the accomplishment of capitalist exploitation and alienation, but also gathers and embodies them.

Nonetheless, Marx is careful to point out the difference "between machinery and its employment by capital," and "the material instruments of production [and] the mode in which they are used."[43] What Marx is suggesting here is that, while alienation and exploitation necessarily attend the capitalist mode of production, they are not the only conceivable result of the technological development of productive forces. "Machinery, considered alone, shortens the hours of labour," says Marx, "but, *when in the service of capital,* lengthens them."[44] There is, then, a certain contingency at play in the impact of technology, the determination of which appears to rely more on the conditions and context of its deployment than on qualities inherent in technology itself. Keeping in mind that productive forces, especially technology, can *never* be considered independently from their embeddedness in the mode of production, Marx nevertheless appears willing to concede that technology can be dispatched in the service of humanity, as readily as it has been used for its enslavement.

Indeed, Marx goes even further and insists that technological development can play a crucial role in emancipating humanity from the alienated social forms of the capitalist mode of production and can contribute to the reconciliation of human beings with their essential nature as freely self-determining producers. In the first place, it was modern industrial technology that "rent the veil that concealed from men their own social process of production."[45] Factory mechanization exposed the industrial labourer for what he or she was – a cog in a vast machine – and revealed the immense productive potential of modern technology to be in stark contrast with the miserable conditions of his or her social existence. It was precisely this contradiction – between the forces and relations of production – that Marx predicted would ultimately undo the capitalist mode of production.[46] Of course, evidence of this contradiction has yet to unleash the revolutionary energy Marx supposed it would, but the point here is that he acknowledged that the role of technology extended beyond exploitation and domination. Minimally, it symbolized the promise of a possible future that looked more desirable than the prevailing condition of most people, and that could be achieved if only they could shed the restrictive economic relationships preventing its realization.

More ambitiously, Marx thought productive technology represented more than a *symbol* – he believed it could also contribute to the *delivery* of a more equitable society. Marx was typically modern, in so far as he firmly believed technology, as the productive form of science, could be enlisted in the relief of humanity's estate.[47] It could do so by fabricating the abundance that, when subjected to egalitarian distribution, constituted the "material premise of communism." According to Marx, the "development of productive forces is an absolutely necessary practical premise, because without it privation, *want* is merely made general,"[48] and he counselled the revolutionary proletariat to "increase the total of productive forces as rapidly as possible."[49] To the extent that technology is numbered by Marx as one of an array of productive forces, it is clearly envisioned by him as among the indispensable material conditions required for progress toward the good society.

Martin Heidegger, Technology, and Being

Martin Heidegger's thoughts about technology were formed in the wake of the Second World War, in the shadow of Nazism and the atomic bomb, so it is not surprising that they pursue the question of the relationship between the essence of technology, and the fundamental essence and existence of human beings as such. However, Heidegger's approach was as much the result of the placement of this question within the trajectory of his own philosophical project as it was an offspring of these particular historical moments. Heidegger's core observation was that the true meaning of a phenomenon was discernible only via its situation in the (correctly understood) relationship between human existence and human essence, between being and Being (*Dasein* and *Sein*), between the ontic and the ontological. "Ontic" derives from the Greek root *on,* or *ont,* for "be." It denotes "of being" and refers to the day-to-day business of *Dasein,* German for "there-being" or "existing in the world." "Ontological" combines *ontic* and *logos,* and so refers to the *gathering* of that business of being into a reasoned, deliberate, expressed account of itself. It is this account Heidegger refers to as

Sein, or "Being." In his view, a study of human practices performed in isolation, without attention to their relationship to the essential Being of the actors, could yield little true or important knowledge about either those practices or those beings. To be meaningful, according to Heidegger, consideration of human activity on the simply ontic level had to be informed by a deeper ontological consideration of that activity's connection to the nature of Being itself – the acts of beings could not be understood completely unless they were seen as forming an account of what it is "to Be."

However, this concern with ontology did not lead Heidegger to metaphysics, the realm where questions regarding human essence had customarily been discussed. Instead, Heidegger maintained that, just as being was significant only to the extent that it was connected to Being, the essence of Being was similarly grounded in the various practices of being. As Heidegger put it, *"fundamental ontology,* from which alone all other ontologies can originate, must be sought in the existential analysis of *Dasein* ... The first priority is an *ontic* one: this being is defined in its Being by existence."[50] Thus, Heidegger's intent was not to jettison the physicality of being human in favour of speculative musing on the metaphysical nature of human Being. Instead, he simply insisted on maintaining a correct view of the relationship between the two: practices of being are significant only to the extent that the question of Being is at issue in them; and a genuine ontology – a sensible account of the essence of Being – must begin with an investigation of the ontic practices of beings.

Heidegger's early work concentrates on establishing the status of the ontic, and on delineating the concrete structures and manifestations of the practices of being.[51] Subsequently, Heidegger turned to the question of the essence of Being, and it is in this period that the essence of technology also became a focus of his concern. In his essay "The Question Concerning Technology," Heidegger begins by affirming that modern technology is rightly considered instrumental, but points out that commonplace assumptions about the nature of instrumentality fail to express the truth about the *essence* of technology itself.[52] Chief among these misleading assumptions is the belief that technology, as a tool or instrument, is neutral – an unfortunate but prevalent misconception that "makes us utterly blind to the essence of technology."[53] Later, Heidegger would put it more bluntly: "modern technology is no tool and it no longer has anything to do with tools."[54]

To overcome mistaken assumptions about the neutrality of techno-logical instruments, Heidegger proposes we re-examine the precise manner by which means are linked to ends in the chain of causality, and he suggests we begin this re-examination with Aristotle. Heidegger presents Aristotle as positing four simultaneous causes of any given thing: the material of which the object is made (*causa materialis*); the form of which the object is an imitation (*causa formalis*); the *telos*, or final purpose, of the object (*causa finalis*); and the agent that directly initiates the effect (*causa efficiens*). So, for example, if we consider a piece of two-inch-by-four-inch lumber, we might say that its *causa materialis* is wood, its *causa formalis* is the standardized formal dimen-sions of a two-by-four, its *causa finalis* is its destiny as a wall stud, and its *causa efficiens* is the sawmill. It is in this fourth cause that Heidegger wishes to deviate from Aristotle's account of causality, because it does not come close to suggesting the breadth or depth of what the sawmill, as a technology, causes. As a tool, the sawmill is the *causa efficiens* of lumber, but, *as a technology*, it is much more. To understand thoroughly that which is caused by technology, Heidegger proposes a fourth cause, corresponding to the deliberate consideration (*logos*) and gathering to-gether (*legein*) of the material, formal, and teleological aspects of cau-sality. Heidegger calls this new, fourth cause *apophainesthai*, which means "to bring forward into appearance."[55] This revision accomplishes two things. First, it implicates human agency in technological effects: only a millwright can consider, deliberate, and gather; the sawmill can only embody or represent the millwright's consideration, deliberation. and gathering. Second, it increases the breadth of that which a tech-nology might cause. More than simply producing a two-by-four, the technology of the sawmill is an "occasioning," a "presencing of some-thing that is present," an "unconcealment," a "bringing-forth" (*poiēsis*), a "revealing" (*alētheia*). The essence of technology, then, according to Heidegger, is to be found in what it reveals: "If we enquire step by step into what technology, represented as means, actually is, then we shall arrive at revealing. If we give heed to this, then another whole realm for the essence of technology will open itself up to us. It is the realm of revealing, i.e., of truth."[56] As a technology, the sawmill reveals the truth about more than just two-by-fours. It reveals the truth about how human beings produce things, and about the relationships they enter into with each other and with Nature in order to do so. Keeping in mind the broader canvas of Heidegger's ontological project, we might say that, to him, the essence of the sawmill reveals the truth about where we stand, or fail to stand, in terms of Being.

Thus, the essence of *modern* technology is to be located in its particular mode of revealing, and Heidegger argues that this mode of revealing is best designated by the word *Gestell,* or "enframing."[57] Modern technology does not reveal through *poiēsis* – it is not a "bringing-forth" of that which is inherent in Nature – but rather by enframing, or delineating the manner in which Nature is to be approached, considered, and gathered. The enframing accomplished by modern technology puts it at odds with the proper essence of Being in two key respects. The first has to do with the way it enframes the relationship between humans and Nature. For Heidegger, the fundamental aspect of Being and its works is their *autochthony,* or rootedness, in the steady and fertile soil of nativity.[58] As he put it, "human experience and history teach us, so far as I know, that everything essential, everything great arises from man's rootedness in his homeland and tradition."[59] Modern technology denies this essential characteristic of Being in so far as its mode of revealing is not a *poiēsis,* or "bringing-forth," of what is rooted in Nature. Instead, it is a "challenging of" or "setting upon" Nature: "The revealing that rules in modern technology is a challenging which puts to nature the unreasonable demand that it supply energy which can be extracted and stored as such."[60] Man's relationship with Nature under the auspices of modern technology ceases to be one of rootedness, and becomes one where Nature is challenged to be a "standing-reserve" (*Bestand*).[61] Heidegger expresses the implications of the technological enframing of Nature-as-standing-reserve quite starkly: "From this arises a completely new relation of man to the world and his place in it. The world now appears as an object open to the attacks of calculative thought, attacks that nothing is believed able any longer to resist. Nature becomes a gigantic gasoline station, an energy source for modern technology and industry."[62]

This privileging of "calculative thought" represents the second dissonant characteristic of modern technology's enframing. Heidegger asserts that, essentially, "man is a *thinking*, that is, a *meditating* being."[63] He differentiates between calculation as an instrumental mode of thinking oriented to effectiveness, and meditation as a contemplative mode of thinking oriented to an appreciation of the authentic essence of Being. The enframing of modern technology subsumes the latter into the former, and neglects the meditative essence of Being.[64] For Heidegger, this represents "*the* danger" modern technology poses for the essence of Being: "The threat to man does not come in the first instance from the potentially lethal machines and apparatus of technology. The actual threat has already afflicted man in his essence. The rule of enframing threatens man with the possibility that it could be denied to him to enter into a more original revealing and hence to experience the call of a more primal truth."[65] Elsewhere, Heidegger, phrases the matter more colloquially when he warns that modern technology could "so captivate, bewitch, dazzle, and beguile man that calculative thinking may someday come to be accepted and practiced *as the only* way of thinking."[66] Just as technology "rips and uproots man from the earth," so, too, does it thrust him into "purely technical relations" that are foreign to his essence.[67] Consequently, in Heidegger's view, the question concerning modern technology is really a question about whether Being can survive technology's assault and, if so, how. Not surprisingly, the stakes he places in the answer are high. "The issue," observes Heidegger, "is the saving of man's essential nature."[68]

Some of Heidegger's views on the possibility of bringing modern technology under human control are alarming, to say the least. He observes that "the essential thing about technology is that man does not control it by himself ... we've found no path corresponding to the essence of technology," and he doubts whether any form of human political system – and he is especially doubtful about democracy – is capable of bridling its juggernaut-like advance.[69] At one point, Heidegger offers the following conclusion: "Technological advance will move faster and faster and can never be stopped. In all areas of his existence, man will be encircled ever more tightly by the forces of technology. These forces, which everywhere and every minute claim, enchain, drag along, press and impose upon man under the form of some technical contrivance or other – these forces, since man has not made them, have moved long since beyond his will and have outgrown

his capacity for decision."⁷⁰ It is this reservation concerning the likelihood of calculating beings making a meditative intervention into the
advance of modern technology that perhaps led Heidegger to his now-
famous observation that "only a god can save us now."⁷¹ Nevertheless,
it would be a mistake to reduce Heidegger's thoughts about the future
of technology to those of a romantic contrarian or despairing pessi-
mist. Despite his concern over the damming of the Rhine, Heidegger
recognizes that "we depend on technical devices, they even challenge
us to ever greater advances."⁷² Despite his description of the autonomous character of technological advance, Heidegger asks who *accomplishes* its mode of revealing, and answers: "Obviously, man."⁷³ And,
despite his identification of the deterministic, constraining aspect of
modern technology, Heidegger affirms that, "still, we can act otherwise."⁷⁴

How are we to make sense of these apparently contradictory positions? What is the source of Heidegger's faith in the face of technological peril? Ironically, its source is the essence of technology itself.
Heidegger adopts a line from the poet Friedrich Hölderlin – "But where
danger is / grows the saving power also" – and makes it his own: "precisely the essence of technology must harbour in itself the growth of
the saving power."⁷⁵ To discover the saving power of technology, we
must set aside despair over its material effects and concentrate upon
its essence. Heidegger suggests that we "look with yet clearer eyes into
the danger" so that we might discover "where the extreme danger lies
– in the holding sway of enframing."⁷⁶ The essence of technology –
and its greatest danger – is not located in this or that particular instance of pollution or disemployment or privacy invasion, but, rather,
in its propensity to enframe the condition of Being, mistakenly, as one
of calculation and rootlessness. Consequently, confronting technology
in terms of the inefficacy of its specific material outcomes – that is, as
a "problem" to be "solved" – serves simply to reproduce the technological understanding of being that makes these effects possible, and
perhaps even inevitable. As Heidegger puts it, "all attempts to reckon
existing reality ... in terms of decline and loss, in terms of fate, catastro-
phe, and destruction, are merely technological behaviour."⁷⁷ Instead,
Heidegger recommends technology be regarded, as one writer has
put it, as "an *ontological* condition from which we can be saved."⁷⁸ We
can be saved from this condition because the essence of modern

technology as an enframing can, with effort, be identified by meditative beings – so long as they concentrate on "catching sight of what comes to presence in technology, instead of merely gaping at the technological"[79] – and think of alternative ways of becoming that are more consonant with the essence of Being. Heidegger is quite categorical with regard to this possibility: "I don't see the position of man in the world of global technology as inextricable or inescapable. The task of thought is to help limit the dominance of technology so that man in general has an adequate relationship to its essence."[80]

Once the enframing essence of modern technology is fully understood, it can be confronted with an account of the relationship between human beings and technology that is more appropriate to the essence of Being. The comportment toward technology that is proper to Being is characterized by Heidegger as a "releasement toward things."[81] It should be noted that the translation here of the German word *Gelassenheit* as "releasement" does not capture Heidegger's meaning adequately. In modern German speech, *Gelassenheit* refers to "composure," "ease," or "calmness," but in older usage it connoted a relinquishing of the material world in favour of offering oneself up to God.[82] "Releasement" means letting go. Thus, Heidegger is suggesting that a correct relationship to technology can be established only in the context of piety. In the midst of our pride about technical mastery, we must be gratefully aware of the limits to human control. Being cannot be mastered – our thinking (*Denken*) of being is a thanking (*Danken*) of Being – and in thanking Being we let beings be. As Heidegger describes it, a releasement toward things is something like a detachment, or critical distance, that prevents an excessive intimacy between human beings and modern technology's enframing essence from taking the place of an intimacy between those beings and the essence of Being. In short, a releasement toward things saves us from viewing the world in a purely technical manner: "We can use technical devices, and yet with proper use also keep ourselves free of them, that we might let go of them at any time. We can use technical devices as they ought to be used, and also let them alone as something which does not affect our inner and real core. We can affirm the unavoidable use of technical devices, and also deny them the right to dominate us, and so to warp, confuse, and lay waste our nature."[83] Pious detachment entails subjecting technical considerations to considerations of Being. For example, if our relationship with technology were characterized by a releasement toward things, we would be unlikely (and perhaps unable) to view the Earth, in which our essential Being is rooted, as standing-reserve.

The second aspect of the proper relationship between human beings and technology is described by Heidegger as an "openness to the mystery."[84] According to Heidegger, "the meaning pervading technology hides itself."[85] Despite the fact that technology – modern or otherwise – touches every aspect of human existence, its complete meaning is never immediately evident; that is, the meaning of technology presents itself primarily as a mystery, the solving of which requires considerable thought. To deny the mystery of technology is to assert the facile mastery that characterizes the destructive mode of enframing peculiar to modern technology. Once people presume to have technology "figured out," they believe themselves free to calculate its deployment as best suits their material preferences, all the while forgetting to contemplate the impact that technology and those calculations might have on their essential Being. Here, by cautioning human beings to remain open to the mystery of technology, Heidegger is asking them to discard pride and, instead, to invoke the humility proper to their essence. He is asking them to recognize that technology's impact on Being cannot be discovered via the calculation of outcomes – which just serves to obscure and escalate this impact – but, rather, only through the appreciation and contemplation of mystery that is appropriate to a meditative being.

Thus, according to Heidegger, "releasement towards things and openness to the mystery belong together. They grant us the possibility of dwelling in the world in a totally different way. They promise us a new ground and foundation upon which we can stand and endure in the world of technology without being imperiled by it."[86] They do so because they create a relationship between beings and technology wherein the latter ceases to be an enframing that, as a setting-upon and challenging-forth of the Earth as standing-reserve, contradicts the rooted and meditative essence of Being. Instead, in the context of a releasement toward things and an openness to the mystery, technology reverts to a mode of becoming that is closer to *poiēsis*. This is a mode of becoming, characteristic of ancient *technē*, which Heidegger describes as "the bringing-forth of the true into the beautiful," and which "illuminated the presence of the gods and the dialogue of divine and human destinings ... It was pious, *promos*, i.e., yielding to the holding sway and safe-keeping of the truth."[87] It is for this reason that Heidegger insisted that only a god can save us from the torrent of modern technology: because only in the presence of a god – an appreciable representation of that which is great and commanding – could human beings be inclined to approach technology with piety and humility, and use it to live well in accord with their meditative and rooted essence, rather than to master and pillage the earth in defiance or ignorance of it. The presence of a god, called forth by the danger inherent in the essence of modern technology, can save us from that danger by thrusting us into a relationship with technology that conforms to the essence of Being.

George Grant on Technology and Modernity

George Grant's deep thinking about technology was shaped by a whirl of influences, and emerges in the context of a comprehensive, trenchant, and politicized critique of the philosophical foundations of modernity. Grant was, variously: committed to the truth of the Gospels and the wisdom of Plato's philosophy; impressed profoundly by Friedrich Nietzsche's laying bare of the emptiness at the heart of modernity; supportive of Leo Strauss's call for a return to the virtues of the ancient tradition; moved deeply by Simone Weil's philosophy and Christian spirituality; a classical conservative; a socialist; and a committed Canadian nationalist. He was also a great admirer of Martin Heidegger, whose thoughts on technology infused Grant's own, although primarily through their somewhat simplified adaptation in the work of Jacques Ellul.[88] These influences were distilled by Grant to produce a potent indictment of the modern technological complex: modern technique colonizes the realms of philosophy, citizenship, and intimacy, and corrodes the virtues of meditative thought, justice, and love residing there. It does so by reducing the imperative to observe and practise these virtues to a mere calculation of comparative advantage.

Grant's critique of the modern disposition is that it undermines our ability to "make true judgments about right action."[89] Modern civilization is more concerned about a plurality of judgments than it is about true judgments, and more interested in free action than in right action; to maintain true judgments about right action, from the modern standpoint, is to deny the historicity of human existence, and to court tyranny. This orientation is manifested in a series of substitutions: that which is valuable or efficient replaces that which is intrinsically good as the primary criterion of worthiness; calculation replaces deliberation about ends, informed by contemplation, as the essence of judgment; and changing what is given to suit one's independently formed purposes replaces the pursuit of harmony with that which endures as the motivation of right action.

This concern with human freedom over human virtue manifests itself most strongly in orientations to Nature and history that Grant identifies as foundational to modern civilization. Free modern individuals regard Nature as "the simply dominated" rather than as "the simply contemplated," and attempt to impose artificial order on what they see as its chaotic contingencies.[90] Modernity's orientation to history has followed a similar path. According to Grant, ancient peoples "simply refused history" by recognizing that virtue was timeless and that eternity was external to history, not an infinite forward projection into it.[91] Judaism and Christianity introduced historical awareness but held that history, conceived of as time-on-Earth, was relatively insignificant and beyond human control: history was a mere ante-room to eternity; human awareness was directed to the prospect of future salvation and peace in the hereafter; and human freedom was perceived as bounded by an unchanging moral order.[92] Secular moderns have seized upon the future orientation of Judaeo-Christian faith, but have dropped the obedience to natural and divine law that accompanied it. Instead, they have elaborated it into a discourse about free, autonomous beings making their own history, and constructing their own moralities, pursuant, not to a beatific and heavenly terminus, but as part of a progress that is ever unfinished.

It is this humanist orientation to Nature and history – an orientation under which "the idea of progress crushes the idea of providence" – that vaults technology to the centre of the human condition.[93] Grant articulated his definition of technology most clearly in the latter stages of his thinking, and it is a definition that bears the distinct marks of

Heidegger's thought on the same subject. Grant thought that the English neologism *technology* perfectly captured the "unity of knowing and making which has determined the modern world."[94] As he put it, "modern technology is not simply an extension of human making through the power of a perfected science, but is a new account of what it is to know and make in which both activities are changed by their co-penetration ... knowing has itself become a kind of making."[95] Grant goes further when he suggests that "knowing has been put in the *service* of making."[96] Here, Grant is arguing that the definitive aspect of modern technology is its encouragement of the belief that knowing the world that exists *despite* man's intervention is less valuable than the facts we create by changing the material world to suit our purposes. Technology is not only the instrument of change, it is simultaneously the *epistēmē* in which the inestimable value of making progress becomes all that we need to know.

Thus, for Grant, technology is not "just a bunch of machines." Instead, it is "a whole way of looking at the world, the basic way western men experience their own existence in the world."[97] Armed with technology of their own making, humanity stands before Creation as an inquisitor, and Nature becomes the *object* of judgment rather than its standard. The modern technological paradigm is based on "the summonsing of something before us and the putting of questions to it, so that it is forced to give its reasons for being the way it is as an object."[98] If we are not satisfied with these reasons – and we rarely are, because, for essentially progressive beings, satisfaction is untenable – technology stands ready to be enlisted in the cause of changing the object to suit our preferences at any given moment. Given its centrality in the operation of the modern spirit, it is little wonder that technology has become a force bearing the marks of autonomy: "It moulds us in what we are ... Its pursuit has become our dominant activity and that dominance fashions both the public and the private realms."[99] To masterful beings who believe their fulfilment lies in making the future as they envision it, or who see progress in altering that which Nature has given, the instruments of that making and alteration take on a special status. They become more than instruments. These instruments – these technologies – become ends in themselves, capable of directing, conditioning, and attracting attention.

According to Grant, liberalism is the political corollary of technology. He describes a technological society as one "in which people think of the world around them as mere indifferent stuff which they are absolutely free to control any way they want through technology,"[100] and defines liberalism as "a set of beliefs which proceed from the central assumption that man's essence is his freedom and therefore that what chiefly concerns man in this life is to shape the world as we want it."[101] Key to the coincidence of technology and liberalism is that both deny the existence of limits that might constrain human action, save the minimal and changeable ones created by people themselves. Both are founded on the belief that liberty is achieved by overcoming or defying necessity, not by living within it. Grant believes that the modern obsession with liberty, so conceived, undermines the capacity for genuine goodness in liberal-technological political communities: "As liberals become more and more aware of the implications of their own doctrine, they recognize that no appeal to human good, now or in the future, must be allowed to limit their freedom to make the world as they choose. Social order is a man-made convenience, and its only purpose is to increase freedom. What matters is that men shall be able to do what they want, when they want ... In other words, man in his freedom creates the valuable. The human good is what we choose for our good."[102] Grant reveals that liberal politics are thus the perfect, and perhaps the only, politics for a technological world. This symbiotic relationship between liberal politics and technology underscores the reality that liberalism is not, as many of its contemporary exponents would claim, a purely procedural constitutional order devoid of substantive preferences and content.[103] Liberalism is a politics of getting-out-of-the-way of technological mastery and the material progress it always promises and sometimes delivers. As such, it is a politics that privileges one conception of what is good – a certain kind of liberty, material progress, and unfettered development of technology – at the expense of a host of potential others. To the extent that technological liberalism purports to be free of a specification of the good, it is a politics in denial.

If this is the first irony of technological liberalism pointed out by Grant, its propensity to generate universal homogeneity under the guise of particularity and pluralism is the second.[104] The concoction of technology, liberalism, and capitalism conjures up an image of an open pluralism in the realm of private tastes, but this pluralism operates only within the horizon of a more general homogeneity: "As for pluralism, differences in the technological state are able to exist only in private activities: how we eat; how we mate; how we practice ceremonies. Some like pizza, some like steaks; some like girls, some like boys; some like synagogue, some like the mass. But we all do it in churches, motels, and restaurants indistinguishable from the Atlantic to the Pacific."[105] This is the case because, despite its private pluralism of tastes, liberal, capitalist, technological society imposes a specifically defined set of legitimate public purposes upon its members. In a liberal society, the only legitimate public purposes are those in which human beings are presented as – if not in fact, then at least in fancy – free and equal makers of their own way.[106] In a capitalist society, the public good is equated with the economically rational, which, in any given instance, is defined by either individual accumulation or corporate efficiency.[107] In a technological society, legitimate public purposes are those that are amenable to technological solutions.[108] In each case, specific goals and standards are valorized, to the exclusion of all others.

For Grant, these respective horizons are complementary to the point of ultimately fusing into a single one: "the universal and homogeneous state – the society in which all men are free and equal and increasingly able to realize their concrete individuality."[109] The problem is that the appeal of the universal homogeneous state is founded on a lie. According to Grant, "pluralism has not been the result in those societies where modern liberalism has prevailed. Western men live in a society the public realm of which is dominated by a monolithic certainty about excellence – namely that the pursuit of technological efficiency is the chief purpose for which the community exists."[110] Modern freedom is the freedom to live one way. Grant has a number of reservations about this modern technological consensus. High among them is his concern that, by excluding all alternative narratives of human excellence, technological modernity deprives us of two accounts in particular – one given by philosophy, the second in revealed religion – that are capable of elevating us beyond the ephemerality and banality of modern existence.[111] In the place of real pluralism, single-minded devotion to progress offers only a "monism of meaninglessness."[112]

In more immediately political terms, however, Grant fears the tight but encompassing circle of technological homogeneity leaves little room for the preservation of localized folkways that run contrary to the imperatives of the technological complex. According to Grant, we live in the midst of "a modernity which at its very heart is destructive of indigenous traditions."[113] Not only are we all equal under the gaze of technological mastery but we are all the same. In this context, ideologies that protect distinction, such as conservatism and nationalism, are nonsensical. Grant saw the consequences of this high-tech melting-pot as especially dire for his home country, an experiment in conservative multi-nationhood, having the misfortune to be proximate to the liberal and technological dynamo that is America. Long before free-trade agreements and the rhetoric of globalization, and long before Canadians would lend a high-tech hand to the American conquest of the cosmos, George Grant lamented that "the aspirations of the age of progress have made Canada redundant. The universal and homogeneous state is the pinnacle of our striving ... our culture floundered on the aspirations of the age of progress."[114] It is not that nationalism is a good in itself – Grant understands that loving the good sometimes makes it difficult to love our own unambiguously – or that human distinctions entail corresponding gradations in basic regard for those who exhibit them.[115] It is simply that, faced with the homogenizing progress of modern liberalism and technology, the maintenance of national or local distinction keeps alive at least the *possibility* of virtue, because one or another of these distinctions *might* harbour those alternative accounts of human excellence excluded by a universally technological society. It is in them that we might encounter spaces where the good can still be loved and defended successfully against the attacks of the instrumentally rational and efficient.

Does Grant's devastating critique of technological society entail a rejection of all technologies? Not necessarily. Throughout his writing, Grant was careful to affirm the goodness of technologies that alleviated human suffering and contributed to well-distributed and genuine leisure.[116] This would seem to indicate that Grant thought technologies *could* be good if they were developed and applied pursuant to good ends. Of course, the goodness of an end can be determined only by using standards existing outside simple technological rationality, standards discoverable by contemplation and revealed in religious devotion. So long as we love *these* standards of goodness more than we value technological progress in itself, the use of technological instruments can be limited, and need not continue down the path modernity has taken thus far. The difficulty lies in recovering this love of the good in an instrumentally rational (anti-philosophic) and secular (irreligious) world that fetishizes the empty pluralism of technological liberalism. More likely is the continued pursuit of technological progress for its own sake, which Grant identifies with the sheer will-to-will depicted by Nietzsche as the desperate nihilism of the modern age.[117] If we can find no good reasons for using (or not using) technologies, other than that technology is *necessarily* good, then we will continue to find ourselves summoned before our machines to give our reasons whenever we ponder the folly of turning them off. And, in this event, the reasons we give are unlikely ever to be convincing.

The Politics of Technology

In many respects, the philosophical positions on technology articulated in the writing of the thinkers discussed in this chapter swirl around and feed off each other. In the foregoing discussion, I have not attempted to trace explicitly these lines of influence, concord, and disagreement. The point here was not to provide a comprehensive genealogy of the philosophy of technology, or to elaborate a distinct "theory" of technology, but, instead, to gather resources for an interrogation of the hope for democracy vested in network technology. Some of the thinkers discussed above – Plato and Heidegger, for example – have fundamental philosophical differences, but each makes a particular contribution to substantiating and developing the hypothesis guiding this investigation: in so far as it combines productive activity with the gathering of significance, technology – especially information and communications technology – says something about what human beings are; what they wish to be; and how they live, or *might* live, together.

This hypothesis derives from three premises supported in the range of thought canvassed here. The first premise is that technology and politics are intimately linked. As Aristotle observed, the instruments of our artful activity are properly subjects of political judgment because they are always oriented to ends that either contribute to, or detract from, a good civic life. Because technology combines *technē* and *logos*, its political impact is not confined to the material world, and to conceive of these consequences as if they were so limited is to ignore the fundamental ontological implications of technologies that weave a particular range of political possibilities into the essential fabric of our humanity. Marx, Heidegger, and Grant understood that technology uses us as we use technology; that technology is not just the motive force changing our external world, but also constitutes our inner world, our mode of thinking about, and caring for, things. Technology affects what we *are*, not just what we *do*. Thus, before discerning, for example, the impact of tele-voting on new social movements or political parties, we must inquire not only about this new technique of voting as an instrument, but also about what this technology makes and says of us as political beings.

At the same time, technology also operates on the field of power, and it is this operation that suggests an encounter with distinctly *political* deliberation and judgment. As Langdon Winner has put it, "the question is ... one about politics and political philosophy rather than a question for ethics considered solely as a matter of right and wrong in individual conduct. For the central issues here concern how the members of society manage their common affairs and seek the common good. Because technological things so often become central features in widely shared arrangements and conditions of life in contemporary society, there is an urgent need to think about them in a political light."[118] Aristotle stipulated that the practical arts of *technē* should necessarily be applied under the direction of the master science of *politikē*. Compounding *technē* with *logos* only enhances the wisdom of this simple teaching. Thus, the character, causes, and outcomes of the "widely shared arrangements and conditions of life" attending particular technologies are political matters requiring judgment in political terms. Failure to undertake this judgment constitutes a surrender to somnambulism, and a resignation to a collective life of trying – but never succeeding – to catch up to our inventions.

The second premise is that the political outcomes of technological adventures are strongly conditioned by the economic, epistemological, and political environments in which they are situated. The mechanized industrial factory was not neutral: it was, as Marx showed, an alienating, de-humanizing, exploitation machine because it functioned to advance and materialize the parochial interests of the capitalists who implemented its design. Not every outcome of a technological application or development is a direct and unambiguous function of its director's intent. Nevertheless, technologies tend to reproduce and reinforce the conditions from which they emerge. For Marx, technologies developed in the midst of industrial capitalism functioned to advance productivity and profits, not to advance safety, health, or the spiritual rewards of work. Grant extended this thinking. To him, technology stood at the convergence of modern progress, liberal politics, and capitalist economics as both trophy and template: a mark of accomplishment and a set of fixed parameters for further advance. He argued that it could hardly have been otherwise so long as these epistemological and material conditions prevailed. The lesson here is that, keeping in mind that technological impact is not wholly determined by intent, sound investigation of the political potential of any new technology requires sober attention to the imperatives and constraints prevalent in the political, economic, and epistemological configuration in which that technology is lodged. In this context we might ask, for example, whether network technologies can be democratic if they are developed and embedded within a fundamentally undemocratic environment.

The third premise emerging from the above survey is that technology conditions political outcomes and possibilities. Regardless of who or what directs it, technology makes certain things possible and other things impossible. This is true on the level of how we perceive the world and our place in it, as well as on the level of possible action. As Heidegger revealed so forcefully, irrespective of the characteristics of this or that instrument, technology *enframes*. For Marx, technology was part of the ensemble of material conditions that produced distinct political formations. Grant agreed, suggesting that technology makes more than just automobiles and missiles and computers; it also *makes us* into a particular sort of political beings.

Marshall McLuhan's famous aphorism applies here: "the medium is the message." Because the medium is *technē* and *logos*, technology crafts and it gathers. Technology both unifies what exists, and constructs new relationships and identities; in every crafting or gathering, there are options chosen and discarded, possibilities included and excluded. Thus, all technological innovation is political. The consideration of technology offered here enables us to ask the following general question: in making and gathering politics through network technologies, which identities, relationships, strategies, power, and politics do we choose or discard? It also suggests a range of more specific questions: What does this technology give to economic practice, and how does economic practice condition its deployment? What is the essence of this technology, and what is at issue in it for the essence of Being? Is this technology subject to political deliberation as to the ends it serves, or is it an end in itself? Does this technology challenge or represent the modern priorities of unlimited freedom, material accumulation, and progress? Finally, is this technology what it appears to be, or does it just stand in for other, more genuine practices and goods? Each of these questions requires an answer before any determination regarding the validity of the hope for democracy in the age of network technology can be made.

networks

· · · · ·

Two out of every three human beings alive today have never made a telephone call, and for every two telephone lines in all of sub-Saharan Africa, there are three on the island of Manhattan alone.[1] Evidently, the label "Wired World" denotes a certain way of defining "the World" better than it explains where the wires actually are, or who uses them. It is important to keep this in mind when discussing network technology. Electronic communications and information networks are not exclusively a technology of the affluent world but, with 90 percent of Internet hosts located in North America and western Europe, they are still predominantly so.[2] Despite this, the discourse surrounding network technology has not been exempt from the tendency of the affluent to regard the world as made in their image and, if not possessing their mode of life, then at least striving toward it. Indeed, it is the very nature of a "net" to encompass and to capture – if it is big enough and fine enough it will encompass and capture everything we need, and therefore everything worthy of our attention. This is an intrinsic consequence of the assignment of *logos* to *technē*, one that is revealed starkly by the name and structure of this particular technology: that which "the Net" does not or cannot gather is extraneous and insignificant. Predictably, the promotion of this technology has been fuelled by exaggerated assumptions about its penetration and reach, and a casual disregard of the material reality of life in many areas of the world. We seem to need constant reminding that the world gathered by network technology is one in which computers and telephones are more a part of everyday life than are hunger and poverty.[3]

· · ·

With this in mind, the present chapter outlines the emergence and basic technical attributes of networked information and communications technologies, by describing their development as instruments, and surveying a range of their relevant applications. This examination of how networks have been crafted, and what it is they make possible or do in technical terms – their qualities as potential *technē* – will prepare the ground for subsequent consideration of their *logos*, or what they gather and say about their users as political beings.

The Incredible Shrinking Computer

Designating the "world's first" electronic digital computer is a difficult task.[4] In 1943 "Colossus," an electronic deciphering machine designed in Britain and based on the theories of mathematician Alan Turing, was put to work breaking Nazi codes.[5] Colossus was simply a machine though, and not a computer, because it did not utilize a stored-program to direct its operations.[6] The Electronic Numerical Integrator and Calculator (ENIAC) was built in a former musical-instrument factory on the campus of the University of Pennsylvania in 1945 by engineer J. Presper Eckert and physicist John Mauchly. It was over two and a half metres tall and twenty-four metres long, weighed twenty-seven tons, employed nearly 18,000 vacuum tubes and 6,000 manual switches, and required so much electricity that, when it was switched on for the first time, the lights of Philadelphia were said to have dimmed noticeably.[7] Its first job was to speed up the calculations (it could perform 5,000 per second) involved in assessing the feasibility of proposed designs for a hydrogen bomb at Los Alamos. The ENIAC machine featured many of the abstract attributes of a computer, including high-speed calculation, programmability, and generality of application, but it lacked the three definitive elements of the modern computer: binary math; a central processing unit; and a general-purpose memory. It used decimal representation and so, in terms of mathematical language, was simply an electronic analogue to mechanical calculators. Because it utilized different processing units for different operations, and separate peripheral "accumulators" to store its instructions and results, programming ENIAC was "a one-way ticket to the madhouse" that involved two days of manually setting and connecting thousands of switches and cables.[8] A computer's programming can be changed and its memory accessed without any substantial physical reconfiguration of its parts. ENIAC was essentially a huge, fast, cumbersome calculator that could perform operations according to programmed instructions, but its "set-up" had to be changed every time those instructions were altered.

In 1944, John von Neumann – a Hungarian-born mathematician who was a central figure in the Manhattan Project at Los Alamos – joined the Eckert-Mauchly Computer Corporation's efforts to improve on ENIAC. It was von Neumann who contributed the ideas of a central processing unit, internal memory, and binary math to what had already been developed in the field of stored programming.[9] Soon, computers bearing these characteristics began to appear. In Britain, a prototype known as the Manchester Mark I ran a factor search on 21 June 1948 and became the world's first operating stored-program computer. Subsequently, the Mark I would be produced and sold commercially by a company called Ferranti – its first customer was the University of Toronto, hoping to enlist the machine in its efforts to design the St. Lawrence Seaway.[10]

Soon thereafter, John von Neumann produced a stored-program parallel processor for the Institute for Advanced Study that became "the paradigm of modern computer design" and spawned many imitators.[11] For their part, Eckert and Mauchly built the Universal Automatic Computer (UNIVAC), a general-purpose, high-speed, big-memory alphanumeric computer designed for the commercial market in the United States. Its first public job was a promotional stunt: it was programmed to predict the outcome of the 1952 presidential election in the United States for CBS television. Pollsters and pundits had predicted a close race, but, with only 7 percent of the returns in, the computer predicted an Eisenhower landslide, giving him 438 electoral college votes to Adlai Stevenson's 93. Fearing the public-relations disaster that would attend a gross mistake, the broadcaster agreed to withhold the initial prediction while the computer was hastily reprogrammed to produce a more conservative estimate. With the machine reconfigured to accord with human expectations, UNIVAC's second guess was broadcast, showing the race as too tight to call. In fact, Eisenhower won a huge victory, on the order of the computer's initial predictions (442 to

89). The original prediction was finally announced three hours after it had been made, with CBS commentator Edward R. Murrow observing that "the trouble with machines is people."[12] Ultimately, forty-six UNIVACs were sold to a variety of industrial interests, and the age of commercially available computers began in earnest.

What, then, is a computer? The word itself combines the Latin prefix *com,* or "with," and *putare,* which means "reckoning." In German, "computer" is *(Be)rechner,* or *Auswerter:* a "reckoner," or that with which one calculates worth, use, or value in numerical terms. "To reckon" is to ascertain the number or amount of a thing through calculation. Nominally, then, a computer is a device *with* which one can *reckon* by calculating. The French word for computer is *ordinateur,* which is linked etymologically with ordering, seriality, and orderliness through the Latin words *ordo ordinis, ordinalis,* and *ordinarius.* As their name suggests, computers calculate and order, but this is not what defines them as instruments. The mechanical calculator, invented in 1623, calculated and ordered, as did the abacus before it, but neither of these was a computer. As an instrument, the modern computer is defined by the three essential attributes brought to its design by von Neumann: memory; binary digitization; and central processing.[13]

A simple calculating machine performs a single mathematical operation and then awaits instruction for subsequent operations. A computer can be instructed to perform a series of complicated operations that, once initiated, proceeds to completion without additional intervention. That is, computers can be *programmed,* and they can be programmed because they can *remember* what they were told to do – computers have a *memory* in which data and instructions can be stored, and from which they can be retrieved.[14] How does a computer remember? It remembers by translating information entered using an input device, such as a keyboard, into a series of electromagnetic charges, and then fixing these to a storage medium, using sensors and polarizers. Storage media have taken the form of magnetized tapes, drums, disks, and chips, with each step down in size typically entailing a step up in storage capacity.[15] Computers use different types of memory for different purposes. "Read Only Memory" (ROM) stores data or instructions in such a way that they can be read, or executed, but not altered or supplemented. The information stored in ROM remains there even when the computer is turned off and, for this reason, a computer's basic operating instructions are usually stored this way. "Random Access Memory" (RAM) stores material temporarily as needed by particular applications, and can be written to as well as read. Software – application programs beyond the computer's simple operating instructions (i.e., word processing) – and the files they create are often stored on disks. These can be either "floppy" disks that are inserted into and removed from the computer, or internal "hard" disks with comparatively greater capacity. Both can be read or written to, remain intact without power, and can be erased at the press of a button. These details aside, for a device to qualify as a computer, its main memory must be housed internally, and it must be capable of remembering both data and instructions, which can be assigned distinct storage "addresses" on the same medium. It is this feature that provides for many of the attributes commonly associated with the modern computer: changes in programming and application without the reconfiguration of hardware; high-speed operation; simultaneous storage and operation of an array of programs; and the rationalization and increase of memory space on the storage medium.[16]

The second definitive attribute of the modern computer is its use of binary digital notation. Calculating machines performed operations using decimal digits ranging from 0 to 9. In a mechanical calculator, the value of a digit is represented by a physical quantity of material corresponding to the assigned value of that digit, and calculation is carried out by the addition or subtraction of physical quantities with designated values. For example, an abacus represents the number 234 with 2 beads in the hundreds column, 3 beads in the tens column, and 4 beads in the ones column. To add the value of 111 units to 234, one extra bead is placed in each column and the result is 345. Calculation can be carried out this way, but very large numbers require large machines to deal with proliferating decimal places, and long series of calculations require time for mechanical operation. Compared to mechanics, the physics of electronics is quick and capacious, but it is difficult to represent electronically the values ranging from 0 to 9 as ten discrete *quantities,* as is required by decimal digital notation and calculation. Electronics work better when each decimal digit is rendered into a series of binary digits, or *bits.* "Binary" comes from the

Latin *binarius*, for "two together," and means dual or paired. A binary digit, then, is one of a pair of only two: it is *either* a 0 *or* a 1. In binary notation, decimal digits are rendered as a string of 0s and 1s: the decimal digit 0 becomes 0000 in binary notation; 1 becomes 0001; 2 becomes 0010; 3 becomes 0011; 4 becomes 0100, and so on.[17] Calculation is then carried out according to an algorithm, which stipulates, for example, that the sum of 0001 (1) and 0011 (3) is 0100 (4). Why is it easier for speedy, capacious electronic components to deal with binary digits than it is for them to handle decimal digits? A bit – either a 0 or a 1 – does not have to be rendered physically as a quantity. Rather, a bit is simply an impulse: a "yes" or a "no"; a positive or a negative; a something or a nothing. More specifically, binary digits can represent the presence or absence of an electromagnetic charge, a charge that can be easily stored, transmitted, and manipulated. Strings of bits, in fact, represent strings of impulses – present and absent charges that themselves represent specific abstract values. Abstract values converted into the physical form of electricity can be more easily stored and more quickly manipulated than those taking other physical forms, such as beads. Electricity takes up less space, and moves faster, than beads do. Thus, a computer is a machine that translates abstractions into the physical form of electromagnetic charges, and vice versa, through the language of binary digits.

Numbers are not the only abstractions that can be converted into bits. Indeed, any finite set of values can be so translated, and it is this that makes the computer much more than just a calculator. The alphabet has twenty-six characters, each of which can be rendered as a string of electromagnetic presences and absences, and represented by strings of 0s and 1s. The letter "A" becomes 1010, or presence (of charge)-absence-presence-absence; "B" is 1011, or presence-absence-presence-presence. Words become strings of strings of bits. This not only allows computers to store, display, and manipulate text, but also to be programmed with relative ease, using common phrases such as "run," "go to," and "if then," which stand in for more complicated strings of bits. As memory increases, so, too, does the number and complexity of programming commands that can be stored as bits and activated by input words. Furthermore, the proliferation of electronic digitization has drastically increased the variety of abstractions that can be input into, and output from, computers. Devices known as transducers make input devices of things like microphones, photographic instruments, and levers, by converting abstracted physical quantities into electro-magnetic impulses. Once converted, things like vibration frequency, light intensity, amplitude, dimension, intensity of pigment, geometric patterns, physical position, pressure, and temperature can all be digi-tized into sequences of binary pulses that can be input, stored, ma-nipulated, and reproduced as output by computers.[18] In this way computers become instruments, not just of simple calculation, but also of music-making, image-drawing, and process-control. Because of bi-nary digitization, a computer can: "read" a list of items and add up a grocery bill; "listen" to a Beethoven piano sonata and instantly replay it backward; "look at" the *Mona Lisa* and immediately reproduce her with a frown; and "sense" our entry into a darkened room and illuminate it more quickly than we can flip a switch.

Where does all this take place? In the "central processing unit" (CPU), the third essential element of the modern computer. The CPU orches-trates and carries out all the computer's arithmetic and logical opera-tions.[19] It is essentially a collection of circuits, or electromagnetic switches, capable of identifying, collecting, dissembling, routing, and assembling strings of impulses. The CPU in a modern computer pro-cesses binary impulses using Boolean algebra. Named for its inventor, the English mathematician George Boole (1815-64), Boolean algebra provides a system of logical procedures for performing operations on binary codes.[20] In this system, binary pairs are subjected to a series of "gates," each of which, depending on the approaching pair's configu-ration, "processes" the pair and passes on the result to the next gate or operation. The three most basic gates in the Boolean system are known as AND, OR, and NOT, and their operations can be illustrated using binary digits as follows: if *both* digits of a pair approaching an AND gate are 1s, the gate passes a 1 on to the next gate (any other combina-tion results in a 0 being passed on); an OR gate will pass on a 1 if *either* digit in the pair approaching it is a 1 (if *neither* is a 1, then a 0 is passed on); a NOT gate inverts 1s and 0s that approach it into their opposites (1s into 0s and vice versa). In Boolean logic, binary pairs are processed by gates, and the results combine in new pairs to be processed at sub-sequent gates until the operation is complete. This all sounds rather arcane, but as one analyst has put it: "Although Boolean algebra contains other operations, AND, OR, and NOT are all you – or a machine – need to add, subtract, multiply, divide, and perform other logical processes, such as comparing numbers or symbols."[21] Boole meant his system to apply to numbers, but it can easily be used to process any kind of abstract binary pairs, including electromagnetic impulses and the absence of them, which makes it the perfect logical system for the computer.

A computer's processing circuitry is, in fact, a vast series of electronic Boolean gates that perform operations on streams of binary pairs of electromagnetic pulse-presences and -absences that pass through them as inputs and outputs. These operations are often labelled "calculations" – perhaps because it is relatively easy to conceive of mathematical calculation operating in Boolean terms – but this is a misnomer, because the system really allows for much more than arithmetic. To illustrate: a child applies leftward pressure on the joystick of a video-game control; the device converts this pressure into a stream of electromagnetic pulses (absences and presences – 0s and 1s); these input pulses are processed through circuitry, or "gates," according to Boolean logic, until an output is produced that is then reconverted into a graphic representation of a spaceship moving across a display screen in exact proportion to the pressure applied initially; these same input pulses are also processed through electromagnetic gates that reconfigure them and output a stream of binary pulses that can be converted into the sound of a spaceship hurtling through the cosmos (if there is such a sound), to be amplified through a speaker system connected to the computer. Thus, computers do more than count; they process electricity and magnetism according to a rigid system of logic whose principles are consonant with the binary nature of these forces, and in so doing produce a vast range of representations immediately accessible to the visual and auditory faculties of the human sensorium through audio and video output devices.

We call a device a "computer" when it performs these operations and processes internally in its central processing unit. Given the mind-boggling number of operations that are involved in eliciting even the most apparently simple representations from a computer, it goes without saying that the circuitry involved in these processes is substantial.

Much of the development of the computer as an instrument has been driven by the pursuit of efficiencies in memory capacity and processing speed: large memories and fast processors that are physically small and inexpensive. As indicated above, early computers processed bits using cumbersome wiring, switches, and vacuum tubes, and storage media took the form of bulky and awkward magnetic cores, drums, and tapes. The UNIVAC computer was fourteen feet long, seven feet wide, and nine feet tall, and used 5,000 cathode-ray tubes and metres of magnetic tape. Its magnetic-core RAM could store about 84,000 bits, its processor could perform about 8,000 operations per second, and it sold for about $200,000.[22] All this changed, however, in the early 1970s, when the tubes, drums, and wires of mainframe computers were replaced by microchips.

The microchip is essentially a programmable, general-purpose integrated circuit of on/off switches inscribed onto the planes of a tiny silicon wafer, using techniques similar to photo-engraving, and its contribution to the realization of computing efficiencies has been staggering.[23] Specifying achievements in current microchip technology is inadvisable, given the dizzying pace of their acceleration, but some snapshots can be illuminating. Today's microchips are typically about the size of a human thumbnail, can feature more than 31 million transistors (i.e., "switches"), and are composed of components measuring about 1/200th the width of a human hair.[24] Miniaturization achieves the dual purpose of increasing memory capacity while confining it to small physical spaces, and increasing speed of operation by minimizing the distance electrical pulses have to travel. It is not uncommon for a contemporary notebook-size computer to have a Random Access Memory with a storage capacity in excess of 128 million bits and an internal-hard-disk memory with an 8-billion-bit capacity.[25] The first microprocessor – the Intel 4004 – could process 60,000 instructions per second. The Pentium microprocessors installed in most new personal computers today are capable of executing 300 million instructions per second. The cost of contemporary computers is expressed typically in term of price per million instructions per second, a ratio that is becoming wider as chips get faster: in 1989, it cost roughly $300.00 for a million instructions per second; by 1995, the rate was $15.00 and declining.[26] The microchip, it appears, has achieved efficiencies of size, cost, speed, and capacity that the early pioneers of computing could barely have imagined.

Indeed, these very efficiencies account, at least in part, for the escalating ubiquity of computers in the daily lives of most citizens in the industrialized world. The microprocessor's speed, flexibility, and inexpensiveness have made it an integral component of most electronic devices, ranging from telephones to clocks, to televisions, to cameras, to automobiles. It is estimated that there are more than 10 billion microprocessors in use worldwide.[27] The microprocessor has also played a key role in the phenomenon of "personal" computing. Early computers were designed for, and employed by, large-scale military and industrial users who had a lot of calculating to do. With the advent of the microchip, it became possible not only to put tiny processors in garage-door openers, but also to put relatively powerful and inexpensive, general-purpose personal computers (PCs) into people's homes and offices. The problem, of course, was that few people knew how to use them or had any idea what they might use them for. In the early 1970s, home-made PCs such as the Altair 8800 were the stuff of electronics hobbyists, but, by 1980, companies like IBM and the upstart Apple recognized potential profits and jumped into the commercial personal-computer industry.[28] All that remained was to develop programming languages, user interfaces, and software applications that would make the computer a useful instrument for the average consumer. There is much detail in the history of these developments, and the discussion of networks I turn to presently constitutes just one aspect of it. Suffice it to say at this point that, since its inception, personal computing has proceeded along a trajectory that has seen the device's sophisticated application to increasingly complex tasks matched by a similarly increasing ease of use. The vast majority of people have little or no understanding of *how* a computer works, but advances in both hardware and software have made sure, at the very least, that they can *operate* one readily – sometimes without even being aware they are doing so.

Computer Networks

In October 1957, the Soviet Union won the space race by launching successfully the first *Sputnik* satellite. A second was launched a month later, and a Soviet probe landed on the moon the following September.

As part of the ensuing flood of research-and-development activity in the United States, in 1958 President Eisenhower directed the Defense Department to establish the Advanced Research Projects Agency (ARPA).[29] One area of ARPA's defence research was communications, under the auspices of the agency's Command and Control Research group, which was founded in 1961 (renamed the Information Processing Techniques Office [IPTO] in 1964). In May 1961, three microwave relay stations owned by the American Telephone and Telegraph Company in Utah were destroyed by explosions, causing severe disruptions in communications, including some used for national defence. Officials suspected sabotage immediately. The bombers were alleged to be members of a group calling itself the American Republican Army, described as "a movement against the telephone company and other big business."[30] Around this time, US scientists became concerned about the "survivability" of the American communications system in the event of a nuclear attack. Calling this "the most dangerous situation that ever existed," a scientist named Paul Baran at the RAND corporation – a think-tank set up in 1946 to preserve and broaden the US operations-research capacity developed during the Second World War – set to work on the problem of designing a communications infrastructure that could remain operational even after a number of its components had been destroyed.[31] In Britain, a scientist named Donald Davies was working on a similar set of ideas, without any knowledge of Baran's work, although with different concerns in mind: "Davies simply wanted to create a new public communications network."[32]

Whatever their separate motivations might have been, both realized the key to designing such a system was to incorporate a high degree of "redundancy" into it. Communications channels are vulnerable and unreliable. In order to ensure that a message will be communicated successfully between a sender and a receiver in the event that the circuit linking them is compromised, it is necessary to provide a large number of alternative links between the two. A limited degree of redundancy was already a typical feature of most communications systems – if conversants in a telephone conversation found themselves "cut off," they simply had to ask the operator to establish a new connection. However, this kind of redundancy is insufficient as a safeguard against a serious breach of system integrity, such as occurs when

Figure 3.1   **Network types**

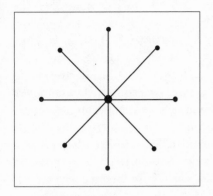

Centralized network
Disabled hub prevents communication between any two nodes.

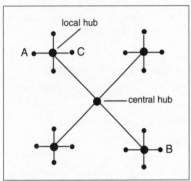

Decentralized network
Disabled central hub prevents communication between A and B; does not affect communication between A and C.

Disabled local hub prevents communication between A and B, A and C, B and C.

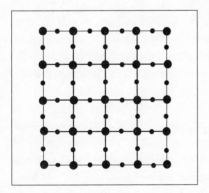

Distributed network
Disabled hub does not prevent communication between any two nodes.

● = *Hub*

• = *Node*
     (party to communication)

Adapted from Baran (1964).

a central hub – a switchboard, for example – is rendered inoperative. Centralized networks where all communications between two parties must pass through a single, central hub are especially vulnerable. The problem is only partially solved by decentralized networks, which allow for localized communication through dispersed hubs in the event of central-hub malfunction, but still cannot provide for communication between parties connected to different local hubs. Effective redundancy requires that communicating parties be connected by a *distributed network* that provides a multitude of potential connecting routes between any two points (see Figure 3.1).[33]

Consider the analogy of a city plan. In a city where all streets radiate from a central plaza, and are connected only by the plaza (a centralized network), a person will be unable to walk from the street he or she is on to the end of another if the plaza is obstructed. Next, imagine a city where satellite plazas, with otherwise unconnected streets radiating from them, are connected to each other by otherwise unconnected streets radiating from a central plaza (a decentralized network). In the event of obstruction in the central plaza, a person would be able to walk from his or her street to another connected to the same satellite hub, but not to a street connected to a different satellite hub. However, in a city whose streets are arranged in a grid pattern with multiple *intersections* (a distributed network), an obstruction at any intersection – or even at a number of intersections – generally will not prevent a person from walking from his or her street to any other on the grid, because he or she will have a number of alternative routes by which to travel around the obstruction. It is possible that multiple obstructions at intersections proximate to either his or her departure or destination point could leave the person stuck, but the more intersections – the larger, more redundant, and densely connected the grid is – the less likely it is that this will be a problem.

Distributed networks provide the degree of redundancy required for secure communication, but they pose other practical problems. Most of these centre on the question of "switching," or the direction of a message along the route between sender and receiver.[34] Traditional centralized and decentralized communications systems, such as those employed in early telephony, utilized circuit switching to get a message from sender to receiver. Circuit switching involves setting up a unique, dedicated electrical circuit between sender and receiver through a series of switches. As long as the parties remain connected, the circuit between them is reserved exclusively for the transmission of their communication, even if they are silent. Circuit switching suffers from two problems. First, it is not conducive to effective redundancy because it relies on the integrity of the initial, unique connection between sender and receiver – once that line is broken (i.e., when a cable, switch, or hub through which the connection is made malfunctions) an entirely new connection must be established. Second, a circuit-switched communication line can transmit only one "call" at a time, because it treats each one as an indissoluble whole. This is not a problem if what is being communicated is a voice transmission, because conversations tend to be fairly continuous and so do not waste much transmission time with silences. Data communications, by contrast, tend to proceed in bursts, often separated by long gaps; dedicating lines for the exclusive use of an entire set of data transmissions from start to finish would entail wasting massive amounts of transmission time. This problem would be compounded, not alleviated, if circuit-switching simply was applied to a distributed network, because to realize the security of redundancy great numbers of connecting routes would have to be reserved in case of failure by the initial circuit. This would mean only a portion of the network's capacity would be available to prospective users at any given time. Thus, in order for communicators to take advantage of a distributed network's redundancy without sacrificing substantial transmission time to underused dedicated circuits, a different switching system is required.

The solution to this problem is known as "packet switching," and it is a technique particularly suited for distributed networks and electronic/digital communication. Assuming the existence of a distributed network, packet-switching works like this: instead of being transmitted as a unified whole, a message is digitized and separated into a number of message blocks or packets, each of which contains about 200 characters; each packet is addressed to the receiver, and headed with instructions for its reassembly in the proper order upon delivery; each packet is also imbued with instructions directing it to follow the most efficient possible route across the distributed network to its destination; once they have arrived at their destination, the constituent packets are collected, reassembled by a processor, and accessed by the receiver; if a packet is lost, the receiving processor sends a message to the sending processor, which immediately dispatches a "twin" to replace the lost packet.

Packet-switching overcomes the problems associated with circuit-switching. First, it is designed to take full advantage of the redundancy of a distributed network: if one potential route is busy or malfunctioning, the packet simply avoids the obstruction and proceeds to its destination by another route. In a distributed network, there is rarely, if ever, a need to re-establish a connection in order for a message to be communicated because *no single connection constitutes the sole path between two points*. This is what gives distributed networks their robust character – communication can continue even when substantial portions of the network are not functioning. Furthermore, packet-switching exploits redundancy without compromising efficiency. Packet-switched channels are occupied only as long as it takes to transmit the packet. Any given message might comprise huge streams of data. If, as with circuit-switching, the channel between source and destination remained dedicated during the transmission of the entire stream as a single unit, the silences between bursts would be unusable. Exploiting these gaps would entail breaking the dedicated connection and disabling the communication. However, if the stream is broken into many packets bearing their own routing and assembly instructions, channels need not be dedicated, and the "silences" between packets can be filled with the transmission of other packets, which need not even be from the same message. The distributed communications network employing packet-switching becomes a lattice of conveyors. This allows the total transmission capacity of the network to be used, by multiple users, at all times. It also makes for easy expansion, because connection to any part of the lattice, or network, facilitates access to routes reaching all other parts of it.

This model of networking and packet-switching is a perfect fit for communicating digitized information. Streams of bits can be more easily broken apart and reassembled after transmission than can, for instance, analog audio signals. Thus, from the outset, distributed network communication has been *digital* communication. It has also been *computerized* communication: the number of operations involved in sending and receiving messages in this way is staggering, and nothing operates on binary digits more efficiently and powerfully than computers. Consequently, bringing distributed, packet-switched, network communication on the above model to fruition has a number of practical requirements: the physical configuration of a distributed network; computers and programming designed to execute the numerous tasks involved in disassembling, routing, reassembling, and storing digitized messages; and the digitization of content. To be realized, the theory of packet-switched communication over a distributed network required computers.

In the mid-1960s, researchers in the Information Processing Techniques Office (IPTO) were growing frustrated with the existing system of utilizing the agency's computing resources. Computers were still very large, very expensive, and few in number, and access to them was localized and in demand. It was believed that the costly isolation and duplication of computer resources could be overcome, and research findings could be shared, if computers at different sites could somehow be linked electronically. Around the same time, responding to the national security concerns alluded to above, Baran's RAND project recommended that the Air Force build a distributed, packet-switching network. In 1966, ARPA decided to fund an experiment, carried out under the auspices of the IPTO, to link its computers, and eventually it would adopt the model proposed by Baran.[35]

Initially, the network was to link the "host" computers at four ARPA sites – the University of California, Los Angeles (UCLA); the Stanford Research Institute; the University of Utah; and the University of California at Santa Barbara – over specially designated telephone lines leased from common carriers. Accomplishing this required the design of special computers to perform all the functions necessary for packet-switched communications between host computers over a distributed network: converting input data into packets; assigning packets with address and reassembly information; sending packets; routing packets sent from another source to another destination; receiving, storing, and reassembling packets; verifying transmission and receipt; and facilitating the retrieval of completed messages by destination hosts. These computers were called "Interface Message Processors" (IMPs), the first of which were built in 1969.[36] According to the early network design, host computers would be connected to IMPs, which would then communicate with one another over telephone lines via a device known as a "modulator/demodulator," or "modem" for short (see Figure 3.2). Modems convert digital information, or bits, into wave signals of either sound or light, which can be transmitted over telephone lines (modulation), and then reconvert these signals back into bits, which can be read by a computer (demodulation). In order to communicate in this way, it is necessary that IMPs connecting hosts employ a common language and grammar: in performing the operations they are designed for, IMPs have to arrange, address, and instruct packets according to a consistent set of rules and procedures so that each machine can recognize and process packets coming from another. This grammar is known as a network "protocol," and the first one was drafted by "an adhocracy of intensely creative, sleep-deprived, idiosyncratic, well-meaning computer geniuses" based at the first four host sites.[37] With four hosts, four IMPs, a battery of modems, and a protocol in place, the four ARPA sites were connected, and the ARPANET became operational in the early days of 1970.

The number of hosts connected to ARPANET, and the network's degree of redundancy, grew about as quickly as IMPs could be built and installed.[38] By the end of 1970, it became clear that what was required was not only an increased *number* of hosts, but also increased *access* to them. A typical host computer could support only four terminals for network access, and these had to be physically proximate to the host itself. It quickly became evident that the network itself was the solution to this logistical problem: instead of limiting access to four terminals physically proximate to hosts, users could access an IMP using a remote terminal connected to it via the network. All that was required to accomplish this was to design a processor – the Terminal Interface Processor (TIP), completed in October 1971 – capable of receiving instructions from remote terminals that could "dial-up" the processor using a modem, and transmit data and commands to the IMP for processing (see Figure 3.3). The first TIP and accompanying multi-line controller could handle up to sixty-three remote terminals simultaneously. This meant a multiplicity of prospective users could access the network and connect with a distant host computer without the mediation of an initial host computer. What is more, TIPs required no user identification before dispensing service – access to

Figure 3.2 **Schematic of distributed, packet-switched computer network based on early ARPANET architecture**

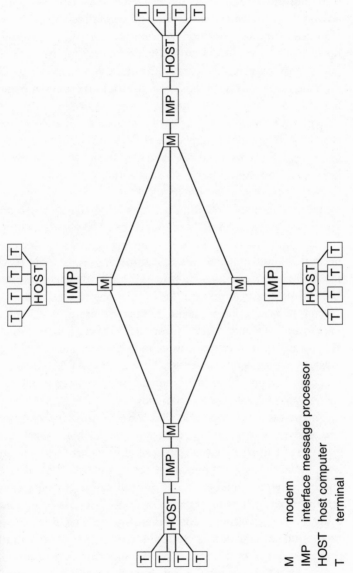

M   modem
IMP   interface message processor
HOST   host computer
T   terminal

files on a host computer might require a password, but access to the net itself did not. Furthermore, all terminal devices connected to a TIP were individually addressable in the same way a host was.[39] Network architecture has changed somewhat, and network service providers have assumed many of the functions previously accomplished via direct dial-up to a TIP. Nevertheless, this early design accomplished the multi-plicity of access, using comparatively simple remote devices, that would ultimately become one of the defining attributes of contemporary net-worked communications technology.

With a functioning, easily accessible network in place, what remained was the development of applications for it. At least part of the initial impetus for the design and deployment of the network was to facilitate sharing of both computing resources and research findings. Accord-ingly, the first two application protocols for the network were designed in the early 1970s with these goals in mind. A remote log-in protocol known as "telnet" allowed users linked to a TIP or IMP to utilize the computing power of then-powerful host computers connected to the network without being anywhere near them. A scientist in rural Cali-fornia, for example, could dial-up a TIP, telnet to the host computer at UCLA, and use it to process the calculations of his experiments. A second application, known as the "file-transfer protocol" (FTP), allowed the retrieval of stored data files from one host computer to another in an instant. These early protocols, in refined forms, remain widely used today. However, soon after the public launch of ARPANET in 1972, it was a third experimental application that would emerge as the so-called "killer app" of early network technology.

The first piece of electronic mail between two computers was sent and received in 1970, and was the result of a "hack": an early version of FTP was modified so that it could not only *retrieve* files from another host, but also *send* a text message from one machine and drop it into another, where it could be read.[40] From the time of its inception, the resource-sharing function for which ARPANET had been designed was eclipsed by its capacity as a messaging system. The expansion of raw processing power in desktop personal computers certainly contributed to the abatement of the need to access huge number crunchers from remote locations, but even as early as 1973 an ARPA study found that 75 percent of all ARPANET traffic was e-mail, despite the fact this was

Figure 3.3 **Schematic of multi-access, distributed, packet-switched computer network based on early ARPANET architecture**

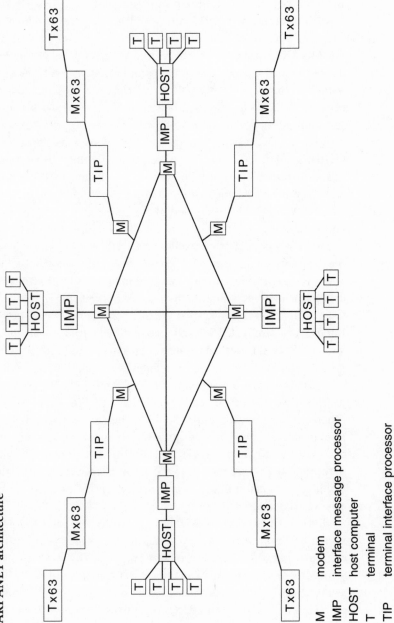

M   modem
IMP   interface message processor
HOST   host computer
T   terminal
TIP   terminal interface processor

an "unofficial" use of network resources.[41] In their final report on the ARPANET experiment, IPTO officials identified "the incredible popularity and success of network mail" as the "largest single surprise of the ARPANET program."[42] Thus, from its earliest days, networking has occasioned an elevation of the communicative capacities of computer technology above its abilities as a calculator; it seems that what began as a calculating device, when made part of a network, became a communicating device.

By 1972, packet-switching networks were being developed in various parts of Europe and Asia, and computer scientists began to investigate the possibility of linking these networks together.[43] It became clear that the same logic that facilitated the linking of computers in a single network could be applied to the construction of a network of networks: a processing computer capable of emulating hosts on the networks being connected would act as a "gateway" through which messages between these networks could pass; and a protocol would be developed to provide for the consistent management of standardized data traffic between autonomous networks employing their own protocols. This protocol was called the "Transmission Control Protocol/Internet Protocol" (TCP/IP) and, in simplified terms, it worked as follows: TCP would break messages into "datagrams" and place them into standardized electronic "envelopes" that could be handled by a variety of otherwise autonomous network protocols (i.e., because the envelopes were standardized they could be delivered despite the idiosyncrasies of their content); IP would then route these datagrams through the gateways connecting the various networks. The refinement of TCP/IP in 1978 was crucial. It meant that "anyone could build a network of any size or form, and as long as that network had a gateway computer that could interpret and route packets, it could communicate with any other network."[44]

And build networks they did. The release of TCP/IP, and ARPANET's switch to this protocol in 1983, prefaced a dramatic proliferation of computer networks linked together through gateways. Many of these were academic research networks in North America and Europe, such as the CSNET for computer scientists sponsored by the National Science Foundation in the United States, and the CA*net in Canada. Others included networks serving users of particular types of computers or programming languages, such as BITNET (the Because It's Time Network – a network of IBM systems), Usenet (a network for users of the Unix programming language that grew to become a distributed network of news/discussion groups), and FidoNet (for MS-DOS users). Local Area Networks (LANs) began developing, primarily at universities for intra-campus communication, and these could be linked, as a whole, to broader networks using a technology called Ethernet. In 1985, five supercomputer centres were built in the United States, and the National Science Foundation built a "backbone" network (NSFNET) to connect them. Regional networks, typically those established by universities and others in the academic and research community, were given exclusive franchises to connect to NSFNET free of charge for two years. In Canada, the first connection to NSFNET was at the University of Toronto in 1988. A year later, NSFNET and its regional subnets replaced the original ARPANET as the main carrier of packet-switched computer network traffic. In 1990, CA*net, which connected Canada's twelve regional networks, established three separate links to the NSFNET backbone.[45] With the establishment of European and Asian connectivity, a network of networks – including the now-famous Internet – was born.

Network Applications

Networks and personal computers grew up together and remain inseparable. Just as networks were beginning to connect increasing numbers of computers on university campuses and at research facilities, small, inexpensive, powerful computers were finding their way onto desktops in homes and workplaces. By 1996, an estimated 32 percent of Canadian households had personal computers, and half of those were equipped with modems.[46] This figure does not include the vast numbers of people who use computers in their workplaces or elsewhere outside the home. As access to computers grew, so, too, did connectivity and the reach of networks. Digital communication can take place over telephone lines, which means that anyone with a PC and a modem is, for all intents and purposes, wired. The advent of fibre optics – which allow digital telephone and computer signals to be transmitted as pulses of laser light rather than as analog signals – and the laying of dedicated, high-bandwidth transmission lines, have opened the floodgates for bits, making network communications faster and more efficient. Today, global data traffic over these latticed lines is growing six times faster than voice traffic.[47] As mentioned before, the Internet is only one of a number of computer networks, but a consideration of its growth suggests the extent of their reach in general: as of January 1999, there were roughly 43 million hosts connected to the Internet, supporting well in excess of 150 million users.[48] By its very nature, network growth is exponential. Projections indicate the Internet will double in size yearly, reaching a predicted 707 million users by the year 2001.[49] Because they consider only the Internet proper, these figures indicate just a thin slice of computer network use.

So what are all these people doing on the networks? In most general terms, they are doing three things: they are producing and gathering information, they are communicating, or they are controlling systems, though the line between these broad categories of activity is sometimes difficult to draw. In this section, I briefly outline the attributes of computer networks as information utilities, as communications utilities, and as systems control utilities, and suggest how, in their most developed form, they collapse these three categories into one.

Networks as Information Utilities

The utility of networks as an information technology derives primarily from their extension of the storage attributes of computers. Networking enables a user whose computer is connected to a network to, at least potentially, access the information stored on every other computer connected to that network. Depending on the size of its memory, a single desktop PC may not be able to store the entire catalogue of the Library of Congress, the unabridged *Oxford English Dictionary*, the *Encyclopaedia Britannica*, Hansard, and the collected works of Shakespeare, along with the various software programs required to browse and read these texts. Even the relatively capacious storage medium of "Compact Disc – Read Only Memory" (CD-ROM) is ultimately limited in the amount of material it can contain. However, a person can access seemingly limitless material if his or her PC is connected to computers that do store this information, perhaps as their only function. In most cases, aside from viewing this information, a person can also copy portions of it, as he or she desires, and store them on his or her own computer for subsequent access. This process of moving information from the memory of one computer to the memory of another via the network is called "downloading." It is important to note that it is not only data of the kind mentioned above that can be downloaded. Software programs of all sorts are readily available for copy via networks, often either for a nominal fee or completely free of charge. In terms that evoke the self-image of network denizens, programs distributed in this way are referred to as "freeware" or "shareware."

Information storage and retrieval take a number of forms on the network. One form popular with early enthusiasts was the Bulletin Board Service (BBS). Typically, BBSs would specialize in a particular interest or geographic area; the BBS host would place text postings of relevant information in the memory of his or her computer, and subscribers would be able to read these, and contribute their own postings, by dialling-up the host computer, using a modem. These became the electronic equivalents of the library notice board, where people could "visit" to offer, solicit, and browse information regarding matters of mutual interest. However, pure BBSs have been more or less replaced by more varied and sophisticated methods of information delivery. Among these is the computerized database, which has come to typify the awesome information-generating and -management capabilities of network technology.

The word "database" is something of a misnomer. "Data" is the plural of "datum," which means a thing known or granted, and derives from the past participle of the Latin *dare,* for give. Thus, data are things taken as *given,* which are more typically rendered in English as "facts." A database, then, would be a collection of data or facts. However, "data" is often used to refer to facts that are unprocessed, thus implying a database could be simply *any* group of facts, however raw and disparate they might be. This does not capture the essence of a database. A database is precisely a collection of things known that have been grouped together for specific reasons or purposes. It is this sort of deliberate processing that turns data into *information* and, somewhat awkwardly, means that a database is a collection of information rather than one of data. According to this definition, a dictionary is a database because it is a collection of information about the words of a particular language, not just a collection of any words whatsoever. It is this quality that renders a dictionary, or any other database, useful. And usefulness, or utility, is a key attribute of databases; they are not just collections of information, they are collections of information *to be used*.

They may not have been referred to as such, but databases have existed in one form or another for quite some time, though they have always been constrained by considerations of their assembly and the requirements of utility. The books and rooms and buildings that house the database known as the library have limitations: their capacity is finite; collecting and arranging the information to fill them is an arduous process; and the navigation required to use them can be difficult. Networked, computerized databases purport to overcome these limitations and make information more easily usable than ever before. For obvious reasons, computers are able to store far more information than previously available storage media. Reduced to strings of bits, the textual, numeric, and graphic information contained in the world's libraries, archives, and documents can be stored tenfold in physical spaces that are too small to be fretted over. The rate at which old data need be culled to make room for new is drastically reduced in the era of the microchip. Most important, when computerized databases are networked, access to this plenitude of resources need not rely on physical proximity to their place of storage. Digitization and networking make it possible for much *more* information to be rendered readily and easily usable.

Computerization and networking also make the collection of data, and the processing of it into information, much easier. Anything that can be digitized – text, images, sound – can be stored in a computer and become part of a database. All that is required is that data be converted into bits and input; at most, this entails physically typing text into a computer somewhere, but it may just as easily mean "scanning" a document in its entirety and storing it as an image. When computers are networked, data can be input at one site and then transmitted for collection as information in a database at another. The ability of computers to be programmed to perform certain operations repeatedly, combined with a network's facility to move bits from one computer to another, also means that digitized accounts of various human activities can be recorded, collected, and organized as information in databases with greater ease than ever before. This surveillance utility is compounded by the increasing penetration of networked devices in daily life. For example, if a bank wishes to construct a database of information relating to the use of its automated teller machines, the fact that these machines are computerized and networked makes this task extremely easy. Prior to networked computerization, constructing and maintaining such a database – about, in this example, how people used *human* tellers – would require a massive, time-consuming, ongoing physical effort that included the collection and processing of mountains of copied documents from a multitude of sites. Today, all this happens in the blink of an eye: a human activity (the management of personal funds) is converted into bits by an input device (an automated teller); these bits are sent via a network to a central computer whose programming categorizes abstractions of the activity represented by these bits (withdrawals, deposits, bill payments, transfers, account balances, time of day); the computer then collects these data with similar sets sent to it from other input sites and, according to its programming, categorizes, analyzes, and stores the results as information in a database profiling the use and users of bank machines. In this manner, networked computing not only allows the easy publication and distribution of already-existing information, but also facilitates the automatic creation of databases full of information resulting from the instant conversion of a variety of human activities into data.

An increasing number of devices serve as input terminals to computer networks; one does not need to be sitting at a computer to send bits careening across networks for registration in the memory of distant computers. Banking machines, cash registers, telephones, electronic door locks, elevators, room sensors, cameras, and numerous other instruments are capable of digitizing inputs and transmitting them to computers for processing. As more and more human activities come to be mediated in some way by networked, computerized devices, collection and storage of information about those activities in databases increase. It is probably safe to say that any human activity mediated by such devices and remotely involving an exchange of resources – commercial or not, private or public, productive or consumptive – is registered in a database in some way. This accumulation of data gives network technology the appearance of being not only a conveyor of information but also a creator of more of it than there has ever been before.

Networked computerization also enables databases to be used more efficiently – if not necessarily more thoughtfully – than they could be previously, primarily due to increased ease of access to them and their contents. Before networks were in place and content was digitized, a person wanting to use a database such as the Library of Congress catalogue would have to travel physically to its site and wade through its substantial card files. Once the records of its holdings were transformed into bits that could be read from distant computers connected to its own via a network, people could "visit" the library and "browse" its catalogue daily without ever leaving their desk. Additionally, specific information in computerized databases can typically be located much more quickly than a card in a catalogue or a document in a pile. The same is true of locating databases themselves. Sophisticated "search" programs enable seekers of information to request that the computer locate databases, and material stored in them, by simply inputting a word or words identifying them (e.g., *Twelfth Night* or "interest rates"). Essentially, this boils down to asking a computer to look through its memory, locate a particular string of bits, and retrieve and display the information accompanying it. When reduced to bits, information can be operated upon using computers, and, when these computers are connected, these operations can be initiated, and their results utilized, at remote locations. Thus, networked computer databases not only facilitate easy collection of large amounts of information, but also make accessing, using, and manipulating this information more efficient. In combination, these effects of networked computing on databases – increased volume, proliferation of information creation and gathering, and efficiency in usability – are largely responsible for the attachment of the label "Information Society" to societies where this technology is prevalent. If we accept that a database is any repository or collection of processed and usable information, it is not difficult to see why this label resonates as readily as it does at present. Networked computers make it easy to construct, maintain, and use databases, and so they have proliferated in electronic form. Databases suggest that, if there is one thing the societies being considered here do not suffer from a lack of, it is information.

Networks as Communications Utilities

Communication is the raison d'être of computer networks. The only reason to connect one point to another is to allow passage between them, and in that way facilitate the sharing of whatever is passed between the sender and the receiver. Computer networks are thus communications media – from the Latin *medius,* for "middle" – in that they are the passage existing in the middle of, or between, connected entities. Technically, computer networks are relatively indifferent to the substantive content of that which passes through them, so long as this content is packaged and configured in a way that allows the medium to accomplish its delivery. They will act as the conduit for anything that can be rendered as bits and separated into packets. This technical indifference to substantive content is a feature of many media, including, for instance, regular mail: the only thing the medium of the post requires for successful delivery is that the content being passed is packaged in an addressed envelope bearing adequate postage – it cares not whether the message inside is written in Hebrew or French. Similarly, the medium of broadcast television is just as capable of sending pornographic images from a transmitter to a receiver as it is able to send images of a puppet show, so long as both were recorded in a manner that satisfies certain purely technical standards. Telephones can communicate grunts and groans as well as dirty language. The point here is that, in most respects, limitations on the substantive content of communications are imposed by considerations extraneous to the medium itself. That being said, we should keep in mind two caveats. First, the technical requirements of a medium still impose *some* restrictions on what can be communicated through it. For example, regardless of the presence or absence of extraneous considerations, a computer network is simply unable to communicate anything that cannot be rendered in the form of binary digits. Second, asserting that media as instruments generally do not limit substantive content does not deny that they influence it as technologies. As discussed in the previous chapter, media have a decisive impact on the environment in which they exist, and so also on the content ultimately produced there.

So what *is* distinctive technically about the communication enabled by computer networks? Aside from the network's predilection for bits, most of what is distinctive about this way of communicating is contained, disappointingly, in the modern vulgarization of natural *fitness* as technological *convenience*. From the Latin *convenientia*, meaning assembly, agreement, or fit, "convenience" has come to denote ease of use or access, and freedom from difficulty. It is the poor cousin of efficiency, which itself is somewhat less virtuous than harmony or genuine fitness. Communication via computer networks exudes convenience because it combines many of the convenient attributes of other communications media, and overcomes their inconveniences. The origins of computer networks remind us, for instance, that this medium of communication is extremely *reliable* compared with other media. Mail disappears, circuit-switched phone conversations are cut off, radio signals are lost amid skyscrapers, and television screens fill up with "snow," but, due to the robust characteristics of packet-switching over a distributed network with redundant connections, computer-mediated communications are rarely lost. Additionally, this form of communication can be both *synchronous* and *asynchronous*. Traditional telephone communication (i.e., prior to the emergence of asynchronous devices such as answering machines and voice-mail) is synchronous, because both sender and receiver must be present for the communication to be successful – I have to pick up the phone and dial your number and you must be there, not busy, and answer for us to communicate. This is not always convenient. Postal communication is asynchronous, in that I can mail you a postcard but you need not be there – either as I write it or even when it arrives – in order to receive the communication contained in it. Should I wish to speak with you right away, though, postal service is not very convenient.

Communications using networked computers can be accomplished both asynchronously *and* synchronously. I can send you a message, perhaps containing reams of bits, which you need not be present at your terminal to receive. It will be stored in your "mailbox" in the server that connects you to the network, and upon reading the message you can download it from this server to the memory of your PC, save it, edit it, produce a hard copy, or delete it altogether. Such asynchronous communications of bits, data, and information can occur between computers that are programmed to communicate even without immediate human initiation. On the other hand, should we wish to communicate in "real time," so long as we are both present at network-connected terminals we can do so using specialized software that directs our terminals to act as graphic telephones. One such service is known as "Internet Relay Chat," where messages typed at the terminal of the sender appear simultaneously on the screen of the receiver, and vice versa. This is far enough away from the audible conversation of telephone communication that it is not charged as such, even over long distances. However, the development of more sophisticated input devices, capable of digitizing voice, for example, moves this type of computer communication closer to traditional telephony, and has phone companies scrambling to find ways of charging long-distance rates for what increasingly resemble telephone calls.

Networks also have had an impact on the cost and speed of communicating frequently, or in large volumes, over great distances. Reduced to electromagnetic bits racing around tiny micro-chips, and to pulses of laser light beaming across fibre-optic filaments, huge amounts of information can be communicated in seconds between senders and receivers separated by continents and oceans. This type of communications suffers the costs neither of transporting documents physically as material over geographic space (as with mail), nor of maintaining a dedicated, exclusive long-distance telephone link (as with fax communications). Bits take up little space, weigh nothing, and move faster than stagecoaches, trains, or airplanes. Thus, especially in the case of voluminous documentary exchanges, computer networks are far cheaper, and far quicker, than comparable communications media. Bits are also easily copied, stored, and erased, which has ambiguous effects on network communication. On the one hand, ease of duplication and storage has made it very simple to maintain and distribute records and copies of communicative activities that otherwise might have been retained or circulated more selectively. Telephone calls can be recorded, letters hoarded, transaction records filed, and copies of these can be circulated to others who might clutter their cabinets with them – but not as easily as bits representing large documents, or long exchanges over time, can be stored on a computer and forwarded to others via a network. On the other hand, ease and inexpensiveness of storage, duplication, circulation, and especially erasure, can also lower the quality and perceived value of many communications. There appears to be a correlation between the ease with which an act of communication is executed and the degree of banality expressed in that act. One can hardly imagine the collected electronic mail of Bill Gates could be as interesting to read as the collected letters of Henry Ford, despite the comparable impact these fellows have had on North American social and economic life. And just as network communications are easily accumulated, so, too, are they easily disposed of. Casually deleting a piece of e-mail does not carry the gravity of deliberately trashing a letter or shredding a document. Paradoxically, the very convenience that has established computer networks as significant communication media also encourages a certain ephemerality in that which is communicated through them.

Another important feature of this medium is that it enables what I shall call *multicast* communications. Typical telephone communications are *bicast,* or occurring reciprocally between two parties on a one-to-one basis. Communications via the media of radio and television are *broadcast,* or non-reciprocal, one-to-many transmissions between a single source and a multitude of receivers. Communication via a distributed computer network can be both bicast and broadcast, as well as many-to-many – hence the term "multicast." Postal communications can also be multicast, but the combination of expense and inconvenience usually succeeds in dissuading all but the commercially interested from using the medium in this manner. However, it is quite simple for parties wired to a distributed network to communicate a message simultaneously to a multitude of receivers who can communicate with that party as part of a reciprocal multitude. On the networks, one person can chat with one other, communicate with scores of others, or receive communications from those same scores with seamless ease. It is this capacity that has contributed to popular images of computer networks as decentralized federations in which every consumer of information is likewise a producer and distributor of it.

The communication capacities of computer networks listed above have combined to elicit a range of actual uses that, along with proliferating databases, have come to characterize this emerging medium. Early file transfer and remote log-in protocols enabled communication of a sort: a user on one computer could "ask" another computer to which his or hers was networked to share some information, or "direct" it to do some work, and these applications persist. However, it is electronic mail that has become perhaps the paradigmatic communicative activity taking place on computer networks. In 1997, an estimated 71 million people were regular users of electronic mail, making it by far the most common application of network technology.[50] In some instances, e-mail is used simply as a souped-up form of regular mail, with messages exchanged between one sender and one receiver at higher speeds, more economically, and with greater frequency than customary with traditional postal service. However, the efficiencies of e-mail have also led to the proliferation of networked "mailing lists" that facilitate automatic circulation of messages to a number of correspondents simultaneously, without requiring senders to produce multiple physical copies of the message being posted. Given that they are created and dissolved daily, it is impossible to estimate accurately the number of network e-mail lists in existence, but they probably number at least in the tens of thousands.[51] Some of these are closed, meaning that subscription to them is limited to people with some sort of qualification (e.g., the personnel of a company, membership of an organization), but most are open and specific to particular topics or issue areas (e.g., labour politics, gardening, baseball). Subscribers to a mailing list automatically receive postings containing information relevant to its subject area in their electronic mailboxes, and can also post information or messages they deem of interest to the list for distribution to all other subscribers. Some lists are moderated, meaning that postings are vetted by a list manager before they are distributed to subscribers; others are unmoderated, with postings immediately distributed to everyone on the list. Subscription, cancellation, management of incoming messages, and distribution of multiple electronic copies are dispatched by software programs that carry out these functions more or less automatically.[52] This software is readily available on the network itself, and those with network access can start a mailing list of their own on a topic of their choosing. Various search engines exist that allow network users to peruse lists of lists to find those they might be interested in. Typically, network mailing lists serve the dual function of circulating information in the form of announcements and news, and providing a forum for asynchronous discussion between subscribers on topics of mutual interest, fascination, or outrage. Information received via mailing lists can be read, deleted, saved, and even forwarded to other individuals or lists, all with the press of a button. The impracticalities of sustaining this type of communication via any other medium are such that, prior to the advent of computer networks, it was effectively impossible to do so in large-scale social organizations.

Newsgroups, or discussion groups, are a communications application of network technology that combines many of the aspects of mailing lists and the bulletin boards described earlier. Subscribers to news/ discussion groups are able to post, read, and respond to messages pertaining to a specific issue area, but, unlike a mailing list, this material is not automatically delivered to participants' mailboxes – users must "go" to the part of the network where the proceedings of the group are stored and employ freely available newsreader software to read them. In this way, news/discussion groups resemble bulletin boards. Where they differ is in that news/discussion groups typically feature more in the way of discussion and debate than announcements or news, much like many mailing lists. A distinguishing feature of news/discussion groups in this regard is their maintenance of a record of topic discussions that can be read for reference by new or ongoing participants. These records – known as "threads" – are made possible by the asynchronous nature of network communication, and the ease with which dialogue transformed into bits can be stored and accessed using computers. The existence of threads is thought to be an inherently inclusive attribute of newsgroup communication because they compensate for the obstacles often facing prospective entrants into conversations already well under way. There is no need for engaged conversants to "go back to the beginning" to bring newcomers "up to speed" in order for the latter to be welcomed as unobtrusive and conscientious participants in a developing dialogue. Neophytes need only review a topic's thread to avoid the pitfalls of redundancy and naïveté. Ease of storage and access also makes these threads a built-in safeguard against misrepresentation and duplicity in discussions, as a record of past statements is readily available to all participants. However, it should be noted that these same conveniences make threads vulnerable to sabotage, a potential problem made more acute by disputes over whether individual participants, group managers, or collective memberships "own" a thread's contents. In 1992, a disgruntled member of a network of news/

discussion groups known as the Whole Earth 'Lectronic Link (WELL) deleted massive sections of discussions he had participated in, rendering what remained nonsensical.[53]

There are many networks that offer, as part of their services, access to a vast array of news and discussion groups, including commercial providers such as America Online (AOL) and the Microsoft Network (MSN). However, an early and prolific network of on-line news/discussion groups known as Usenet was distinctly non-commercial, and a brief consideration of its operation illuminates much of the character of this mode of communication. Usenet encompasses several thousand newsgroups, covering a broad range of topics, though not all of these groups are carried by every network service provider. These groups are separated into categories, some of which are global, while others are geographically specific. The seven major global categories in the Usenet hierarchy are: computers (comp.); network news (news.); recreation (rec.); science (sci.); debates (talk.); society (soc.); and miscellaneous (misc.). There are also minor categories for business (biz.) and so-called alternative (alt.) topics. Geographically specific categories generally mimic national and regional divisions. Thus, there are categories for Canada (can.), British Columbia (bc.) and the other provinces, and even cities, such as Halifax (hfx.). Within these broad categories are several hundred, even thousands, of subcategories or topics that, when combined, denote the subject matters of various newsgroups. So a designation like rec.arts.sf.books indicates a discussion group dedicated to the recreational arts of science fiction literature, while can.politics.socialism might be the electronic meeting place of a nostalgic cabal reminiscing about the lost art of left-wing politics in Canada. The creation of new groups is left up to the democratic decision making of users: proposals are posted to a special site, and after a period of open discussion Usenet users vote on whether the proposal merits the establishment of a new discussion group; if the proposal is successful a "newsgroup control message" informs Usenet carriers that the new group is official and available for distribution.[54] According to some observers, "it is this global co-operative effort concerning the establishment of new newsgroups that is at the heart of USENET."[55] This gives the impression that one is free to talk about whatever one wishes to talk about on a network, provided one can find enough others to talk with. This is true to a certain extent, but it should be kept in mind that the establishment of discussion groups on commercial network service providers is not necessarily so organic, and responds to other priorities besides simple topicality. Nevertheless, once established, these groups provide forums for the consideration of a wide range of topics among discussants connecting across vast geographic expanses, and constructs an ongoing record of their deliberations according to rules often determined through some manner of consensual process. It is for these reasons that they typify many aspects of the communicative utility of networked computers.

The Web and the Collapse of Information and Communication

Up to this point, I have been treating the information and communication utilities of network technology as separate categories. In some respects, this misses the point of the technology entirely: *network technology encourages the collapse of the distinction between information and communication.* To illustrate how it does this, we will consider a network known as the World Wide Web (www, or "the Web") that, in many ways, best exemplifies the paradigmatic features of network technology.

The Web was formed in 1992 by Tim Berners-Lee at the European Nuclear Research Centre as a computer network that would facilitate a new way of delivering information, and by 1994 it had eclipsed previous network delivery systems in its volume of traffic.[56] Information on the Web is organized as a massive, searchable database of "pages" existing, not in just a single computer somewhere, but in all of the computers linked to the network via Web servers. This means that the Web is not simply a means of accessing remote databases, but rather a network *that is itself a giant, expanding database.* Every Web page is assigned a distinct address that can be accessed using a navigator, a software program typically available free of charge as an incentive for network use. Pages are often grouped by their creator into a linked series, or Web site, and prefaced by a homepage. Using a navigator, users input the Uniform Resource Locator (URL) address of the page or site they wish to visit and request that the location be opened; within seconds the contents of that site are displayed on the user's computer screen. With a flick of the wrist, these contents can then be downloaded onto the user's computer for storage, alteration, distribution, or other use. Should they not have the URL address of the site they are looking for, users can locate it by entering keywords into one of a number of search engines accessible by their navigator. These search engines maintain large databases of sites on the Web, and can present users with a listing of sites pertaining to the entered keywords, or allow browsing by subject area.

Most of the applications discussed in this chapter so far have been strictly text-based, wherein the information being generated and consumed, or the communication being enacted, starts out as words or numbers, is transformed into bits, and ends up as words or numbers again. Web pages, on the other hand, are *multimedia* because they can communicate information in the form of sound, and graphic, pictorial, or even moving images. Rich image and complex audio files are heavy with bits – it takes more os and is to get a computer to produce the sound of Jimi Hendrix's guitar and an image of him setting it on fire than it does for it to spell out "Wild Thing" – and their fidelity is contingent on both the capacity of the connection and the sophistication of the user's equipment, but these technical constraints are on their way to being overcome. What is important to consider here is the following: the advent of the Web has meant that computer networks have progressed beyond the point of being just another way to deliver words and numbers into or between people's homes and workplaces. They are also capable of delivering sound and moving images – just like television.

The Web, however, differs from television in a number of key respects. For example, unlike television, the Web is an asynchronous medium. To watch a TV show, the consumer must be present in front of a television set at the time of broadcast. The advent of programmable video-cassette recorders (VCRs) has altered this somewhat, but not to the point of making television a truly asynchronous medium. While a viewer need not be in front of the set bodily, *some* receiving presence *at the time specified by the broadcaster* is still required for successful reception. If you forget to program your VCR, you are out of luck. In contrast, the content of Web pages is stored, and awaits a "visit" by a user at his or her convenience, rather than according to the schedule established by the source. Second, because it is carried over a distributed network that enables many-to-many communications, the Web is a multicast medium, unlike television, which broadcasts signals to multiple receivers from a single, central source. Thus, because the instrument that mediates the *consumption* of content – the personal computer – is the same instrument that mediates its *production*, the Web is said to be home to as many *potential* producers as consumers of network content. One cannot produce a television show and broadcast it using a TV set, but one can produce a homepage that is accessible to anyone else connected to the Web using a regular PC.

Most important, the Web distinguishes itself from television in the apparent dynamism of the interface between it and its users. Watchers of television have a limited array of options as to the manner in which they can consume the content being offered to them. In addition to being subject to the time constraints imposed by synchronous broadcast alluded to above, television viewers are generally confined to consuming the material being presented in the manner intended by its originator. A person watching the news must wait for the sports coverage; he or she cannot skip ahead. At best, the viewer can change the channel, but neither remote controls nor even VCRs are capable of the nimbleness or flexibility of information delivery made viable by the Web. This is the case because information delivered via the Web is presented in the form of *hypertext* (a misnomer, because it applies to images as well as text), which "links" Web documents to any number of other Web documents. The author of a Webpage can build these links into his or her document, using a relatively simple programming language known as "Hypertext Markup Language" (HTML). Navigator software then enables viewers of this information to jump easily from one document to another along these links.

To illustrate how this works, consider the example of the Web site maintained by the Canadian Broadcasting Corporation surrounding the 1997 Canadian federal election.[57] To access the site, a user enters the URL and asks his or her navigator to open the location. On the PC screen appears the homepage of the CBC election site, displaying a welcome message and a list of topics including: results; riding profiles; sights and sounds; platforms; news archives; political links; and input. Each of these topics is highlighted, by either underlining or colour, and this indicates that each is what is known variously as a "pointer," a "hotlink," or a "hypertext link." What this means is that the URL address of other pages is built into the homepage appearing on the user's screen, and to access the contents of any one of these pages, he or she simply moves the mouse appliance attached to the computer until it places the cursor on the desired topic, and "clicks." Automatically, the user's navigator opens the location of the new page – which itself may be festooned with more pointers – and displays its contents on the screen. The user has no need to ascertain, search for, or enter the URL of the new destination page because his or her navigator software, in conjunction with a document written in HTML, does it. Thus, a user wishing to peruse the profile of the electoral riding in which he or she lives simply opens the CBC election homepage, clicks on `riding profiles`; clicks on his or her `province`, clicks on the relevant `riding`, clicks on `past results`, clicks on `candidate profiles`, and so on. A person clicking on `sights and sounds` will be presented with a gallery of still images on which he or she can click to view a full motion and sound excerpt from a leader's speech. Click, click, click, and the user has bounced around in the midst of a substantial amount of information. Another click or two and he or she can review discussions occurring at the `input` pointer, or similar discussion lists at the party pages linked to the CBC site, and add his or her own commentary or questions. All of this can be downloaded onto the user's computer for storage and future reference.

This example only scratches the surface of the capacities of hypertext. Ted Nelson originally defined hypertext as "nonsequential writing – text that branches and allows choice to the reader, best read at an interactive screen."[58] The prefix "hyper" comes from the Greek word *huper,* for "over" or "beyond." Indeed, it is the ability of hypertext to link users to information *beyond* the particular site at which they began that gives the Web its distinctive power as an information and communications utility. Not only can hypertext link users to related documents managed at the same site, or even connected to the same Web server, but it also has the capacity to link them to documents and sites connected to the Web by any server, any place in the world where there is network access, all with a click. So, in the example I have been using, a user looking for campaign information can: begin at the CBC election homepage; click on `party platforms` and get a list of them; click on the `Green Party` and get their homepage with a list of related resources;

· · ·

click on the `Alternate Press News Service` and get a homepage that lists like-minded organizations; click on `Corporate Watch` and get a list of current topics; click on `Industrial Development in Indonesia` and get a profile of North American companies doing exploitative business there; click on `free fax` and have a template of a letter demanding the cessation of one of those companies' operations in Indonesia appear on screen; type in his or her name and anything special he or she might wish to add; and click on `send` to drop it in the CEO's electronic mailbox. After the second click, the user had already passed *beyond* what the CBC might have explicitly intended – in seven short clicks, the user went from wondering about party platforms to assailing the ramparts of multinational capitalism. And with seven more rapid clicks on the "[<BACK]" button of his or her navigator, the user will have returned to the party platforms, and will perhaps check if they say anything about investment in Indonesia. If the user can stand it, he or she might click on the `sound bite` of the prime minister of Canada announcing to the press what a great boon for Canadian industry his latest trip to Southeast Asia was.

This protracted example serves to illustrate the ways in which the Web represents something of a culmination for computer networks as instruments. In the first place, the Web accentuates the capacity of networks to enmesh with one another and become networks of networks in an exponentially increasing matrix of connectivity. On the Web and when hypertext is being used, documents themselves *become networks,* in the sense that they are the media through which connection is established to other networks of information and communication. Second, the Web places a vast amount of information and a broad communications capability at the fingertips of its users in a manner that differs considerably from previous methods of delivery. I am referring here to the much-ballyhooed "interactive" nature of network engagement. Not only are users able to chart their own course through the reams of available information using hypertext links, they are also, in many cases, able to input information when visiting a site and so affect what is happening "there." Interactivity may take the form of simply choosing one hypertext link instead of another, downloading data or software, participating in the operation of a program running at a remote site, contributing to the information being gathered there, or having a discussion with other visitors to that site. It is in this way that computer networks, particularly the Web, erase the distinction between information production, information consumption, and communication. They do so by reducing these activities to the same thing: the movement or exchange of bits. In the regime of the network, data become useful information only when they are rendered in binary language for easy movement from one site to another. And, in this regime, communication shares less with community and commonalty that it does with commodities (useful articles of exchange; from the Latin *commodus,* for "with measure" or "convenience") and commuting (interchange of one thing [usually payment] or place for another; from the Latin *commutare,* for "with change"). In societies where computer networks are the ascendant medium, information and communication not only occur simultaneously, but also collapse into the single category of exchange. Thus, while they are portrayed as media of *interaction,* computer networks might be described more accurately as media of *transaction.*

Networks as Control Utilities

In discussions about network technology, the word "control" is regarded with great suspicion. This is the case primarily because it chafes against the image of this technology preferred and promoted by a good many of its purveyors and users. Control, they imagine, is the residual preoccupation of an era that technology has rocketed past. As one writer has observed, "the dominant theme of modernity has been to control chance through certainty."[59] Max Weber's account of iron cages of rationality, and Michel Foucault's insights into panoptic discipline, support this characterization and suggest its technological aspects. The denizens of the 'Net do not challenge these views of modernity. Instead, they take them as proof that computer networks are a technology that frames an *exit* from modernity: contemporary network technology *unleashes* chance, *negates* certainty, and so, in its very essence, *defies* control. Networks, it is presumed, are a *post*modern technology in that they are inherently pluralistic, decentred, anarchic, disembodied, immaterial, libertine, and radically democratic. If control is a concern in this environment, it is so only to the extent its appearance signals a brutal and artificial instance of domination

fundamentally at odds with the nature of the medium itself, an impo-
sition that turns the medium into something other than what it *really*
is. As the title of a popular book about networks insists, they are thought
to be, by nature, *out of control*.[60]

There is much to contend with in this line of argument, and I will
return to it in ensuing chapters. In the present context, however, I wish
only to point out that those who try too hard to believe that networks,
as *"hyper*media," are essentially *beyond* control neglect to consider the
many ways in which they are used *for* control. "Cyberspace," one of the
popular euphemisms for computer networks, is a play on "cybernetics," the theoretical field of automated communication feedback and
control pioneered by Norbert Wiener in the late 1940s.[61] The word
"cybernetics" derives from the Greek *kubernētēs*, for "steersman," or
the pilot of a ship. "Control" originates in the medieval Latin verb
contrarotulare, which combines *contra* and *rotulus* – "against the rolls"
– and refers to the comparison of performance against information
inscribed on rolls of paper, which served as official records before computer printouts and databases. In this sense, control denotes not only
the power of direction and command – of steering – but also processes
of verification, regulation, and adaptation. These, it should be recalled,
were explicitly built into early packet-switched, distributed computer
networks. Indeed, it has been argued that the rise of computer technologies is directly linked to attempts to gain control over the explo-
sion of social, economic, and material change in the nineteenth and
twentieth centuries. In this respect, James Beniger refers to the development of computer technology as part of the "Control Revolution,"
which "represented the beginning of a restoration – although with increasing centralization – of the economic and political control that was
lost at more local levels of society during the Industrial Revolution."[62]
Industrialized economies had become too large, complex, and technically mystifying to allow for centralized – let alone local – stewardship.
According to Beniger, the restoration of control required the construction of an infrastructure in which "the twin activities of information
processing and reciprocal communication" were made "inseparable
from the process of control." Central to this relationship was "the con-
tinual comparison of current states to future goals," and "the comparison of inputs to stored programs."[63] If one were looking for a recipe

that would produce computer networks, one could do much worse than Beniger's list of information processing, reciprocal communication, data inputs, and stored programming. He goes on to illustrate comprehensively how computerized information and communications technologies have been brought to bear as instruments of control in the realms of mass production, distribution, consumption, and bureaucratic organization.

This suggests the information/communication and systems control utilities of computer networks are not inherently opposed; in fact, they are complementary. In any system, control involves "purposive influence toward a predetermined goal."[64] Enacting such influence requires the maintenance of information about the goal desired, about the behaviour and interaction of inputs to the system designed to reach it, and about the relationship between the two. It also requires the communication of this information to the controlling agent. A factory, for example, is a system for producing automobiles. In order to control this system, information about the goal (e.g., the number of automobiles to be produced, their design and engineering specifications), and about the interaction of inputs contributing to its realization (e.g., raw materials, energy, labour) must be constantly communicated to a controlling agent capable of comparing combined inputs to desired outputs and altering the former to realize the latter. Such information gathering and communication can take simple forms in simple systems. In a home, a mindful parent can control the system of household maintenance by posting a list of chores on the refrigerator, having children check off listed items as they are accomplished, and altering the system of incentives and sanctions to induce completion of the tasks. However, as the scale of modern social and economic enterprises escalates, so, too, do both the complexity of the systems developed to control them, and the information and communication requirements of those control systems. As Beniger writes, "because both the activities of information processing and communications are inseparable components of the control function, a society's ability to maintain control – at all levels from interpersonal to international relations – will be directly proportional to the development of its information technologies."[65] Networked computers meet ably the control requirements of our system-saturated society and economy, due to their ability to store vast amounts of information in the form of bits; communicate that information automatically, reliably, and quickly in great volumes; and use that information to perform complicated operations according to stored programming designed to achieve a wide array of specific goals.

The penetration of networks as control utilities proceeds in tandem with the systematization of contemporary social life at numerous levels. At the highest levels, computer networks are the fabric of a "Global Information Infrastructure" (GII), designed to facilitate control of an economic system whose reach presses the limits of planetary expanse. When US vice-president Al Gore – a noted promoter of the so-called Information Superhighway – says that a GII is necessary to facilitate "stewardship of our small planet," he speaks the language of a steersman in cyberspace; and he means to say that computer networks are a particularly suitable replacement for instruments (such as national governments) whose utility as control devices has been vanquished by the complexity and scale of globalized economic systems.[66] Likewise, as the complexity and system saturation of everyday domestic life increases, access to computer networks is presented as an essential need, and the ability to use them an essential life-skill. In this way, computers and networks become an indispensable instrument in the control of productive, consumptive, educational, or family life. Consider the following account of the features of the "wired homes" being built in "Canada's Premier Interactive Community" in suburban Toronto: "Built with a high speed ATM network infrastructure, the Stonehaven West community allows residents to take advantage of a range of services and applications in the home, the community, and around the world ... Network Ports located in different rooms of the house will contain 'plugs' for telephones, fax machines, personal computers, televisions, and stereos connecting all communications appliances to the In Home Network. The In Home Network provides an interface for home automation systems including security, environment controls, lighting, appliances and emergency services."[67] Evidently, life in the suburbs is not what it used to be and, in terms of systems control, what's good for the global economy is good for private homes. Networks not only enable the control of various systems within the home, but also facilitate controlled integration into the greater system of community life by providing an interface with external networks. Thus, designers of the computerized telephones in wired homes remind us: "It's a bank. It's a shopping mall. It's a restaurant. It's a weather report."[68] And there is seemingly no end to the systems-control benefits of network technology mediated through in-home appliances: for "parents with kids," it "makes day-to-day routine at home easier by eliminating some time consuming errands and shopping trips, and eliminates the hassles of bringing kids along"; while, for "empty nesters," it "lets people pay their bills exactly when they wish to, regardless of bad weather or other transportation issues" and makes it "easy to monitor finances as often as desired."[69]

The point here is that the restriction of the reach of network technology to its use as an information and communication utility, narrowly conceived, is decreasing. Networks are not just the preserve of those who wish to access information quickly and easily, or who wish to chat with one another with more convenience than afforded by telephone or postal service. Because of their peculiar attributes as instruments for the exchange of information in the form of bits, networks have also become a crucial systems control utility, deployed across many levels of contemporary life. And because they are networked, these levels of systems control commingle: the activities of a person using a computer network to exert control over his or her highly systematized home and working life (e.g., by banking via the Internet) become inputs in the control regime of a larger system oriented to its own ends. One person's transfer of funds is another institution's input information to be communicated to its own controlling agent, whether it be a human systems manager or an algorithm. Certainly, the embeddedness of systems has always been a feature of complex societies, but networked computers appear to be an instrument capable of achieving this integration with greater efficiency, and invisibility, than was possible in their absence.

From Eniac to the Web

Computers and networks are complicated instruments that seem to become more and more sophisticated on almost a daily basis. What I have presented above is a description of the rudimentary attributes of these instruments by concentrating primarily on their origins, in order to encourage a basic understanding of what it is they are and do *as technical instruments.* Regardless of how advanced its gadgetry has become, a computer is still a central processing unit and a memory that, according to stored programming and using Boolean logic, performs operations on abstractions digitized in the binary language of os and 1s, or the presence and absence of electromagnetic charges. The coincident sophistication and simplification (i.e., as computers become more complicated technically, they become easier to use) of this process has not changed its basic nature, but it has resulted in the deepening penetration of computers into daily life. The proliferation of digitization, the miniaturization of microchips, the arrival of the PC, the creation of a range of input devices, and the development of a panoply of programming and applications have made computing a nearly ubiquitous element of living and working in the so-called First World. It is not hyperbole to suggest that there can only be a very few people in North America, Europe, and parts of Asia whose productive, consumptive, financial, domestic, or leisure activities are not mediated in some way by these instruments. Indeed, the opposite claim would be far less credible.

Computers and computerized devices thrive on bits and, like blood between humans, bits can be passed between these instruments. The medium by which this passage of bits occurs is a network of networks – a matrix of multiply redundant connections between computers across which packages of bits can be sent reliably, in massive volumes, at remarkable speeds, and with considerable efficiency. This network of networks has become an instrument in its own right, and in this chapter I have presented a range of its applications as an information, communications, and control utility. As an information utility, networks effectively extend the capacities of computers in general, providing ready access to vast stores of information. Of particular note in this regard is the network's facility as an instrument for the construction, maintenance, and use of voluminous and complex databases accessible from remote locations. As a communications utility, computer networks provide for reliable, asynchronous, recordable, quick, inexpensive, multicast exchanges. This array of conveniences has produced a range of communications practices that are distinctive to computer networks, including electronic mail, mailing lists, and news or discussion groups. Even more distinctive are the applications that have developed on the network known as the World Wide Web, which effectively collapse the distinction between information and communication. The Web – with its enmeshed lattices of hypertext links, point-and-click interface, ease of navigation, multicasting, and multimedia delivery – gives rise to the prevailing image of computer networks as interactive media through which the practices of information production, consumption, and communication pass as undifferentiated streams of bits.

This ultimate instrumentality of computer networks – their propensity to encourage the rendering of a wide range of human activities into the exchangeable and therefore manageable form of bits – is amply illustrated in their utility as instruments of systems control, the final application discussed in this chapter. Whether it is tracking averages on the world's stock markets and automatically issuing risk models, or allowing Mom to log on to the company's mainframe and word-process from home while she takes care of the kids, networked computers have emerged as a crucial instrument in maintaining control over the complex demands and outputs of an increasingly system-saturated existence. The irony is that, despite their promises of convenience, proliferating computer networks contribute to this saturation and, in many ways, increase its demands. Even more curious is the fact that networks make the source of these demands difficult to identify, by weaving information, communication, and control into a seamless web that appears before (or between) us, as does water to fish. Networks are fast on their way to becoming invisible media of transaction, wherein value is a function of the exchangeability of bits. This sort of discussion, though, takes us well beyond the consideration of computer networks as artful instruments, which has been the subject of this chapter. The suggestion that networks are more than instruments – that they also constitute environments, or places in which we carry out the practices of living – invites consideration of their *logos*. It invites us to consider computer networks *as a technology:* not just what is said about or via networks, but also what networks say about us, about how we live and wish to live as a political community.

the political economy
of network technology 1:
the mode of production

. . . . .

In a networked society, the real powershift is
from the producer to the consumer, and there is a
redistribution of controls and power. On the Web, Karl
Marx's dream has been realized: the tools and the means
of production are in the hands of the workers.

DERRICK DE KERCKHOVE, 1996

KARL MARX WAS a historical materialist, not a dreamer, which suggests
his investigations into the effects of technology were conducted with
his eyes open, rather than closed.[1] We can only assume he would have
approached network technology similarly, understanding that – even
if this technology harboured within it a potential for non-exploitative
relationships – the technology itself would be less determinant of its
own ultimate impact than would the mode of production in which it
was situated and deployed. The aim of this chapter and the next is not
to put words about network technology into Marx's mouth; it is, instead, to consider some of the questions about network technology that
he might have asked. Thus, before rushing to the conclusion that networks represent the virtual shovel with which the grave of advanced
capitalism will be dug, we should consider the following: What is the
mode of production in which network technology is embedded and
what does it give to that mode as a force of production? Who owns
network technology and what is the nature of that ownership? What
characterizes the relations of production, consumption, and exchange
in this mode of production and how does network technology affect

these relations? And, finally, what are the effects of network technology on working life, the crucial activity through which people express what they are as species-beings? The present chapter addresses the role of network technology in the capitalist mode of production and its utilities as a force of production. Chapter 5 discusses the relationship between network technology and emerging regimes of work, consumption, and exchange.

Still Capitalism after All These Years ...

The information society is a capitalist society. Despite their suspected potential for a new style of politics, computer networks have not revealed so much as even a hint of how they might form the infrastructure of a *fundamentally* reorganized (i.e., non-capitalist) economic life. This should come as no surprise, given that network technology has developed, by and large, squarely within the context of the capitalist mode of production: "PCs and networks may well be useful for politics; but the form of the network and the structure of computing equipment is determined first and foremost by the needs of the state and capitalist corporations."[2] Nevertheless, a belief persists that somehow the advent of networks signals or has precipitated a profound economic shift. There are a number of related factors at play in this perception. One is the belief that the cooperative academic research culture, which flourished in the early days of network development and featured a liberal approach to information exchange, is somehow representative of the economic reality of the technology on a broader scale.[3] Similarly, many observers assume that the non-commercial, communitarian spirit of many electronic bulletin boards and discussion groups infects the technology in general, leaving other network applications unable to inoculate themselves against it.[4] Both of these beliefs are based on the experience of cultures that are becoming increasingly marginal to mainstream network application and use. Granted, marginal cultures are important sources of change and often bleed alternatives into the mainstream. However, in this case, the communitarian cultures on the fringes of network life are not primarily *economic* cultures, and their participants are not customarily linked to one another in economic relationships. Members of academic research networks and non-commercial on-line communities compare findings, exchange recipes, tell secrets, and maybe even strategize for revolution, but they do not produce goods, they do not sell products, and they are not each other's bosses or employees. It is hard to imagine that alternative forms of sociability on the margins of network use represent a fundamental departure from or challenge to the economic logic in which this technology has developed and is embedded. As one observer has put it, "to the extent that these technological enthusiasts believe these technologies can override the logic and power of capital, there is little evidence to support such a view."[5]

There are also more properly economic bases upon which it is argued that network technology is part of a profound change in capitalism. On the one hand, it is observed that the ascendancy of computer, network, and other digital technologies in the economies of wealthy nations has introduced a new cadre of elites, populated by whiz-kids such as Bill Gates of Microsoft and Steve Jobs of Apple Computers, who have struck it rich on the heels of their technical innovations, and who now wield considerable influence in the new economy. On the other hand, it is asserted that the escalating penetration of network technology into nearly every facet of economic life has altered the functioning of the economy in innumerable ways: entirely new industries have been born; the distinctions between existing industries and enterprises have been blurred; the production, marketing, and distribution of goods has been reorganized; and work has changed. Take, for example, the following testimonial as to the inherent pluralism of the information economy: "The point is that the Digital Revolution will be shaped not just by what John Malone, Ray Smith, Bill Gates and the other road warriors do with their corporate strategies, but by the decisions of Safeway and A&P as well."[6] Similarly, the growing presence of service, knowledge, and entertainment industries in North American economies is often presented as signalling a shift in power from steeltown to tinseltown.[7] If it is true, news that economic control in the digital world might not be monopolized by telecommunications behemoths may be somewhat comforting, but the redistribution of a bit of clout to include massive retail chains and entertainment conglomerates hardly constitutes a revolution in any serious sense of the word. Doubts about the accuracy of suffixing "information" with the word "revolution" are not new. Numerous such critiques were levelled at the "information society" thesis in the late 1980s.[8] The dubious character of this usage is revealed when its advocates admit that "the Digital Revolution [is] a curious revolution if ever there was one [because] its target is not the levers of political or economic power but rather the dials on your television and buttons on your PC."[9] If it has nothing to do with a change in who – or, more specifically, which *class* – manages the levers of political and economic control, then the emergence of digital network technology cannot be considered revolutionary.

It is important to understand the nature and depth of change that network technology has brought to the operation of capitalist economies. However, before trying to do so, one must establish what has clearly remained the same – namely, that these changes have all occurred squarely within the parameters of an economy that remains capitalist, perhaps more capitalist than ever before. The boardrooms of the new information industries may be populated by new faces, managing the sale and distribution of new goods and services in new ways to people doing new jobs with new tools, but these are still capitalist bosses, enterprises, commodities, and workers. Our appreciation of the endurance of capitalism should not be distracted by the new gadgetry with which this economic system has, once again, secured its longevity. If it is a question of which exerts greater influence on the other, I think it is fair to say the capacity of capitalism to determine the deployment of network technology far outstrips the likelihood that network technology can, or will, independently *transform* capitalism.[10] That being said, network technology is intimately involved in a striking *enhancement* of existing capitalist relations and processes. Ellen Meiksins Wood puts it as follows: "The old Fordism used the assembly line as a substitute for higher-cost skilled craftsmen and to tighten the control of the labour process by capital, with the obvious objective of extracting more value from labour. Now, the new technologies are used to the same ends: to make products easy and cheap to assemble, to control the labour process, to eliminate or combine various skills in both manufacturing and service sectors, to replace higher with lower wage workers, to 'downsize' workers altogether – again, to extract more value from labour."[11] In so far as they reproduce and extend the logic of capital accumulation, computer networks are unexceptional instruments, and follow in the footsteps of nearly every communication and information technology that has preceded them.[12] Nevertheless, there are complexities to be considered when it comes to the relationship between capitalism and networks at the present juncture. Wood is correct when she observes that "this isn't just a phase of capitalism. This *is* capitalism."[13] However, it is still advisable to explore the particular economic attributes and functions of network technology within the current capitalist configuration.

Ownership of the Means of Power

If the information society, of which computer networks form the infrastructure, is a capitalist society, that means it is a class society in which a minority exerts power over a majority by virtue of its control of economic and other social resources. In classical Marxist phrasing, the attribute that simultaneously draws the line dividing these two classes and confers power on one of them is "ownership of the means of production."[14] For Marx, the first step toward understanding the politics of capitalism involved coming to grips with who owned the means of production, and the character of this ownership. As will be elaborated upon in the next chapter, computer networks are much more than simply means of production: they are also means of control, means of labour, and means of exchange and consumption. In societies where economic livelihood colonizes most other aspects of human existence, networked computers are, comprehensively, means of power. For some, this quality of the technology has rendered traditional Marxist categories incapable of contributing to a sufficient appreciation of the politics of capitalism's current configuration.[15] However, I would suggest that, far from invalidating Marx's injunction to investigate the matter of ownership, the expansive role computer networks play in structuring economic and social power in advanced capitalist societies simply increases the importance of doing so.

It is important to disabuse oneself of the facile notion, expressed eloquently in the epigraph of this chapter, that because a great many people own personal computers connected to networks, the "means of production" are "in the hands of the workers" and that, consequently, we are in the midst of a dispersal of social and economic power. Such a claim is equivalent to saying that ownership of a hammer and saw – which, admittedly, represents a marginal increase in power – extends to a carpenter the ability to manifest structural changes to the economic parameters and conditions within and under which he lives. Hammers and saws are, of course, only part of the picture. Like hammers and saws, as tools of work, personal computers simply grant their owners connectivity to a regime of wage labour to which they must be wired (chained?) for their survival, but that affords them little in terms of "ownership" of the complex means by which their existence – and the privileged existence of their bosses – is secured. Means of production encompass far more than simply tools of the trade; means of power, even more. Owning a personal computer may be preferable to relying on one belonging to an employer in so far as it reduces dependency *somewhat,* but it hardly confers the power to eliminate broadly the conditions by which wage-labourers are dependent on wage-providers.

Similarly, it is at least a mistake, and perhaps disingenuous, to equate the opportunity networked computers provide for individuals to "produce" content and transmit it via networks with control over the processes of production, labour, exchange, and consumption that together constitute the field of economic and political power in advanced capitalist societies. Certainly, no single capitalist controls this entire field either, but the control capitalists enjoy *as a class* also certainly precludes significant control over these domains by working individuals. Much will be said in the following pages about the complex role networked computers play in each of these domains. One hopes that this discussion will substantiate the rather obvious point I am making here: wage-earners and other working-class people own computers; they do not run the global capitalist economy. It reflects a profound misunderstanding of the economic dimension of network technology to assume that someone using the Internet in his or her spare time to post details of the Nike Corporation's labour practices, or to document the pathologies of trade liberalization, has power equivalent to, or even remotely challenging, that of Nike's Phil Knight or the capitalist state. Networks might provide those who use them with an interesting and potentially significant political tool; they do not alter the users' class position.

This misunderstanding is related to another that also requires clarification before a discussion of the ownership of the means of power in a networked world can be undertaken, and that is the mistaken belief that *computer networks cannot be owned.* A number of factors combine to produce this curious assumption, including popular beliefs about the uncertain status of intellectual and electronic property, leftover mythologies about the costlessness of network access, and the conviction that a network comprises solely the proliferation of independent people at its terminal points, and not the computers and wires that connect them. The fact is that the cables and wires that carry packets of bits, the switches and processing equipment that regulate this traffic, the computers and other devices that send and receive it, the software programs that instruct the entire process, are all owned and paid for in one form or another. Indeed, there are network resources that are public and there are those that are distributed freely to consumers. This does not detract from what should be obvious realities: every aspect of network technology has been designed and built with human labour; its construction and maintenance cost money; it is bought and sold; it is a source and instrument of profit. Networks and their constituent elements are owned as surely as highways and streets and alleys and driveways, and the automobiles that traverse them, are owned.

This is most clearly the case with the numerous private proprietary networks that exist within and between enterprises in order to facilitate the movement of data and information in the form of bits. A company that operates and maintains an intranet or local-area network to connect its employees to one another and to relevant information owns and controls that network. Networks that enable electronic data interchange and other transactions between large transnational corporate and financial actors are private, and owned either by a consortium of the actors using them or by an outside service provider who sells exclusive, secure access and maintenance to them. Networks available for public use are owned in a variety of ways: infrastructural transmission media are owned by telephone, cable, or satellite companies; access points and technologies (i.e., servers, service providers, "on-ramps") are owned and made available by various private and public interests; collections of services, tools, software, and content are developed, owned, sold, and distributed by commercial and non-commercial interests (i.e., the Microsoft Network, America Online, Netscape, Yahoo!); and network sites and terminals are owned by the proprietors of the computers in which they are housed. While it may be true that no single interest or person owns, for example, the Internet wholly or entirely, it is also true that every constituent part of the network is owned and controlled by an actor of some sort. Consequently, it is important to determine what the attributes and character of this ownership are, especially in regard to the ostensibly public networks that mediate an increasing range of everyday human activities.

As detailed in Chapter 3, computer networks arose from publicly funded defence research in the United States, and were brought to fruition via the establishment of government-supported "backbones" administered by arm's-length agencies, which mediated bit traffic and provided connectivity either for free or at highly subsidized rates. These origins have contributed to the now antiquated but persistent notion that networks are completely public (and, therefore, "ownerless") and completely free. However, as network use began to extend beyond the rather selective communication of academics and computer enthusiasts, and as the various commercial and industrial utilities of the technology became apparent, the nature of government involvement in its development changed somewhat. In the first place, governments in both Canada and the United States continued to make massive investments in the development of network infrastructure. In 1993, the US National Information Infrastructure (NII) program allocated $2 billion a year for the construction of an "information superhighway."[16] In Canada, the federal government has committed in excess of $100 million over six years to the Canadian Network for the Advancement of Research in Industry and Education (CANARIE). Established in 1993, CANARIE is an industry-led consortium whose objectives include the development of "broadband high-speed networks" and the "commercializ[ation of] cutting edge technologies, products, applications and services in information technology."[17] The 1998 Canadian federal budget included $205 million earmarked for a program called "Connecting Canadians" that ostensibly seeks to expand access to computer technology in neighbourhoods and schools, but also includes $55 million for a new high-speed optical network touted as "the foundation of our digital economy for the new century."[18] This is in addition to massive government purchases of information technology and the opening-up of general infrastructure-renewal funding to network-related projects. Total annual federal spending on network-related technology has been estimated in excess of $3 billion.[19]

However, despite this considerable investment, the 1990s have also witnessed a massive retraction of government ownership, operation, and regulation of network and other telecommunications technologies. Combined with massive public subsidies for infrastructure *construction*, the vacation by government of *ownership* of this resource constitutes what more than one observer has characterized as "perhaps the largest liquidation of public property in the history of capitalism."[20] Between 1984 and 1995, the privatization of telecommunication interests worldwide reached a value of $105 billion.[21] For the Internet – which is exemplary in this regard because of both its origins in government and the continuing widespread perceptions of its status as a free and public utility – the decisive moment came in 1995, when federal US funding for the non-profit, publicly administered National Science Foundation Network (NSFNet, the Internet's initial backbone) was terminated and the infrastructure was sold to America Online (AOL), a commercial service provider. AOL, along with other private network service providers, operates regional network access points, connectivity to which is sold to local Internet service providers, who likewise sell access to individual, corporate, and institutional network users. Canada's CA*net was commercialized in 1997. As demand for network connectivity increases, more and larger players are entering the network service market, including telephone and cable companies hoping to cash in on the traffic that heretofore merely paid a toll to travel over their lines.[22] The key point, however, is that the Internet – like most other networks of significance – may have been developed as a public resource using public money, but it is now owned privately and operated commercially, albeit in the midst of continued government infrastructure subsidization. Nicholas Baran has written that, when it comes to computer networks, "the government is essentially just another commercial customer."[23] It would be perhaps more accurate to say that government is a customer who not only pays cash on the barrel, but also contributes mightily to the construction and maintenance of the barrel itself.

This means that network computers have more or less mimicked the development of communications technologies that preceded them: publicly funded pioneering and early development followed by privatization, continued subsidization, and limited regulation once the profitability and viability of the new technology has been established.[24] At this point, regulation is reduced to determining which private interests will be allowed to capitalize on the emerging market, a process that typically has more to due with the relative strength and influence of the actors involved than with the public interest.[25]

That the governments of both Canada and the United States are committed to the private and commercial elaboration of network technology is beyond doubt, as is readily apparent in the policy directions and legislative activity of both governments in this area during the 1990s. In 1993, the US NII *Agenda for Action* stated as its first principle the conviction that further development and control of network infrastructure should be the responsibility of private investors in an unfettered market. In 1996, the US Congress and president passed into law a new Telecommunications Act confirming this principle. As Robert McChesney has described it, "the overarching purpose of the 1996 Telecommunications Act is to deregulate all communications industries and to permit the market, not public policy, to determine the course of the information highway and the communications system."[26] In concert with the Communications Act of 1995, this legislation "guarantees that the eventual information highway based on the interactive telecomputer will be a thoroughly commercial enterprise with profit maximization as its founding principle."[27]

Similarly, Canada's 1993 Telecommunications Act has been described as a "handing off of responsibility to market forces,"[28] a milestone in a long deregulatory trajectory institutionalizing a "'powershift' away from the public sector towards the private sector."[29] This liberalizing approach to network technology was endorsed unambiguously in the final report in 1995 of the Canadian Information Highway Advisory Council (IHAC), which recommended that "highway network and new infrastructure should be left to the private sector, and the risks and rewards of the investment should accrue to the shareholders"; and "the provision of the Information Highway facilities across the nation must be driven by existing or potential market demand."[30] The council recommended that government limit its involvement in this sector to that of ensuring an investment-friendly regulatory environment, acting as a "model user," and encouraging the development of interoperable standards. In all respects, the council recommended that the government develop the liberal spirit evident in its 1993 Telecommunication Act and allow it to animate the exploitation of network technology.[31] The Canadian government confirmed its endorsement of this approach in its response to the IHAC recommendations in 1996, and took steps in this direction that included dropping domestic-ownership requirements in the broadcast industry to the same level as those in the telecommunications sector – a reduction from 33 to 20 percent.[32] In its second "final" report, IHAC applauded such deregulatory efforts by the government as helping create a "competitive" environment for technology development, and urged more of the same.[33]

The terms used to describe the ownership of computer networks as means of power include: "private," "commercial," and "lightly regulated." Customarily, when these three in combination refer to media ownership, they signal the applicability of a fourth: "concentrated." Indeed, the history of mass media in Canada and the United States has been one of varying but substantial degrees of concentrated ownership. This was obviously true of monopoly and near-monopoly ownership of common-carrier media such as telephone and postal services, in which the privileges of concentration traditionally were mitigated by substantial degrees of regulation. However, concentration of ownership has also occurred in less-regulated, privatized areas of the Canadian telecommunications sector, where the number of companies fell from 183 at the beginning of the 1980s to just 62 in 1990, despite a growth rate in this sector (8.6 percent) that outpaced that of the rest of the economy (3 percent) over the same period. Of these 62, a mere 9 accounted for 83 percent of all revenues.[34] Concentrated ownership has also been a dominant feature of relatively less-regulated, content-providing, broadcast media such as newspapers, radio, cinema, and television. As Mary Vipond has documented in the Canadian context, "in all the media, concentration and conglomerate ownership is increasingly the norm. A few large corporations whose names could be listed on the fingers of both hands control a very high proportion of our mass communication outlets."[35] There is little to indicate that this dynamic will not be repeated in the case of network-technology ownership. Indeed, there is much to suggest that – contrary to beliefs that everyone will be an equal producer of network content in a disaggregated network media environment – network technology will further concentrate ownership of primary communication and information media.

As computer networks began reaching critical levels of deployment in the mid-1990s, the worldwide telecommunications and information-technology sectors underwent what has been described as a "rash of mergers," with merger and acquisition transactions increasing by 57 percent in 1995 alone.[36] It should be kept in mind that this flurry of amalgamation occurred in sectors that *already* featured a high degree of ownership concentration. Two factors have contributed to this escalation. The first, as Peter Golding has explained, is the incentives arising from the complex of privatization, globalization, and deregulation that characterizes the current telecommunications market environment: "Information technology companies sought energetically to achieve the critical mass necessary for competition in the international market, and especially sought the synergy that association with the newly deregulated telecommunications companies would allow."[37] A second factor contributing to the concentration of ownership in this sector inheres to network technology itself, in so far as its facility with the movement of digital bits renders vertical and horizontal integration easier to accomplish.

Vertical integration occurs when a single enterprise owns and controls operations involved in every aspect of the dispensation of its products. Thus, a vertically integrated media conglomerate such as Disney develops an idea for a film; produces it; markets it; distributes it; screens the film in its theatres; produces, manufactures, distributes, and retails its soundtrack and related promotional merchandise; develops and produces advertising for these products; and places this advertising and these products in its amusement parks and in sports arenas where professional teams owned by Disney play. Digitization, and the availability of computer networks through which bits can be manipulated and exchanged, simply makes it easier to transfer materials, and coordinate productive, distributive, and operational activities, between previously distinctive enterprises. Network control utilities such as these are discussed in greater detail below, but it should be noted here that one of their most significant results is likely to be increased vertical integration of ownership in communications media generally. These utilities also make it easier for non-media interests to take advantage of the opportunities for horizontal integration – the concentration of ownership *across* rather than *within* distinct sectors – presented by a relaxed regulatory environment that fails to discourage concentrated cross-ownership of communications media. Examples of horizontal integration in the communications sector include General Electric's ownership of the NBC television network, and Westinghouse's ownership of CBS, in the United States. These horizontal integrations of media and non-media interests precede the widespread use of computer networks, but it is certain that the development of this new technology will enable enterprises owned across sectors to be integrated and coordinated with even greater ease than before.

Despite the pro-competition rhetoric contained in the IHAC final reports in Canada and the NII *Agenda for Action* in the United States, it appears likely that the combination of a privatized, deregulated business environment and a technology offering highly efficient control utilities will lead to competition in the telecommunications and media sectors only between actors of a particular sort: a handful of large, vertically and horizontally integrated companies capable of competing in a globalized market with other global, vertically and horizontally integrated enterprises. As John Malone, chair of the highly integrated media conglomerate TCI, has boasted: "Two or three companies will eventually dominate the delivery of telecommunications services over information superhighways worldwide. The big bubbles get bigger and the little bubbles disappear."[38] Interestingly, in June 1998, AT&T Corporation purchased Malone's TCI for $32 billion in a bid to gain control of the so-called last mile of the US communications infrastructure: the delivery of local telephone, long-distance, Internet, and cable-television service into residences all via a single cable.[39]

If this bubble scenario is plausible (even remotely), it is likely that the Microsoft Corporation will be one of the big bubbles, and the case of Microsoft provides an illuminating example of how computer networks enable and even encourage the increased concentration of ownership in the information technology and communications media sectors.[40] Between its early MS-DOS and more contemporary "Windows" software packages, Microsoft products make up close to 95 percent of the operating systems used in personal computers (PCs) today. That means whenever a PC is turned on somewhere in the world, it is overwhelmingly probable that a Microsoft program executes the basic operations of that machine. This dominance extends through a variety of other software areas as well: Microsoft has captured an 80 percent share of the word-processor market; as of March 1997, 87 percent of software developers in the United States were developing programs for Microsoft platforms, and 53 percent were doing so using Microsoft's Visual Basic programming language. Despite this already substantial degree of market penetration and 1997 revenues in excess of $11 billion, from 1994 to 1997 Microsoft spent nearly $5 billion on mergers and acquisitions. These have included the purchase of interest in a wide range of companies involved in a variety of activities, including the development and sale of: competing operating systems; desktop applications; server operating systems; databases; vendor training and professional services; multimedia Internet standards (i.e., audio and video streaming); Web-site development tools; network financial and commercial transaction applications; on-line news media; interactive entertainment; Internet-access services; and cable and satellite technology. A more specific sampling of Microsoft's investments reveals the character of its integrationist strategy: in 1997, the company invested $3.3 billion in the Teledesic system of Low Earth Orbit digital transmission satellites, $1 billion in the Comcast cable company (which includes a home shopping network), and $450 million in WebTV, an interactive network-access device; in 1995 Microsoft invested $500 million in MSNBC, an on-line news service; and in 1994 the company purchased an undisclosed stake in the Hollywood studio Dreamworks SKG, along with a $30-million interest in Dreamworks Interactive. Pointing out that during this period Microsoft has been "acquiring strategic technologies at a rate of over one per month," Nathan Newman has determined that "there is a very real possibility of Microsoft becoming an unprecedented financial and technological colossus bestriding more markets and industries than any monopolist has ever aspired to dominate."[41]

More remarkable than Microsoft's predatory appetite is the relationship between the peculiar character of network computer technology and the dynamics of concentration. As has often been pointed out, the real desire of a media monopolist such as Microsoft is not simply to control products and content but, more important, to control the standards and protocols of future product development. The combination of widespread digitization and the fact that an expanding array of human activities are mediated by devices connected to each other in proliferating networks makes such control easier to secure. As Newman points out, "the nature of high technology makes each individual market inextricably linked to other markets through a combination of software standards, training skills, development tools and physical architecture that must all be able to work in combination. The key to the economics of network technology is that products and markets do not stand alone in these high technology markets but, instead, reinforce one path of innovation versus any alternative path."[42] If a company enjoys market dominance in one area of computing technology – say, for example, desktop operating systems – there are substantial incentives for that company to attempt to use its control in that market to leverage dominant positions in others. This is accomplished primarily by extending the control of standards afforded by dominance in the first market to enforce the compatibility of products in the second. The enmeshing of computing devices into networks simply raises the stakes and escalates the potential rewards of such gambits. For a company like Microsoft, the proliferation of networks represents, not a market out of control, but, instead, an opportunity for further market colonization.

Microsoft's recent attempts to use its monopoly position in the operating-system market to leverage similar dominance of the market for Web browsers provide a case in point. Because of Microsoft's overwhelming dominance of the operating-system market, computer manufacturers and retailers have found that, in order to remain competitive, they must sell new PCs with the Windows (and previously MS-DOS) operating system already installed. Hoping to achieve a similar market position for its Web-browser software by piggy-backing it on its already ubiquitous operating system, in 1996 Microsoft began to demand that, as a condition of their licence to install Windows, PC manufacturers also install Microsoft's Internet Explorer Web browser. Unlike the monopoly position it enjoys in regard to operating systems, Microsoft faces competition in the browser market. However, if Microsoft is successful in its plan to exclusively "bundle" its own browser with an operating system that already enjoys nearly total saturation of the PC market, the future for this competition is limited. This is a situation that has not escaped the attention of anti-trust investigators, although the technicalities of digital monopolies have proved vexing. In response to charges of unfair monopoly leveraging, Microsoft has insisted that its operating system and Web browser are, in fact, one inseparable product – the bits that form the browser cannot be removed without disabling the bits that comprise the operating system – and so "unbundling" them is impossible without destroying the product upon which the company has built a legitimate monopoly.[43]

It is this technical attribute of computers and networks that – more than any psychopathology of predatory, monopolist appetites – explains why the ownership of media constructed upon this infrastructure is likely to be as, or more, concentrated than in other media and technology sectors. In the first six months of 1998, there were 3,700 corporate combinations in the United States, valued at $626 billion. This figure is twice that for the same period in 1997, and exceeds the twelve-month totals for each of the previous four years. Massive telecommunications mergers have been cited as the driving force behind this increase.[44] Ownership in this sector concentrates not just because of one man's megalomania, but also because of the technical specificities and apparent imperatives of network technology. Operating systems, browsers, software applications, and the content they deliver, in their digitality, are decreasingly identifiable as completely distinct products, and the ability to enforce concentration-conducive standards through existing market presence is considerable. As McChesney has written, "the nature of digital communication renders moot the traditional distinctions between various media and communications sectors."[45] Secondarily, as computer networks grow to form a mesh from which contemporary economies cannot be extricated, the imperatives of interconnectivity and interoperability – the need for the pieces, both software and hardware, to fit together seamlessly – demand increased attention.

There are two ways of ensuring that the various complex elements of the digital infrastructure work together effectively. The first is for the state to develop and enforce common technical standards capable of integrating the diverse activities of a panoply of service, product, and content providers. The second is to allow a reduced range of private interests to develop market-wide standards on their own via their pursuit of the economic rewards of monopoly-level market shares. To a certain extent, these options are mutually exclusive. Referring to Microsoft's owner, one writer has observed: "The problem is you can't have less government interference and less Gates at the same time."[46] Engaging the first option runs the risk of standardizing technicalities before they have been optimized, disincentivizing research and development, and, thereby, preventing further technological advance – a proposition that, in the current discursive climate, borders on the unthinkable. Perhaps more seriously, rigid, state-enforced technical and industrial standards run afoul of the privatized, minimally regulated and commercial priorities described above as characteristic of liberal approaches to this technology and its ownership. More consistent with these is a commitment to allowing capitalism and capitalists to sort out network standards on their own, despite the likely consequence of increased ownership concentration, at least to the point that the technology and its standards stabilize.

Thus, the ownership of network technology is resolutely capitalist in character: it is private; it is only moderately regulated; and it is acquisitive, accumulative, and commercial. As with other communications media and information technologies situated in this economic context, the ownership of networked computer technology is concentrated and, due to its peculiar properties, it is likely to be increasingly so. This means that the ownership of network technology, like the ownership of most other means of power in a capitalist society, is class based and located squarely in the hands of a powerful minority. These stark realities stand in marked contrast to the rhetoric of democratization and pluralism that has accompanied this technology. As will be discussed below, ownership is not the only area in which network technology's marriage to capitalism prevents it from living up to the revolutionary promises people have been eager to make on its behalf.

Perfecting Capitalism and Networks as Control Utilities

Network technologies figure prominently in capitalist economies that, unfettered by popularly legitimate alternatives, are racing to perfect themselves. Bill Gates, the world's richest man, owner of Microsoft, whose Windows and MS-DOS operating systems run nearly all the world's personal computers and who hopes to extend that dominance to the networking sector, describes perfect capitalism as "friction free."[47] According to Gates, "capitalism, demonstrably the greatest of the constructed economic systems, has in the past decade clearly proven its advantages over the alternative systems. The information highway will magnify those advantages."[48] What does it mean to say that perfect capitalism is "friction free" and what role do digital networks play in this perfection? In basic terms, capitalism is a system driven by the accumulation of private wealth in the form of profit, and the distribution of economic value, commodities, and resources via the mechanism of market exchange. Friction is a physical or mechanical phenomenon pertaining to the resistance caused by the rubbing of one body against another. Friction slows things down. There are a number of sources of friction – things that slow down the amassing of profit and the exchange of value – in imperfect capitalist economies: the need for human labour to transform providence into value; the physicality of moving value in the form of a commodity from the site of its production to the site of its consumption; the regulation of market exchanges; and the redistribution of value and wealth as enforced by law, to name but a few. Specifically, things such as the maintenance of a stable workforce, the spatial constraints of markets, taxation, and regulation all rub capital the wrong way: though they are, to some degree, unavoidable, they all cause friction in the form of costs that slow down the accumulation of profit. If profit accumulation is deemed good, then we can say that capitalism is "better" when it faces the fewest possible obstacles in the pursuit of this, its primary end. The perfection of capitalism can thus be described as proceeding in exact proportion to the rate at which sources of friction are minimized or eliminated.

Networks are becoming a ubiquitous factor of daily life at the very moment capitalist economies are globalizing and privatizing on a massive scale; that is, network technology is maturing in the midst of a concerted effort to minimize and eliminate as much friction as possible in the operation of capitalism. Globalization is a euphemism that captures a number of phenomena centred on the dismantling of national conditions that proscribe finance practices; capital ownership; and the manner in which goods and services can be produced, distributed, and marketed. Capital flows – in the forms of trade, foreign direct investment, and financial and monetary transactions – have been

referred to as the "primary enzymes of global capitalism," with liberalization, deregulation, and privatization comprising the "three engines
of globalization."[49] Liberalization allows enterprises to move their capital
and operations to locations that offer the best competitive advantage
(i.e., disciplined, low-paid labour; favourable taxation structures;
minimalist regulatory regimes) and to sell the goods they produce, or
the services they provide, in those and/or other national settings without being subjected to prejudicial market conditions (i.e., tariffs, sur-
taxes, subsidized domestic competition). Deregulation is an essential
enabling condition of trade and finance liberalization, and privatization
is an unavoidable consequence of both, in so far as they involve a massive retraction of the state's public role as a regulator of economic activity and redistributor of resources. If states wish to remain competitive
(i.e., attractive homes for capitalist enterprise) in a liberalized environment, and wish to secure reciprocal access to lucrative foreign markets
for their domestic industries, then interventionist public instruments
such as taxation, regulation, labour laws, industrial policy, and resource
redistribution pursuant to social welfare must be used sparingly. The
dismantling of the welfare state is thus often viewed as an intrinsic
element of globalizing capitalist economies. Scholars in the Group of
Lisbon describe the features of a ten-year program adopted by the German government in 1993 as typical in this regard: "reduction of public
expenditure, in particular social security; financial, fiscal, and other
regulatory incentives to promote private investments; reduction of income taxes and corporate profit taxes; stabilization and reduction of
wages; further privatization of telecommunications; reduction of the
role of trade unions; and the relaxation of environmental regulations
that could impinge on the competitiveness of German firms."[50] As these
authors observe, "at the core of the dismantling process is the conviction that the more labor costs are cut and related social benefits are
reduced, the better will be the country's competitiveness and effectiveness in fighting unemployment."[51] Nevertheless, while they have apparently stimulated rising levels of economic growth, the dynamics of
globalization have also contributed to the structural entrenchment of a
global crisis of unemployment.[52]

What have networked computers to do with this set of phenomena?
Bill Gates believes incomplete information exchange between buyers

and sellers is the principal source of friction in capitalist economies, and he is confident that digital networks will alleviate this problem as they become "efficient electronic markets that provide nearly complete instantaneous information about worldwide supply, demand, and prices."[53] However, network technology contributes to the perfection of capitalist economies more because of its proficiency as a control utility into which information and communication collapse than because of its discrete communicative and informative utilities. As discussed in the previous chapter, a control utility is one that communicates information about goals, inputs, interactions, and outcomes to the controlling agent of a system, who can subsequently alter the system's operation. Networked computers, it was argued, are a highly effective control utility because of the speed at which they can process and communicate vast amounts of complex information, and because of their ability to perform complicated operations automatically, according to stored programming designed to achieve a wide array of goals. The desirability of computer networks as systems-control utilities is enhanced by the fact that individual networks mesh easily to form *networks of networks,* which in turn serve the control needs of even larger, more integrated, systems. The globalized, privatized capitalist economy is a system-rich environment and it is pervasive. As Wood writes, "this is the period when capitalism itself has become for the first time something approaching a universal system."[54] Capitalism has become universal not only because of its geographic reach, but also in the sense that "its social relations, its laws of motion, its contradictions – the logic of commodification, accumulation, and profit-maximization [are] penetrating every aspect of our lives."[55] An expansive, complex, deeply penetrating system presents formidable control requirements. Computer networks have formidable control capacities. It is this complementarity that has allowed computer networks trafficking in indifferent bits to emerge as a lean substitute for slower, bumpier control devices such as the state, which traffics in cumbersome regulations and laws: networks are the ideal control utility for "friction-free" capitalism.

In a world in which economics has superseded politics as a science of judgment, worthiness is rendered a function of value. In previous eras, everything of value in a capitalist economy (property, capital, labour, commodities) and the processes involved in exploiting this value (production, distribution, marketing, consumption) could either be reduced to, expressed in, or mediated by dollars and cents. In the era of universal capitalism – globalized, privatized, friction-free, perfect capitalism – these values and processes are all represented or mediated by strings of binary code. Bits are the currency of universal capitalism: a thing must be reducible to bits; otherwise, it cannot be very valuable. Thus, if capitalism has penetrated every corner of our existence, it is, at least in part, because computer networks have carried it, in the form of bits, effortlessly to and from these corners. As one particularly enthusiastic cyber-capitalist has put it, because of the reach and penetration of network technology, "there is no place to hide."[56] Economic life – the only life in a universe of capitalism – is becoming increasingly digitized in nearly all of its aspects. Why? Because bits moving over networks are subject to very little friction. Profit-seeking and accumulation suffer less "rubbing off" of value when conducted using the currency of bits and the utility of networks than was the case during previous capitalist configurations, in which currencies and control utilities were subject to greater friction. If the elimination of friction is the motivation for a digitized economy, then network infrastructure is its primary enabling condition. Canada's Information Highway Advisory Council refers to the "highway" for bits that is network technology as an "'enabler', because it changes the techno-economic paradigm and expands the productive capacity of the economy's resources."[57] Just as real highways, railways, factories, machines, bank vaults, cash reserves, radio, magazines, television, and a spatially concentrated workforce formed the infrastructure of the imperfect, industrial, capitalist economy, digital networks undergird perfecting, friction-shedding, information capitalism.

Networks and Production

As economies have been digitized, the *mode* of production has remained distinctly capitalist, but the practices of production have become increasingly mercurial due to the control utilities of networks. The impact of networks on capitalist production extends well beyond the automation of labour within factories that represents the climax of the Fordist/Taylorist model.[58] Trying to capture its decisiveness, at least one observer has referred to the shift in production design as "Gatesism," although it is not clear that Mr. Gates is as responsible for the genius of this model as he is for the technology that has made it possible.[59] Indeed, there is no shortage of pithy labels circulating to describe various aspects of productive practices in which networks play a central role: flexible manufacturing; just-in-time delivery; mass customization; lean production; total quality management; the virtual corporation; agile competitiveness; and process re-engineering, to name but a few.[60] In order to define the role of networks in facilitating the perfection of capitalism, it is necessary to understand the changes captured by these euphemisms.

Flexibility – the ability to bend without breaking – is perhaps the definitive attribute of successful enterprises in the rapidly perfecting capitalist economy. To be flexible in an accelerated capitalist economy is to be able to adapt swiftly and decisively to changing market conditions, in order to exploit these conditions to one's own advantage. As it is defined in one early articulation, "flexible specialization is a strategy of permanent innovation: accommodation to ceaseless change, rather than an attempt to control it."[61] Two important fields in which contemporary capitalist enterprises must exercise flexibility if they hope to maximize profits are location and production operations. In an environment of liberalized trade and investment agreements, and of retracted state intervention in economies more generally, flexible corporations locate their production facilities in areas where conditions – labour costs, taxation structures, regulatory regimes – are most likely to yield high returns on investment. Furthermore, as governments race each other to the bottom in an effort to minimize disincentives to capital investment, flexibility means being able to move operations around to sites of least friction. The threat of easy mobility or "flight" also allows capital to extract concessions from governments in jurisdictions where operations are already located. The control utility of networks is a crucial factor in making this type of flexibility possible: "With instantaneous world wide communications it is theoretically as easy to control a factory in Asia as it is to control one right next door. The ease with which production can be integrated around the world will provide companies with greater flexibility than ever in selecting plant locations."[62] The operation and control of a productive enterprise is a complex task involving the exchange, coordination, and integration of massive amounts of input, process, and output information (orders, specifications, coordination, inventory, pricing, accounting, delivery, etc.). Such control is difficult enough to exercise when all aspects of the enterprise's operations are physically centralized, in close proximity to system administrators, and stationary. It would be nearly *impossible* to exercise control efficiently and profitably in a decentralized, highly mobile enterprise using pre-network media of information storage, retrieval, and communication. However, "the technologies now exist to synthesize geographically dispersed knowledge, information, operational facilities, and expertise. Furthermore, these technologies are becoming increasingly robust, widespread, and comprehensive."[63] These technologies are computer networks. The relaxation of market restrictions in most national economies has given capitalist entrepreneurs a motivation for overcoming these limitations, and network technologies – by enabling the translation of systems-control information into bits that can be stored in vast quantities, retrieved effortlessly, and communicated instantly and reliably – have provided them with a means of doing so.

Another type of flexibility prominent among capitalist enterprises in the current environment is flexibility in their productive processes and products, whether they are primarily manufacturers or service providers. In this context, "flexibility refers to the capability of an organization to move from one task to another quickly and as a routine procedure."[64] Under the Fordist manufacturing model, profitability required the rationalized, repetitive production of mass quantities of standardized goods marketed to a mass of consumers whose needs and desires were homogenized and stable. The flexible manufacturing model is based on the assumption that the pay-off of mass production and consumption patterns has been replaced by the profitability of specialization and customization. The discourse of the current episode of capitalism insists that markets have fragmented into a plurality of highly differentiated consumers demanding low-price, high-quality goods and services that conform perfectly to their particular, though ever-changing, requirements. Consequently, enterprises must be prepared to be flexible specialists, able to reconfigure their productive systems quickly and seamlessly to meet demands for a proliferating variety of customized products with short demand cycles. Flexible specialization has spawned a number of subsidiary models. Pioneered in Japan, "just in time" manufacturing and supply refers to a system capable of producing customized finished goods, parts, or subassemblies in limited quantities in a very short time, thus eliminating the need for manufacturers to maintain costly, and vulnerable (i.e., to obsolescence), inventories of goods.[65] Closely related is the "lean production" model whereby enterprises seek to eliminate any source of waste and inertia that might decrease their "fitness" for rapid resource redeployment and operational reconfiguration.[66] Customarily this involves a rationalization of labour, physical plant, and on-site resources.[67]

To operate in this environment, capitalist enterprises must be "agile." Agility has been defined as "a comprehensive response to the business challenges of profiting from rapidly changing, continually fragmenting, global markets for high quality, high-performance, customer-configured goods and services."[68] A key component of agility is the productive responsiveness and customization made possible by sustained, detailed, and direct communication between producer and consumers. The Fordist mass- production model was serviced by a market of consumers who either were, or had been, standardized, because standardized goods were the only sort of goods that could be produced on a massive scale in a controlled process. The new styles of production are based on a shift from the assumption of a very large and stable market to a market that, having reached the limits of its growth, is nevertheless capable of perpetual and rapid reinvention. To profit in this market, agile enterprises must be responsive, and responsiveness requires highly integrated systems of design, production, marketing, and distribution. In some cases, this has been accomplished by a blurring of the distinction between these functions *within* organizations. More striking, however, is the emergence of what is known as the "virtual" company or corporation, in which "complementary resources existing in a number of cooperating companies are left in place, but are integrated to support a particular product effort for as long as it is economically justifiable to do so ... it is increasingly easy to integrate design, production, marketing and distribution resources distributed around the world into a coherent 'virtual' production facility."[69]

The move to "agile," "mass customized," or "virtual production" is occurring across a wide range of service and manufacturing industries.[70] Specific examples are legion: a recent study of the virtual economy provides a descriptive list of more than 100 major international corporations who have adopted agile production methods.[71] The list is not nearly exhaustive, but it is representative. In 1994, the Ford Motor Company integrated all of its operational activities, dispersed among thirty countries, into a single system of "centralized supplier relations and purchasing, true worldwide production scheduling, and integration and assimilation of front-end and back-end operations."[72] This is made possible by the coincidence of a liberalized, global market environment in which national economic distinctiveness is not a factor, and a technology – computer networks – that allows an operation of this scale and complexity to be controlled effectively. AT&T relies on a Global Information Systems Architecture "to standardize its business manufacturing systems worldwide ... to improve its ability to compete in all telecommunications markets in whatever part of the world they may be ... using physically distributed resources and bringing about a reduction in product cycle time and cost."[73] In Mexico, the Grupo Azteca soft-drink company has linked its twelve bottling plants to its 180,000 distributors using satellites and computers, including hand-held terminals that enable individual delivery drivers to conduct "real-time, on-site tracking of store-by-store inventory and product-by-product sales rates."[74] In large retail operations throughout North America, quick-response point-of-sale technology reads product bar codes, then relays information to central computers that track inventory and highly specified sales information, automatically dispatching orders to manufacturers and distributors for out-of-stock or fast-selling merchandise.[75] In Canada, Journey's End Hotels uses a computerized, networked databank to file detailed information on guest preferences regarding accommodations and services, making it possible to obtain instantly customized reservations at any location.[76]

Agility is not only about "the integration of the internationally distributed facilities of a single company into a truly coherent global production resource," but also entails strategic, synergistic alliances among erstwhile competitors.[77] Efforts to perfect capitalism are literally festooned with such alliances of convenience, but perhaps most interesting is the way in which networks facilitate them. Key to the formation and success of these ventures is the establishment of "cross-industry computer network and database systems that ... provide instantaneous access to detailed information about the capabilities of hundreds of thousands of companies, about the cost and availability of their expertise and facilities, and about their terms for participating in collaborations."[78] Numerous such networks exist, including Strategis, the Enterprise Integration Network funded by the Advance Research Projects Agency of the US Department of Defense, CommerceNet, TYMNET, SWIFT, and SITA.[79] Networks provide enterprises looking for temporary strategic allies with an instrument for accessing information about potential suitors readily and reciprocally. Just as important, however, is that the cooperative efforts of these allies be able to interface with ease. Put crudely, different conventions of greeting – the Japanese bow, the French kiss, the Canadian handshake – can make it difficult for people to be introduced to each other without friction; a standardized greeting exercised through a common medium makes getting together much easier. This is what networks provide for enterprises whose competencies, products, and services may vary greatly. When most aspects of business operation are expressible in the standardized form of bits and packaged according to standardized protocols, the operations of one enterprise easily interface with the operations of another. When a series of bits registers at a check-out counter as a sale, they are immediately understood by a manufacturer's computer as an order for another pair of jeans and, in turn, by a supplier's computer as a request for an additional bolt of denim. No one is bowing while someone else is trying to kiss him or her on the cheek. Networks that act as efficient, reliable, widespread media for the exchange of standardized bits make it infinitely easier for otherwise distinct enterprises to "shake hands," roll up their sleeves, and get to work.

In 1993, Statistics Canada conducted an extensive survey of technology adoption in Canadian manufacturing.[80] The results were telling. The study found that 81 percent of all manufacturing shipments in Canada were carried out by establishments using at least one advanced production technology. Nearly 60 percent of all shipments came from plants utilizing at least five such technologies. Roughly half (46 percent) of all manufacturing shipments were carried out by companies using flexible manufacturing systems in the actual physical assembly and fabrication of goods. Most strikingly, nearly three-quarters (73 percent) of the goods manufactured in Canada were produced by firms using some combination of programmable process controllers, local-area data networks, intercompany computer networks, computers for controlling factory-floor operations, computers for manufacturing resource planning, and networks for supervisory control and data acquisition. The study also found that the percentage of shipments from companies using ten or more such technologies in their productive operations increased by 15 points between 1989 and 1993, and was expected to grow by another 18 points by 1995. Significantly, manufacturing industries in which over 50 percent of shipments originated in companies using these technologies included resource extraction and refinement, food-processing, clothing and textiles, printing and publishing, transportation equipment and machinery, synthetics, woodworking and furniture making, to name but a sample. All of which led the authors of the survey report to conclude: "The computer-based revolution in manufacturing technologies has been widely felt in the Canadian manufacturing sector."[81]

Network technology, especially at these high levels of penetration, is what makes the new production paradigm, in all its various guises, possible. Computer networks did not create the globalized, privatized economy; they do, however, make it possible to exploit this economy. Flexible production and mass customization are based on speedy, but controlled, processing of systems information, a requirement that the movement of bits over networked computers is particularly suited to meet. In the current configuration of capitalism, the privatization of markets has become synonymous with "change." Networks are control utilities, but they are not deployed to control change – change is clearly in the interests of capitalist enterprises looking for regenerative markets and salutary production environments – but they do allow these enterprises to control their own operations efficiently and profitably in the midst of what is presented to the world as a maelstrom of irresistible dynamism. Networks are the essential technology for those "agile" and "virtual" enterprises that are "thriving on change and uncertainty" in the era of perfecting capitalism.[82] The intimate relationship between computer networks and the new way of making profit in what remains an old capitalist mode of production is clear: "For opportunistic collaborations, whose lifetimes are defined by the profitability of the market opportunities they were created to exploit, electronic integration reduces the burdens of ownership of human and physical resources. It lowers the threshold of profitability of a new operation. It promotes networks of mutually profitable cooperative relationships with other companies among which necessary resources are dispersed."[83] If this is what network technologies do *for* capitalists, Marx would ask us, then what do they do *to* those burdensome human resources whose lifetimes are defined largely by the work they do, and the manner in which they carry out the activities of consumption and exchange? It is to these questions that I turn in the next chapter.

the political economy of network technology 2: work, consumption, and exchange

. . . . .

IN 1848, MARX OBSERVED that "the bourgeoisie cannot exist without constantly revolutionizing the instruments of production, and thereby the relations of production, and with them the whole relations of society."[1] The preceding chapter considered the role of network technology as a control utility in a capitalist mode of production striving to shed friction and achieve maximum flexibility on a global scale. While the vicissitudes of transnational production regimes certainly have an impact upon the everyday lives of people whose existence they frame, such impact is generally felt most acutely in the commonplace activities that constitute the routine exercise of economic life: work, consumption, and exchange. Here, too, network technology contributes to structuring relationships and practices that are synchronized with the accelerating rhythms of capitalism more generally. A consideration of these relationships and practices, and the role of networked computers in them, will complete our examination of the political economy of this technology.

Networks and Work

In Marx's view, the ontological impact of any technology is felt primarily via its insinuation into human labour practices, the activity through which people express their essential being. As a force of production, a technology's effect on working life is directed by the mode of production in which it is deployed. With this in mind, Marx determined that, while it increased productivity and profits, the technology of industrial machines also served to complete the alienation already experienced by people working under the auspices of the capitalist mode of production. Machines accomplished this by further distancing workers from the capacity to determine freely the conduct, conditions, and fruits of their labour. In the automated factories of industrial capitalism, Marx surmised that workers were transformed "into a living appendage of the machine."[2] What have they become in the virtual economy of a globalizing capitalist system striving to perfect itself?

If alienation is the flip-side of profit in a capitalist economy, then it is fair to say that network technologies have contributed to the perfection of both. David Noble, a keen student of the relationship between work and technology throughout history, expresses the matter starkly: "In the wake of five decades of information revolution, people are now working longer hours, under worsening conditions, with greater anxiety and stress, less skills, less security, less power, less benefits, and less pay. Information technology has clearly been developed and used during these years to deskill, discipline, and displace human labour in a global speed-up of unprecedented proportions."[3] This view is, of course, in marked contrast to the one advanced by those who see in network technology an exciting opportunity for individuals to customize their working situations – "reinventing themselves - possibly more than once"[4] – in the interests of maximum flexibility. However, with few exceptions, it would appear the control utility of network technology has been deployed so as to minimize the friction attributed to the necessity of human labour in the production of surplus value. There are a number of ways in which digital networks have been deployed with this goal in mind, and while some individuals have managed to squeeze benefit out of technological upheaval in their working lives, for workers, *collectively*, networks have represented a step back in terms of relieving the general alienation and exploitation to which they, as wage labourers in a capitalist economy, are systematically subjected.

The Disappearance of Work

According to Marx, one result of capitalism's incessant revolutionizing of the instruments of production is that "all that is solid melts into air."[5] Indeed, network technology has enhanced the ability of capitalists to make work disappear, contributing to what has been called the "jobless" growth and future of free-market societies. There are some who believe eliminating labour and its attendant costs is the raison d'être of digital networks: "from the individual employer's perspective, the real purpose of technological innovation is labor displacement as a vital component of reducing costs."[6] According to Stanley Aronowitz and William DiFazio, "computer-based technology *inherently* eliminates labour" and, citing forecasts by the US Bureau of Labour Statistics that show that high-tech jobs are likely to replace only 50 percent of lost manufacturing jobs, they conclude that "high technology will destroy more jobs than it creates."[7] Jean-Claude Parrot, in his dissenting minority report to Canada's Information Highway Advisory Council, points to skyrocketing global unemployment, including 35 million people out of work in the OECD countries, and affirms that "the accelerating pace of technology is among the reasons for this persistently high unemployment."[8]

Nevertheless, there are also those who see in network technology the prospects of an upsurge in employment, including the authors of a report that was central to the deliberations of this very same advisory council: "Technological change both destroys and creates jobs. However, history demonstrates that the use of new technologies, including Information Highway technologies, has led to productivity increases and created new employment, often in whole new industries, which has more than offset the initial job losses."[9] On the basis of "extensive studies" they refer to but do not cite, these authors conclude that "Information Highway development and use will have a net positive effect on employment."[10] It is certainly the case that unemployment in the United States, the leading country of the network revolution, is at its lowest level in decades, and that European countries lagging behind in terms of the development of this technology also suffer comparatively higher levels of unemployment. It is also the case that many of the jobs fuelling employment in the United States are low-paying, temporary, and insecure. Predictions about the impact of network technology on net employment thus divide rather neatly along the lines of ideology and interest: left-wing observers and representatives of the working people upon whom technology is inflicted tend to see joblessness on the horizon; right-wing analysts and representatives of the corporations who wield the power of bits predict a dynamic future full of opportunities for those who are ready for them.[11]

If the strict causality of network technology in relation to net unemployment is as difficult to establish at a macro level as these divergent opinions would suggest, then what can we determine about the effect of this technology on work? First of all, it is apparent that unemployment has settled in as a structural element of many economies at the very moment that computer networks have become *infra*structural in those same economies. In Canada, the rate of unemployment has congealed in the 1990s at around 10 percent, a figure that does not include the doubling factor of those who have ceased looking for work, or those who are employed only partially and/or temporarily.[12] Second, it is certain that technological change centring on the introduction of computer networks has contributed greatly and directly to the specific displacement of large numbers of workers across many sectors. Many of these lost jobs are "replaced," in statistical terms, by jobs in other sectors or firms that have arisen to meet or suit the needs of a networked economy. However, the stabilization of the unemployment rate in Canada at the relatively high level of roughly 10 percent would seem to suggest a perpetual and consistent lag in the overall movement of workers from technologically *destroyed* jobs into technologically *enabled* ones. Even those who are confident that networks will increase total employment recognize the problem of "mismatch" and "transition" for displaced workers.[13] The point here is that, even if jobs eliminated by network technology are eventually replaced by jobs "elsewhere" in the economy, the fact of their elimination is more significant in the lives of the people who held them than is their replacement with a job for somebody somewhere else. This being the case, to discern the impact of networks on working life as regards employment, it is less important to determine how *many* jobs are created relative to those eliminated than it is to understand the *manner* in which existing jobs vanish at such an alarming rate.

Jobs disappear into bits mediated by network technology in nearly every sector of contemporary economies. In manufacturing industries that retain domestic production facilities, networks contribute to job elimination because they complement the regimes of process control necessary for competitiveness in an increasingly perfect, global capitalist economy. Mechanization introduced physical force into the

· · ·

production process on an unprecedented scale, but people were still required to operate the machines. Automation increased the speed and volume of production, but limited individual factories to a narrow range of repetitive processes. Changing these processes required great effort, and coordinating them among various enterprises or facilities required the exercise of complex administrative control by working human beings. The networked computerization of production makes it "flexible," "customizable," and "responsive" because, when process information is reduced to the standardized, universal form of bits that can be exchanged almost instantaneously and with considerable reliability across vast distances in great volumes, the need for friction-ridden human labour in systems administration and control is greatly reduced. The trend of replacing human workers whose jobs involve some control over production processes with networked computers has manifested itself in numerous work areas, ranging from resource extraction and processing, construction, textile and garment manufacture, to the manufacturing end of railway, telecommunications, aerospace, and automotive industries.[14] In the recession of the early 1980s, worker layoffs were the order of the day in the Canadian auto industry. A study by the Canadian Auto Workers found that up to 80 percent of these "downsizing" plants purchased computerized process-control technology in the late 1980s, rather than hiring back their laid-off employees.[15] Similarly, in 1995, Statistics Canada reported that large-scale enterprises with more than 500 employees were three times as likely to adopt multiple (i.e., more than five) advanced manufacturing technologies than were small companies with fewer than twenty employees.[16] The reason for this is that network technology, unlike real workers, meets the large-scale manufacturers' desires to be flexible, agile, and virtual: "to respond quickly to changing customer needs ... to become the most efficient and lowest cost mass producer, with components outsourced to suppliers worldwide ... manufactur[ing] customized products in shorter runs at the low cost of high-volume, mass produced products."[17]

Significantly, the disappearance of work into networks occurs not only in factories, but also in any type of work where decisions and skills can be reduced to the standardized and easily controlled form of bits. For example, in the construction industry, networks make it possible for a number of standardized components fabricated en masse at a variety of remote and disparate locations to be selected, ordered, dispatched, then quickly assembled on site, with a fraction of the human labour (and, thus, at a fraction of the cost) required to build a house from the ground up. As described in a recent report on technology in the building trades, "the 'systems' approach sees the building process not as a series of tasks which need to be completed sequentially on site, but rather as the integration of numerous components at one location as quickly as possible ... Such systems are increasingly applied to entire building projects."[18] Furthermore, "the systems approach to building will obliterate the existing trades structure."[19] Work, jobs, and even entire trades are obliterated by networks because this technology facilitates the reorganization of production to drastically minimize human labour and human process control. Entire walls, complete with hung doors and windows, electrical wiring, paint, and trim, are mass-produced and easily altered to meet changing specifications in a centralized factory that requires less human labour relative to output than a traditional building site. These walls can be priced, ordered to specifications, matched with other components from other manufacturers, dispatched to remote locations, and paid for, all using the technology of computer networks. When the finished wall arrives at its destination, there is no need for framers, carpenters, electricians, drywallers, bricklayers, or painters: all that is needed are a few good men to follow the instructions on the box and complete the assembly. This is how work disappears into bits. Certainly there will be jobs at the wall factory, but not the *same* jobs, probably not filled by the *same* workers and, at any rate, not *as many* jobs as there were on the construction site prior to application of networks to the production process there.

The dynamic pervading the Group of Seven (G-7) economies features a steady decline in overall manufacturing employment, and a concomitant rise in the proportion of people working in service industries, and this is especially true in Canada and the United States.[20] Between 1989 and 1992, 20 percent of all manufacturing jobs in Canada were lost.[21] By 1995, service industries accounted for no less than 70 percent of Canada's overall employment and gross domestic product.[22] Figures such as these often give rise to the nostrum that while work is being lost in the manufacturing sector due to (among other factors) the penetration of network technology, it is simultaneously being created in the service sector, particularly in those industries related to the operation, maintenance, and deployment of networks and related services. Indeed, it is true that some job growth has occurred in high-technology service areas such as software development, network-systems engineering and administration, and computer graphics.[23] However, these gains tend to be offset by massive employment reductions in other service areas, such as government, banking, finance, insurance, retail, and communications. In January 1994, of the 108,000 people who lost their jobs in the United States, half had been employed in the communications, banking, finance, and insurance industries.[24] Furthermore, it is important to keep in mind that the technological innovations eliminating employment in the manufacturing sector are simultaneously being deployed in the service sector, compromising the latter's ability to simply absorb the losses experienced in the former.[25]

In fact, the service industries are particularly vulnerable to the kinds of job disappearance resulting from networked computerization described above in relation to manufacturing. Most service-industry jobs involve people working on, dealing with, distributing, and exchanging information in one form or another.[26] As information is reduced to bits that can be stored, processed, and exchanged via networks and the devices connected to them, the human labour required to deliver services that rely on this processing, storage, and exchange is minimized.

Instances of the effective replacement of service workers by networked computers are myriad and increasing. One area where costly service workers stand to be replaced by cheaper and more efficient network technologies is in the check-out line at the supermarket, where digital point-of-sale systems eliminate the need for cashiers altogether. One system, now in widespread use in Europe, allows customers to scan their purchases into a hand-held device as they select them from the shelves. When their shopping is complete, the device totals their purchases and the customer pays this amount at an express cash register. The data from the device is then automatically input into the store's main computer system for accounting and inventory control. The eight cashiers previously required to process items and customers at an efficient rate are replaced by a fleet of pocket calculators and a single change clerk. A second system goes one step further, allowing shoppers to act as their own, unpaid, cashiers: customers scan and weigh their items (under video surveillance to deter fraud) at a self-serve check-out and, when presented with a total, they make payment using credit or debit cards at an automated teller machine. The advertised benefit of the first system is that it promises to save a customer with twelve items (who accounts for 60 percent of the transactions in a grocery store) approximately two and a half minutes. The second system – because it still requires mass scanning (by customers who may be slower at it than trained cashiers) at the point of purchase rather than at the point of selection – does not promise substantial customer time savings. Here the attraction seems to be solely the system's propensity to eliminate human labour costs.[27]

The employment reductions enabled by network technology extend beyond lower-level service jobs at the supermarket: "as technology is already substituting human labour in almost all routine banking, insurance, tourism, administration and social services, technological improvements and progress will in the future affect high value-added services, especially business services."[28] Automated banking computers that facilitate the processing and exchange of information about money and accounts eliminate the need for human bank tellers; networked customer service kiosks with multilingual touch-screens eliminate the need for front-line personnel in the delivery of many government services. In the mid-1990s, the Canadian federal government installed 4,200 Human Resources Development kiosks across the country, replacing hundreds of Canada Employment Centres and their staffs.[29] These examples are somewhat crude but they are to the point: most jobs that involve the simple processing or exchange of information – a task that defines many service occupations – can be done faster, more efficiently, and more cheaply by a networked computer than by a human being.

The service sector has grown in advanced economies primarily because the labour-cost savings enabled by network technology promise a high return on investment in that sector. Remarking on the relative increase of service-sector investment vis-à-vis manufacturing, Aronowitz and DiFazio have observed: "There is no shortage of capital, only shortages of capital investment in economic sectors associated with production, which, compared to other options, cannot deliver the maximum possible return."[30] The service sector promises high rates of return because, even more so than with manufacturing, its products and processes can be controlled using networked computers, which are far more flexible and far less expensive than working human beings. The irony, of course, is that the same technology fuelling the dramatic growth of the service sector is also responsible for "cannibalizing" a great many of the jobs in that very sector. This conundrum is the result of internalizing the orthodoxy that holds that a healthy employment picture results from industrial competitiveness – whether in the resource, service, or manufacturing sector – and that competitiveness requires the adoption of flexible, friction-free network technologies. The pathology of this orthodoxy is expressed well (albeit unintentionally) by the authors of the key report to Canada's Information Highway Advisory Council: "sectors will be able to increase employment only if they can increase their output in global markets. Increasing competitiveness entails, among other things, the use of new process technology, *which is largely labour-saving*."[31] According to this logic, increasing employment requires the adoption of technologies that eliminate employment. Such doublespeak perhaps makes sense of phrases like "jobless growth," but it also illustrates the manner in which work transformed into bits tends to disappear into networks.

Work that disappears into networks at one place often reappears, in a reconstituted form, in other places. This phenomenon is known as the "migration" or "deterritorialization" of work, and it is particularly prevalent in low- to medium-skill service occupations and industries.[32] In cases where labour cannot be completely eliminated by network technology, it can very often be moved to geographical locations where it costs less. A recent report prepared for the International Federation of Trade Unions recounts a number of cases where the presence of network technology allowed the relocation of work to areas where wages and rates of unionization are low, and workplace regulations are minimal.[33] Data processing is one field that has largely been relocated "offshore" to take advantage of comparatively low wage rates, with the Philippines ranking first in the market for remote data entry. In 1992, data-entry clerks in the United States received an average of $65.00 for 10,000 keystrokes, while workers staffing the 2,000 keystations in the Philippines received between $4.00 and $6.00 for the same volume of work. The islands of the Caribbean (English-speaking) have also provided a profitable home for this type of work, where young workers, most of them female, enter data for wages ranging from as low as $0.80 per hour in Jamaica to $2.88 per hour in Barbados. The same work would compel an average of over $8.00 per hour if it were done in the United States. Remote data entry for a fraction of the cost that would be incurred in a domestic facility staffed by domestic workers is made possible by networks that carry data in the form of bits back and forth from the site of its entry to central databases. However, despite their role in securing higher profit margins, even these "offshore" jobs are vulnerable to disappearance, as developments in digital scanning and optical character recognition threaten to reduce the jobs of 100 data-punchers to a single job loading forms into a machine that feeds a scanner that "reads" the data and sends them to designated databases via computers connected in a network.

"Call centres" represent another example of how networks enable the deterritorialization of work. A call centre is a remote facility where numerous operators process telephone requests for sales or service using computers that are connected by networks to a centralized database or system. Computer sales and technical support, airline and hotel reservations, insurance claims and inquiries can all be handled in call centres serving international markets: a customer calling a local number for SwissAir in London, England, may be answered by an operator in Bombay who can sell him a ticket, provide schedule information, and reserve rental cars or accommodations with applicable discounts all via her network connection to the company's computer system.[34] Call-centre work requires somewhat more education and communications/language skills than data-entry work, which makes the location of call centres in the lowest of low-wage zones infeasible. Nevertheless, call centres will locate in areas that afford the greatest prospects for profitability due to low labour and operating costs while still providing workers who are qualified. Thus, developed areas looking to bolster their economies – such the Republic of Ireland, the so-called Call Centre of Europe – have managed to attract call-centre operations by making themselves profitable investment environments.[35] In Atlantic Canada, the government of former premier Frank McKenna of New Brunswick established a strategic partnership with Northern

Telecom and New Brunswick Telephone to develop a high-capacity network that facilitated the location of nearly 100 call centres in that province by the mid-1990s. Included among these was a call centre for Purolator couriers that – replacing eighty separate offices across Canada – processes pick-ups, traces packages, and handles ordering, billing, and general inquiries via a single 1-800 telephone number. Many of these calls can be handled by the system automatically, without the caller ever speaking to a live agent. Banks, trust companies, facilities-management firms, and appliance service providers are among the enterprises who have located call centres in New Brunswick.[36] Along with developing a network infrastructure capable of carrying digital traffic, the New Brunswick government has offered subsidies, forgivable loans, training guarantees, low workers' compensation rates, exemption from payroll taxes, and labour costs significantly lower than other Canadian locations as incentives to enterprises shopping for call-centre locations.[37]

The location of call centres in underdeveloped or stagnant locales can obviously provide much-needed employment boosts in those areas. The question here is whether the deterritorialization of work enabled by networks and symbolized by call centres renders employment more precarious generally. Put bluntly, digitization makes it as easy for a call centre to move out of an area as it was for the operation to move into it. The deterritorialization of work is one aspect of what is often described as "capital mobility," the ability of capitalists to locate their enterprises in areas of least friction and greatest profitability, and to quickly relocate them as new incentives arise. While this has engendered the appearance of some work in desirable locations, it has meant the disappearance of work in others – in particular, wherever working people have managed to extract from capitalists a closer approximation of what their labour is actually worth, and a financial commitment to the health of their communities. Two phenomena have enabled this work-eliminating mobility, one political and the other technological. Politically, a liberalized, more perfectly capitalist environment of trade, commerce, and investment has diminished the penalties capitalists suffer in moving their operations to locations promising the least friction and greatest returns. It is this same framework that has forced governments to race each other to the bottom in attempting to accommodate the profit requirements of capital. Technologically, the conversion of a great deal of productive, commercial, service and finance activity into the form of bits, and the development of a network infrastructure through which these can be exchanged, processed, and controlled, have enabled capitalist enterprises of many sorts to take full advantage of this environment. When capital moves – because it *wants to* in a liberalized environment and because it *can* when networks are everywhere – work disappears. As I have argued above, it either disappears completely, or it moves far away from where the person who *did* the work *has done* the work. This, in most cases, is tantamount to the work disappearing altogether.

The De-institutionalization of Work

Not all work has disappeared into computer networks, but a great deal of the work that remains is increasingly mediated by them – in Canada, nearly half of all workers use computers on the job[38] – and a large portion of this work is what we could call "de-institutionalized." In previous eras, "work" meant a full-time job or occupation; it meant steady, reliable employment and remuneration; and it often meant a "place" where that work was done. Much of the work that is being maintained or created in the era of network technology fails to reach these standards: "Work on the fringes – part-time, temporary, term-contract, self-employment, and other 'contingent' work – represented the bulk of new job growth through the 1980s."[39] A 1997 study prepared by the Canadian Council on Social Development reported that, in a time of declining full-time employment, part-time work had grown to represent 18 percent of the jobs in Canada.[40] Of these part-time jobs, 60 percent are in clerical, sales, and service occupations, although part-time work in managerial and professional jobs is also growing at break-neck speed. Work on the fringes is generally bereft of fringe benefits, lacking in security, non-unionized, and poorly remunerated. Forty percent of part-time workers earn less than $7.50 per hour, and fewer than one-fifth of them are covered by occupational pension, medical, dental, and paid sick-leave plans.[41] One-third of all part-time employees work irregular hours and receive earnings that vary from one week to the next; 28 percent of part-time workers in Canada describe their jobs as "non-permanent" and therefore insecure (compared with the one-tenth of full-time workers who describe their jobs this way).[42] Fully half of all part-time employment in Canada is involuntary, with 50 percent of part-timers wishing to work more hours than they already do. Finally, while one-third of the part-time workers in Canada are men and women under the age of twenty-five, and 17 percent are men aged twenty-five and older, half the people working under the conditions described above are women over the age of twenty-five.[43]

Networks figure prominently in enabling the shift to this kind of work, examples of which can be found across office, retail, and service occupational categories in both public and private enterprises. In school boards, hospitals, libraries, municipal governments, insurance agencies, accounting firms, financial houses, and medical clinics alike, there has been a concerted shift from full- to part-time employment; from permanent to contract positions; from full to partial or no benefits; from salaried remuneration to piece-work. In 1985, a comprehensive study conducted by the Canadian Union of Public Employees of more than 2,000 of its locals in the public sector found that 40 percent had experienced decreases in full-time positions and corresponding increases in part-time positions. In over 60 percent of those locals where more than 50 full-time positions were lost, significant computer-related change had occurred at the worksite. Computer-driven technological change was also prevalent in roughly 90 percent of those locals that had experienced increases in part-time positions.[44] Causality in this situation would, of course, be difficult to prove. Nevertheless, it is safe to say that the arrival of computer networks has at least coincided with changes creating a condition wherein, as Aronowitz and DiFazio put it, "the 'meaning' of work – occupations and professions – as forms of life is in crisis."[45]

The rise of so-called telework has contributed greatly to this process of de-institutionalization. "Telework" refers to work performed "at locations other than the traditional workplace for an employer or client, involving the use of telecommunications and advanced information technologies as an essential and central feature of the work."[46] Teleworking generally takes place in the home of the worker, although there are a number of possible arrangements within this scope. The combinations of variables include: location (single or multiple); status of worker (self-employed, freelance, contract, or employee); type of remuneration (piece, salary, or wage); degrees of voluntarism (i.e., does the worker have a choice between working at home or working in the office); on-line versus off-line time; and the mixture of tele- and on-site work. Whatever the particular arrangement, telework is a form of labour that is growing as networks proliferate and the cost of personal computers decreases. In 1995, Statistics Canada estimated the number of teleworkers in Canada to be 300,000.[47] By 1997, that number had reached 650,000, with another doubling expected by the turn of the century.[48] In the United States, the number of teleworkers hovers around the 20-million mark.[49] The trend toward increasing telework as an alternative to traditional jobs has led several labour organizations – including the International Trade Union Federation and the International Labour Organization – to formulate wide-ranging recommendations concerning the regulation of this form of labour practice, and it was precisely the inattention to this issue that prompted the lone dissent from the final report of Canada's Information Highway Advisory Council.[50]

For some, teleworking provides a degree of flexibility that allows them to manage more creatively the demands of work in the course of their daily lives. This is the case for teleworking executives and professionals: who can choose whether and when they wish to work at home or at the office; who determine their own daily schedules; who do their work off-line and connect to the company's network only intermittently; who increase their leisure by decreasing physical commuting time; who are salaried (or whose wages and overtime rates are protected by collective agreements); who have the resources to create a workspace in their home that is safe, healthy, and distinct from their private or social space; and who work under their own supervision. Unfortunately, this does not describe the conditions under which all, or even many, teleworkers pursue their livelihoods. Most people for whom telework is the only work are employed in either low-level administrative/managerial tasks, or clerical, sales, and service occupations.[51] Of this second category – the workers most vulnerable to the various pathologies of teleworking – the majority are women, trapped in a homeworking situation that has been described as "a female-dominated work ghetto."[52]

The potential pathologies of telework can be categorized into issues of status and issues of site. Issues of status pertain to the ambiguous or tenuous status that many teleworkers experience in relation to their

· · ·

employers. Most teleworkers are employed on a contingent basis; that is, they are part-time, limited-term contract, piece-workers who are not considered full "employees" of the firm they are working for. The absence of employee status is accompanied by an absence of proper employment rights, protections, and benefits.[53] Many teleworkers are considered "self-employed" and therefore responsible as individuals for their own working conditions; ineligible for benefits and training offered to full-time employees of the firm; exempt from collective agreements; and unprotected by codes regarding, for example, notice of termination, unpenalized absences due to sickness, and holiday pay. Taken together, these matters of employment status suggest that, while the teleworking arrangements enabled by computer networks provide increased flexibility for employers, they also mean increased insecurity for teleworking employees. It is for this reason that trade union calls for the protection of teleworkers characteristically include demands to clarify the status of teleworkers as employees eligible for the same rights, protections, and benefits as conventional on-site workers.[54]

Whether they are traditional firm employees or pseudo "self-employees," teleworkers are vulnerable to exploitation primarily because their worksite is physically removed from the central location of the employer's operation. The pathologies of site take a number of forms. In the first place, teleworking often entails a "shift in the capital costs from the employer to the workers ... [whereby] workers are expected to absorb the costs of doing business."[55] Overhead costs, including equipment and supplies (computers, modems, furniture, software, telephones), utilities (heating, lighting, electricity, phone service), training (courses and manuals), and maintenance, are, in many cases, borne by teleworkers themselves.[56] Given the limited resources of many teleworkers, this burden often translates into an unhealthy, ill-maintained, and potentially hazardous worksite, where the purchase of suitable computer equipment often takes precedence over considerations of ergonomics, lighting, and ventilation. The decentralization and dispersal of telework sites makes it extremely difficult to inspect and monitor working conditions in them. Typically, worksites in the home are not even separated physically from general living and social spaces, with networked terminals set up in bedrooms, dining areas, or family rooms. Executives and professionals who choose to do *some* of their work at home can usually afford to set up and maintain decent "home offices." Teleworkers who have no choice but to do *all* their work at home and on-line typically cannot afford such a luxury, and work instead in "silicon work cells" that also happen to be their kitchens.[57] This obliteration of the distinction between a worker's home and working life has profound consequences when one considers that the majority of teleworkers performing clerical and service work are women who, in working at (low-)paying jobs, are already under pressure for neglecting their duty to perform unpaid maternal and domestic work. It is not without reason that teleworking has been forwarded as a possible "solution" to the day-care crisis.[58] When performed at the same site where cooking, cleaning, and diaper-changing waits to be done, network-mediated telework enables women to exceed even Aristotle's designation of their utility: they can be, simultaneously, unpaid domestic managers as well as poorly paid but economically necessary wage slaves.

Trade unions have sought to protect women and other teleworkers from this sort of exploitation, but the difficulty of doing so illustrates another subset of pathologies pertaining to considerations of site. Workers who are physically dispersed are hard to organize, are less likely to be informed about the situation of their fellow employees, have difficulty reporting abuses, are more subject to the psychological effects of isolation, and are unable to realize the social benefits of association with peers in the workplace.[59] As one teleworker in Toronto, who processes pizza-delivery orders in her bedroom, using a telephone and networked computer terminal, for $7.00 an hour, has described: "[I'm] totally isolated from everybody. You have to be a loner to be able to work alone at home ... you have no social life ... It's easy to get lazy and not get dressed. Sometimes my husband comes home from work and I am still in my housecoat."[60] Teleworkers who are isolated in their homes are also less likely to be aware of opportunities for advancement, and develop neither the connections nor the skills necessary to take advantage of these opportunities in the unlikely event they *are* aware of them. Thus, many teleworkers are effectively ghettoized, and this ghettoization makes them particularly vulnerable to abuse.

The most common form this abuse takes is overwork that is not properly compensated. People whose entire job consists of home-based telework often put in far more hours in a day than their contemporaries in on-site work settings, particularly if their remuneration is a lump sum or based on performance rather than an hourly wage or a salary. In order to increase meagre incomes, people who are paid twenty-five cents per call – the rate the Canadian firm Pizza Pizza paid its nonunionized teleworkers in 1995[61] – will try to maximize the number of calls they process by extending their hours; people who are paid a lump sum for a designated amount of data-entry work will try to accomplish it in as concentrated a time as possible (i.e., twelve hours in one day rather than in two six-hour days) in order to move on to the next contract. The pressures are similar for people who work a full day on-site *and* have computers and network access at home: in attempting to meet the increased workload demands created by staff cuts in both the public and the private sector, many employees find themselves putting in eight hours at the office, and a few more logged-in to their home computer after supper. A recent study surveying more than 6,000 public-sector workers in the Ottawa region found that employees with appropriate technology in their homes worked an average of 2.5 hours more per day than did their non-wired cohorts, with many of these overtime hours being classified as "informal," and therefore noncompensable.[62] There is no doubt that some people choose and prefer to work at home. However, we should not overestimate the "freedom" of this choice. As the union representing many of Canada's civil servants reports, "for the most part, PSAC members have worked at home to complete the heavy workload [assigned] during the regular work day."[63] It could be that a number of eviscerated public services in Canada continue to exist simply because network technology allows them to be delivered by unpaid labour. Again, it should be stressed that many of these employees are women. For these workers, network technology means a paid eight-hour shift in the workplace, followed by an unpaid five-hour shift of cooking, laundry, and child care at home, and a final, unpaid three-hour shift of typing and logging reports at their personal computer. Despite the promised domestic benefits of working at home, for many female teleworkers "the double day becomes an endless day."[64]

The Degradation of Work and Control over Workers

The union representing teleworkers in the Canadian federal civil service has concluded: "Telework is a low-wage, low-capital cost, employer initiative that serves the employer agenda of 'more for less' but does little to provide a healthy alternative to workers' individual needs for flexibility and more leisure time."[65] If this is the case for unionized workers in the public sphere, one can only imagine that the situation for non-unionized and private- sector workers is even worse. In either case, what seems clear is that the cost savings enabled by network technology are seldom realized in the form of increased leisure or remuneration for workers, and more often are converted into value for the employer – increased profits in the private sector, staff/cost reductions in the public sector, and increased control in both.

This assessment stands in stark contrast to the glowing portrayal of the impact of network technology on work presented by employers and their supporting cast of enthusiasts. For these, teleworking offers employees the chance to customize their working situations to match their own personal preferences and needs. Similarly, it is argued that networks provide the infrastructure that employees in service and manufacturing occupations require in order to exercise the high degree of autonomy and responsibility required of them in the supposedly dispersed and flattened hierarchies of flexible, just-in-time, quick-response operations. In this view – often associated with what is known as "total quality management (TQM)" – "teams" of employees committed to specific projects, as well as to the overall efficiency of the operation and the "total quality" of its products, replace strata of middle managers in exercising limited discretion over micro-process decisions. Though it is often dressed in the seductive garb of worker self-management, this approach is properly described as a function of the firm's need to respond quickly to the changing and particular needs of customers in a dynamic marketplace, the demands of which render traditional, hierarchical, cumbersome, bureaucratic management structures impractical and costly. As Gene Rochlin has written,

> what the computer transformation of business and industry has done is to maintain the appearance of continuing the trend toward decentralization, to further reduce the visible hierarchy and formal structures of authoritarian control while effectively and structurally reversing it. Instead

of the traditional means of formalization, fixed and orderly
rules, procedures and regulations, the modern firm uses
its authority over information and network communications
to put into place an embedded spider web of control that is
as rigorous and demanding as the more traditional and visible hierarchy. Because of its power and flexibility, the new
control mechanism can afford to encourage "empowerment" of the individual, to allow more individual discretion and freedom of action at the work site, and still retain
the power to enforce the adjustments that ensure the efficiency of the system as a whole.[66]

Advocates of "going flat" stress that "one cannot underestimate the
role in this of the computer,"[67] which connects "teams" not only to
customers, resources, and the production process itself, but also to
central executive-decision makers who are no longer able to see or control their operations through the gaze, and by the hand, of middle
managers. Thus, while many employees in this situation perform the
added, uncompensated work of *managing themselves,* they are rarely
afforded the more valuable power to *control their work.*

If teleworking leads to flexibility for workers, it does so only for those
few who already have the power and resources to conduct this work on
their own terms; and if the flattening of organizations into teams using computer networks has led to increased perceptions of worker autonomy in some firms, it has done so only incidentally and exceptionally.
In general, the increased control that flows through networks into the
hands of employers is drawn directly from the decreasing control most
workers exercise over their working lives. Hopes, and even perceptions,
of democratization have typically accompanied the introduction of new
techniques and control regimes in the workplace, only to be dashed
when these become embedded in the logic of capital. As Rochlin observes, "the democratizing phase is just that, a transient phase in the
evolution of the introduction of new technology that eventually gives
way to a more stable configuration in which workers and managers
find their discretion reduced, their autonomy more constrained rather
than less, their knowledge more fragmented, and their workload increased."[68] From this perspective, it is apparent that the motivation for

adopting these technologies is not simply increased *profits,* but also an increase in the *power* of capitalists vis-à-vis working people. This is ironic, given that much of the rhetoric surrounding the "flattening" of organizations into "quality"-conscious "teams" promises a reversal of the Taylorist separation of head and hand that characterized advanced mass-production regimes. Instead, the colonization of working life by network technology has achieved exactly the opposite in adding to, rather than subtracting from, the overall power of capital in the labour process, primarily by degrading many work activities to unskilled, often decentralized, forms of bit-manipulation that can be centrally monitored.

Drawing on the work of E.P. Thompson and Harry Braverman, Aronowitz and DiFazio point out that, historically, capitalists have tried to control the labour process not only to increase profits, but also to maintain control over workers themselves.[69] Workers have traditionally maintained what little control they have over work through their mastery of a particular skill, and the task capital has repeatedly set for itself has been the reduction of even this minimal control by eliminating as many aspects of skill from the labour process as possible, especially where doing so was coincident with overall gains in productivity and profit. In some cases, this was accomplished crudely by smashing craft unions, but, more regularly, it has taken distinctly *technological* forms. Skill entails the unity of head and hand, thinking and doing, expertise and execution, knowledge and ability. A skilled person both *knows* how to fashion a thing and is *able* to do so. The development of a skill requires instruction, study, and practice, and so its mastery is often a relatively exclusive achievement – a true skill *excludes,* by its nature, those who have not studied and practised it. One who masters an exclusive skill is customarily referred to as having gained a craft. The word "craft" comes from the Old Norse *krapt,* for "strength," which correctly signals that the possession of a skilled craft often brought with it a certain amount of power. Craft, though now largely an anachronism, is linked to what are known contemporarily as "trades" and "professions."

The mechanization that drove the Industrial Revolution ruptured the relationship between craft and work, and Fordist mass assembly lines completed their divorce. Work in factories devolved to the execution of physical operations that required more efficiency than skill. The system of craft destruction reached its zenith in the 1920s, when Taylor's scientific-management scheme completely separated thinking and doing, subjecting workers to "one best way" of accomplishing rote tasks.[70] Edward Andrew has pointed out that "scientific management eliminated the waste of workers thinking about work."[71] The beauty of Taylor's system was its efficient production not only of goods and profits, but also of workers' powerlessness. To the extent that mechanization (and, later, automation) combined with scientific management to deprive workers of the control afforded by skill and craft, it also granted control over *workers* to capitalists in degrees that matched their control over production.

Automation has remained a viable part of both manufacturing and service delivery, but Taylorism in its traditional form is not well suited to the needs of lean, flexible enterprises facing dynamic demands for customized products with very short cycle times. Employees in agile enterprises require relatively greater latitude in "thinking" than did their forebears on the assembly line. However, allowing workers to think – even within highly circumscribed parameters – restores at least some of their power, thus removing them partially from the control of their bosses. Accordingly, capitalist firms have required a new means of effective control in a labour environment in which workers individually and collectively could not be divested entirely of independent volition. Luckily, the perfect instrument was available in the same means that gave rise to the problem in the first place: "What could not be accomplished by Taylorism was finally achieved by computerization."[72]

There is considerable debate as to whether networked computers are part of the trajectory of de-skilling that has more or less defined the relationship between technology and work. For some, the new regimes of networked, flexible production, and the computerization at their heart, "create" new skills and promise to sustain a need for them.[73] Paul Adler, for example, asserts that because of technology "the long-run evolution of the quality of jobs is ... overwhelmingly positive."[74] Others argue that, "as a result of technologically induced job polarization, an overall de-skilling is becoming a permanent feature of the labour force."[75] Research in Canada has indicated that neither of these positions is correct. John Myles has found that, "in the 1980s, measurable skill change has been modest" in aggregate terms.[76] However, he does find evidence of an emerging skills polarization *within* white- and blue-collar working-class occupation categories; for example, "between high and low skill clerical workers, high and low skill blue collar workers, high and low skill sales workers and the like."[77] Their disagreements aside, each of these assessments neglects to consider the possibility that network technology has altered not just the *presence* or *absence* of skills in various types of employment, but, rather, the very *status* and *character* of skill. As Aronowitz and DiFazio have argued, perhaps what is at issue with network technology is not whether one set of skills is being replaced by another, but, rather, the *end of skill* itself, as it is customarily conceived.[78]

The typical proclamation that we are on the cusp of a new era of highly skilled employment reads something like this: the new economy is based on networked computers; computers need people to operate them; as computers are complicated devices, the people operating the computers will need knowledge; as knowledge is an important part of skill, the new economy will bring with it the return of skilled jobs. What this scenario fails to illustrate is that the kind of knowledge required of "networkers" actually represents the final obliteration of skill from working life, rather than its return, and so also an alarming escalation in the overall disempowerment of working people. The word "skill" comes from the Old Norse *skil*, which also translates as "distinction." A crucial feature of the knowledge or expertise element entailed in a skill is its distinction or relative exclusivity. The spiritual satisfaction derived from the mastery of a craft is evidence of the intrinsic *worth* of a skill, but what makes a skill *valuable* is the fact that not everyone has it. The value of a skill is a function of its marketability: in a town full of coopers, a person who makes barrels is not very valuable, but a cobbler is quite dear. Accordingly, skilled workers will be more powerful in a market featuring a plurality of skills than they will be in a market where a single set of abilities is generalized. Such a plurality of skills can be assured where there is a substantial variety of materials requiring mediation or work for purposes of use and exchange, and where investment in learning to be skilful with one material generally precludes investment in learning another. Genuine, powerful skills are thus not only exclusive, but often specialized: the demands of cultivating a knowledge of woodgrains distract the cooper from mastering the subtleties of leather; and the cobbler's fine hand for stitching must avoid the splinters that plague the clumsy dilettante in a woodshop. In a market economy – even one where work is organized in factories – the cobbler is powerful not simply because he or she has the skill to make shoes, but also because the cooper does not, and vice versa. Marx's utopian view of non-specialized labour – wherein "it is possible for me to do one thing today and another tomorrow, to hunt in the morning, fish in the afternoon, rear cattle in the evening, criticise after dinner, just as I have a mind, without ever becoming hunter, fisherman, shepherd or critic"[79] – applies only *after* the capitalist division of labour has been abolished. While this condition still persists, collective action and individual, specialized skills are the workers' only defensive resources.

In the networked economy, the skills required of many workers are not peculiar to wood or iron or leather, and they are far from exclusive or empowering. The knowledge required to be a so-called, but misnamed, skilled worker in the information age is not specialized, distinctive, or pluralized. Instead, it is standardized and generalized in a manner that mimics the endless and indifferent streams of bits to which it uniformly applies. "What is being replaced, or displaced," writes Rochlin, "is not the direct and tactile involvement of the worker, but tacit knowledge, in the form of expert judgment of the state and condition of operation of plant, process, activity, or firm, and the range of experiential knowledge that allows flexibility in the interpretation of rules and procedures when required."[80] When governments and industry leaders join in counselling young people to hone their skills for a place in the new economy, their message is utterly transparent: learn to use computers and networks because *everyone* will be pushing bits in the new economy.

And people are learning. The combination of a massive penetration of computer education in the school system, and a growing ubiquity of computer interfaces in most aspects of daily life means most people acquire computer "skills" whether they like it or not. And when computers grow more sophisticated, they become easier to use, further driving down the entry costs for admission into the new proletariat. It is true that some jobs, such as high-level programming and design, will remain "highly skilled," but the vast majority of computer and network-mediated occupations will be designated as "semi-skilled," a clever euphemism for "unskilled." What pass for "skills" today – things like the ability to use a word processor, construct a Web site, search the Web, and send e-mail – are simply the undifferentiated *lingua franca* of the new economy. If these are "skills," then so is showing up on time for a shift at the factory and knowing what to do when the whistle blows for lunch. If using some form of computer is something everyone can do, then it is not a skill, and its generalization across the working population is a sign of accelerated working-class disempowerment. In acquiring these pseudo skills, people disempower rather than empower themselves, because, when they can use a computer, they become indistinguishable in a competitive, universal market of similarly attributed workers, with no exclusive skills and no power – replaceable, like bits, and often *by* bits. The option, of course, is to refuse to develop a "working knowledge" of computers, and join the ranks of the unemployed.

The degradation of skilled work is one way in which networked economies facilitate increased control and power over workers by their capitalist bosses; the enhanced monitoring capacity afforded by these technologies is another. The question of network-mediated surveillance will be treated more broadly in the next chapter, but it is important to consider specifically here the proliferation of electronic monitoring in the workplace. As discussed above, networked computers make possible a massive centralization of control over nearly all aspects of productive systems and operations, including human labour. Along with the conversion of a great deal of work into the processing of bits, networks enable the centralized, remote, and often automated supervision of the activities and behaviour – also rendered into bits – of workers who are involved in that processing. And while computerized surveillance has certainly become the norm in network-mediated occupation categories, it is also increasingly being used to monitor people working in jobs that otherwise have little to do with the exchange of bits. For example, in 1997, the US company Net/Tech International began marketing a "hand washing documentation system" called Hygiene Guard. The system is designed to enforce the washing of hands after toilet use by workers in hygiene-sensitive occupations (i.e., food and health workers). Workers wear "smart badges" encoded with personnel identification numbers, which are activated by an infrared signal upon entering the lavatory. A monitoring unit in the washroom's soap dispenser then records whether the employee pulls the soap lever and uses the sink for at least fifteen seconds. Failure to do either results in the automatic registration of the employee's "infraction" by a centralized management computer.[81]

What is important to note here is that network technologies make it feasible to monitor people as well as processes, workers as well as work.[82] As Susan Bryant points out, the degree of surveillance enabled by computer networks is both more *ex*tensive and more *in*tensive than was possible under previous modes of work-process, and worker, control. As the hygiene system described above suggests, computerized surveillance is more *ex*tensive because it can be applied to a growing number of occupational categories and to a broader range of activities within those categories, some of which have little to do with productivity. It is more *in*tensive for two reasons: first, it facilitates the detailed surveillance of behaviour and performance at the level of *individual* workers, as well as providing aggregate production data; second, surveillance mediated by networked computers, unlike surveillance by human supervisors, has the capacity to be virtually constant. As Bryant observes, "new information technologies enable employers to gather and analyze highly detailed performance-related data, not just about the work, but about each individual worker – in many cases on a minute-by-minute basis and often without the employee necessarily being able to detect the watching."[83]

This intensiveness has led many analysts to draw fruitful analogies between contemporary techniques of workplace surveillance and either Jeremy Bentham's design for, or Michel Foucault's discussion of, the panoptic prison.[84] A key attribute of a properly functioning panopticon is that those being supervised maintain a *perception* of constant surveillance, which induces them to be self-supervising, even when the gaze of the bosses or jailers is not, in fact, directed at them. The power of the panopticon is thus primarily cunning and discursive: it often derives from what is actually an *absence* of supervision, even though its persuasiveness relies on the physical architecture of the surveillance system. Significantly, contemporary networked surveillance systems are somewhat more panoptic (i.e., "all-seeing") than the panopticon, in so far as the supervision they enable is often constant *in fact*, and not just in perception: "in some applications of [networked] workplace surveillance the gaze is not merely an unverifiable possibility but is actually a constant and continuous certainty."[85]

It is very difficult to determine the number of working people who are subjected to electronic monitoring on the job, especially given the reluctance of private organizations to divulge the extent of their practices in this regard. Conservative estimates place the number of workers subjected to electronic surveillance of one kind or another at up to 20 million in the United States alone.[86] A poll conducted in 1997 by the American Management Association found that nearly two-thirds of US employers: record employee voice-mail, e-mail, or telephone calls; review employee computer files; or videotape workers on the job.[87] However, perhaps the best way to understand the operation of the electronic surveillance of workers is simply to consider the range of ways in which it is carried out. Typically, there are two kinds of electronic workplace surveillance, although they are not mutually exclusive: performance monitoring and behaviour monitoring.

Performance monitoring occurs where the bulk of an employee's work is rendered in digital form, mediated directly by a networked, computerized device, and so immediately recorded and measurable in strictly quantitative terms. There are numerous examples of this sort of surveillance in contemporary workplaces. The most common is perhaps the counting of keystrokes relative to elapsed time by data-entry clerks of all descriptions – an updated version of the now antiquated "words per minute" measurement of stenographic typing. Data clerks and word processors across a range of industries are typically expected to log between 11,000 and 13,000 keystrokes per hour, and can expect sanctions when they fail to reach this standard.[88] In supermarkets, the same networked scanning technology that enables computerized inventory management and eliminates manual price-entry also simultaneously records the rate at which the cashier (who enters a personalized employee code into the terminal before beginning) processes items, provides change, and moves on to the next customer. In call centres, where computers and telephones are integrated into a single system (i.e., for airline reservations), data recorded both in the aggregate and for individual operators include: total number of calls handled in a given time frame; average duration of calls; elapsed time between calls;

average sales per call; the number of rings before a call is answered; and time away from the workstation. At Bell Canada, the Traffic Operator Position System (which replaced the old "plugs and sockets" system of call routing) can instantaneously deliver to management up to seventy-six distinct pieces of data on an individual operator's performance.[89] As indicated above, surveillance in these cases is constant and relentless, not intermittent. The key to each of these forms of monitoring is computer networks, which not only act as the instrument with which the actual work of the employee is done, but also accomplish the simultaneous, total, and automatic observation, recording, and analysis of every facet of his or her work activity. As increasing numbers of occupations come to involve more interaction with computerized data-recording and process-control devices of one sort or another – such as transport and delivery workers who log their handling of shipments and goods, machine operators whose equipment is integrated in a networked control system, stock clerks who enter the merchandise they handle into hand-held computer interfaces – so, too, do increasing numbers of workers become subject to this sort of surveillance.[90]

Networks also make it possible to monitor the behaviour, directly work-related and otherwise, of employees whose work is less characterized by routine bit handling, or less measurable in terms of the speed and quantity of data they process. Andrew Clement refers to this as "fishbowl" – as opposed to "sweatshop" – surveillance.[91] Digital telephone networks allow employers to record and listen to employee conversations, ostensibly to gauge the quality of customer service being offered; call accounting systems also record detailed information about the time, duration, destination, and cost of employee telephone calls. Similar systems exist, and are even more easily implemented, for the monitoring of electronic mail and Internet use by employees.[92] A standard feature of network management software enables managers to "pull up" the screen of any employee computer connected to the network at any time for direct observation, which is apparently necessary to track the progress of documents and files as they are passed from one employee or department to the next. Finally, in many, particularly large, otherwise "self-supervised" workplaces, card-swipe security systems require employees to pass a digitally encoded "smart-card" through a reader connected to a central database in order to move past various points in the physical site. Such systems deliver security without the bother of numerous keys or the menace of physical guards; they also allow the automated collection and analysis of information, including the time an individual employee enters or exits the building, when an employee arrives at or leaves his or her floor, when he or she enters and exits the lunch-room, and how many times he or she uses the washroom and for how long.[93]

Trade unions and the federations they belong to have consistently insisted on the legislation of regulatory standards on workplace surveillance of this kind, and have also bargained – with limited success – for protections against abuse in their collective agreements.[94] Nevertheless, with few exceptions, statutory and regulatory protection for workers against undue electronic monitoring has been hard to come by. The Criminal Code of Canada and the Electronic Communications Privacy Act in the United States make it an offence for employers to listen in on workers' private conversations carried out by telephone or other electronic means. However, in both cases, employers are permitted to monitor non-personal conversations and, in the Canadian case, are exempted from the prohibition on eavesdropping if the employee has an expectation that his or her conversations *might* be monitored. Of course, neither of these provisions addresses the broad range of electronic workplace surveillance that extends beyond employee "conversations." The Canadian Labour Code is basically silent on the issue of electronic workplace surveillance, and while most governments in Canada have legislated limits on the use of private information gathered about citizens by public authorities, these protections do not pertain explicitly to issues of workplace monitoring, and are not binding on private- sector employers. This difficulty is compounded by one of the ironies of network-mediated workplace surveillance: while networks make it easy for employers to collect reams of information about their employees, the practice of monitoring is itself difficult to monitor because private enterprises enjoy proprietary rights over the data they gather. Thus, private corporations enjoy a level of immunity from surveillance that their employees decidedly do not. In fact, in its final report, Canada's Information Highway Advisory Council observed that, with regards to workplace surveillance and privacy, "the private sector is virtually unregulated."[95] In line with its overall preference for a market-based approach to network technology deployment, the council recommended the development of "national *voluntary* standards," which nevertheless allow individual sectors and organizations "the flexibility to determine how they will refine their own codes."[96] The council recommended this approach – which, as detailed below, has more or less been adopted by the Canadian government – in the context of a general consideration of *personal* privacy (i.e., medical records, consumer and credit information, vital statistics) and only incidentally suggests that these flexible voluntary standards be extended to include workplace surveillance practices.[97]

This general orientation is consistent with the approach taken in the Canadian government's privacy legislation introduced in 1998.[98] The Personal Information Protection and Electronic Documents Act (Bill C-6), is presented as "an Act to support and promote electronic commerce" and only incidentally contemplates the issue of workplace surveillance and privacy. The legislation will apply initially to the federally regulated private sector (federal works, chartered banks, telecommunications and broadcasting firms, airlines, interprovincial transport firms) and will subsequently be extended to cover commercial activities conducted by the private sector. Adopting the fair information principles contained in the Canadian Standards Association's "National Standard for the Protection of Personal Information," Bill C-6 basically replaces voluntary with mandatory compliance.

The legislation applies to the collection, use, and disclosure of employee information collected by the specified organizations, but it does not prohibit the practice of electronic workplace surveillance.[99] The most significant provision in this regard requires employee consent before any "personal information" about them can be gathered, used, or disclosed, but it is not at all clear that this provision will have a significant impact on workplace surveillance. It is highly unlikely, for example, that firms would hire prospective employees who decline a general request for consent regarding the collection and use of information regarding their workplace behaviour and job performance. Additionally, the legislation contemplates a number of exceptions to the general consent requirement. For example, consent is not required when "the collection is clearly in the interests of the individual and consent cannot be obtained in a timely way."[100] It is not unthinkable that firms could make a compelling case that performance monitoring is crucial to the health of the firm and therefore "clearly in the interests of the individual." Furthermore, Bill C-6 stipulates that "the way in which an organization seeks consent may vary, depending on the circumstances and the type of information collected," and that while "an organization should generally seek express consent when the information is likely to be considered sensitive ... Implied consent would generally be appropriate when the information is less sensitive."[101] It is not clear exactly which of the myriad types of information firms might collect about their employees would be considered more or less "sensitive," or who gets to decide the level of a given piece of information's sensitivity. This would seem to indicate that a whole range of workplace surveillance practices may require only implied consent, the sort of consent a worker "implies" by merely being there and doing his or her job.

In this environment, it is not surprising that companies and other organizations have been quick to capitalize on the competitive and power advantages to be gained through the widespread use of detailed electronic workplace surveillance. It has been estimated that in the United States, up to 80 percent of workers in the telecommunications, banking, and insurance industries are subject to computer- and telephone-based electronic monitoring.[102] What are the effects of electronic surveillance of this breadth and intensity on work and working people? As Clement concludes, it is difficult to describe current monitoring practices as anything other than "degrading and an affront to human dignity."[103] Implicit in the constancy of network-mediated surveillance is the assumption that every minute of an employee's working time should be spent in activity that is maximally productive. The consistent and demonstrated results of this demand structure are increased levels of job stress and an increased vulnerability – on the part of workers engaged in routinized tasks where every keystroke counts – to repetitive-strain injuries such as carpal tunnel syndrome, tendinitis, and other muscular/skeletal conditions.[104] Significantly, in many jurisdictions repetitive-strain injury and stress-related conditions are not covered by workers' compensation agreements and, even where they are, such claims are difficult to prove.

In social terms, the impact of computerized surveillance is similarly undesirable. In electronically monitored workplaces, relationships between workers become strained as time for social intercourse is constrained, and space for unsupervised interaction disappears (either because electronically supervised workers can be physically dispersed or because sites of interaction are monitored). The same dynamics make the process of unionization more difficult in electronically monitored workplaces.[105] The availability of up-to-the-minute, quantified, performance ratings coupled with productivity incentives thrust individual employees either into comparison and competition with each other, or into competing "together" as "teams" to reach performance standards that reflect the employer's, but not necessarily the workers' own, interests. In either case, some combination of embarrassment, sanction, demoralization, ostracism, insecurity, and feelings of inadequacy is the result for workers whose performance does not "measure up."[106] Similarly, interaction between workers in service occupations and customers or clients is diminished as electronic surveillance of service provision increases: cashiers are less likely to engage in casual pleasantries with customers in their check-out line if the time between item registrations and transactions is being closely monitored. Relationships between workers and their supervisors are also altered significantly by network surveillance technology, as either employees become "automatically" supervised by computers and software they can neither see nor talk to, or their human managers find themselves limited in their discretion by the combination of unambiguous performance data provided by the monitoring system and strictly enforced company-wide standards. In short, when it comes to the sorts of work that are susceptible to extensive and intensive levels of electronic surveillance, monitoring technology manages to remove from working life all those forms of social interaction that make these otherwise routine and repetitive jobs more pleasant, dignified, and bearable.

Given the significance of work in the everyday life and self-image of most human beings, the degradation of work wrought by networked computer surveillance technology does not bode well for the extension of healthy social behaviours and relationships into the broader public or civic sphere. Traditionally, the workplace has been not only a site of sociability, but also a site of socialization. As Clement writes, "implicit in the use of any device that monitors the activity of an adult is the presumption that he or she is not entitled to be left alone."[107] The only people we cannot leave alone are those who are without the capacity for good and reasoned judgment. To the extent that the surveillance function of networked computers in the workplace socializes people to accept their incapacity to exercise good judgment independent of omnipresent supervision, it is difficult to imagine how anyone could advocate letting these same people loose in the public sphere as genuine democratic citizens. In this context, it is also difficult to comprehend those who insist that network technologies constitute the infrastructure of a democratic renaissance. Menzies has argued that, "as the context for their work is digitized, people are being systematically stripped of their capacity for human involvement and judgment."[108] Similarly, Clement suggests that "participating fully in a democratic society requires practice in exercising initiative and thinking for oneself. Close surveillance stifles independent effort by promoting excessive conformance to norms established by higher-level authorities."[109] The question that remains is whether those driving the increasing penetration of network technology into working life, who simultaneously express hope for a renewal of democracy, have failed to understand the connection between degradation at work and the inability to exercise democratic citizenship, or whether they simply understand it all too well.

Means of Consumption, Instruments of Exchange: Networks as a Commercial Technology

Work does not constitute the whole of a modern person's engagement in economic life – people not only produce, they also exchange and consume. Historically, technologies mediating production, exchange, and consumption have been relatively distinct. Factory machines could produce automobiles, but they could neither expedite a car's sale nor incite its purchase; hard currency facilitated the exchange of goods, but it could not squeeze rivets or fabricate needs; television has come closer to integrating these three functions by producing need-ridden consumer masses in the form of audiences, and by selling these to advertisers. However, traditional television remains limited as a medium of exchange because it does not accommodate the ordering of goods or exchange of payment for those goods independent of the adjunct technology of telephony. It is possible, of course, to purchase goods displayed on the TV screen; however this cannot be done through the television itself. To buy the latest in home exercise equipment, one must take the extra steps of telephoning in an order, and remitting payment via credit card or through the mail. The great utility of net-worked computers lies in their integration of production, consumption, and exchange – the properties of which are increasingly reducible to bits – into activities carried by a single medium. Just as networks have reduced friction in the processes of production, so, too, they stream-line the economic practices of consumption and exchange.

In the discourse enveloping the proliferation of network technol-ogy, this gathering of production, consumption, and exchange into a single medium is referred to as "interactivity," and is promoted as per-haps the definitive attribute of this medium. Network guru Nicholas Negroponte gushes: "interaction is implicit in all multimedia."[110] Net-works are thought to be interactive in so far as, through them, people are able to input directions that have an effect on the information the network delivers. Thus, attributes – such as hypertext; customized search engines; synchronous and asynchronous transfers (i.e., com-puter and video conferencing); the ability to download, alter, and up-load documents; and multiple-user software applications – are all seen as paradigmatic interactive features of networks. Interactivity of this sort is on the cusp of graduating from a value to a virtue in the network age. As Gaëtan Tremblay writes, today there is "an undisputed claim regarding the positive value of interactivity" wherein "interactivity is good in itself."[111] Implicit in this "unconditional valuing of interactivity" are subsidiary claims equating interactive technology with non-hierarchical relationships of equality, symmetry, mutuality, and recip-rocal creativity.[112] Most important, interactivity is meant to invoke that sense of participation that is indispensable in societies whose self-image is democratic.

However, as I suggested in Chapter 3, interactivity misnames the essential activity mediated by networked computers that, on their own, can offer little in the way of equality, mutuality, and participation. *Inter*action connotes reciprocal action between or among actors. Net-works do allow the people they connect to interact in some ways and, in a minimal sense, networks allow users to act on data while those data are being presented to them. However, the suggestion that human users interact with networked computers entails the worst sort of anthropomorphism, in so far as it makes a person of a machine made by people. The fact is, even when its use involves ongoing choices by the user, a networked computer remains an instrument, and a human being cannot *inter*act with an instrument that is inanimate. Actions require commitment, and instruments are committed to ends – intended or otherwise – only by the direction of their users. The use of an instrument to achieve an end should not be confused with interacting with that instrument. One does not interact *with* networked computers and their programs, though it is possible to facilitate interaction *through* them.

This being said, I would suggest that interaction is not the main type of action facilitated by networks. Robust interaction requires a set of enabling conditions that are richer than simple access to a neutral and efficient medium – conditions such as shared norms, goals, and resources that are external to the medium itself. The enabling conditions of robust, generalized interaction are not consistently present in the environment in which networks are situated. On the other hand, networks are well suited to facilitate *trans*actions, which are defined as exchanges performed *across* a connective medium, and which require little else for their completion besides connected parties and a reliable medium. A transaction is not as demanding as an interaction, and networks do appear ready to mediate the former by enabling the movement, carriage, and exchange of binary digits, none of which is enough to support the latter. As noted previously, at this point network transactivity – even when mislabelled as interactivity – shares more with commodities, commuting, and commerce than it does with communication and community.

This characterization is made manifest in the emerging role network technology plays in the economic activities of consumption and exchange. Typically, network enthusiasts have been quick to applaud this aspect of network media: "digital technology will turn every home computer into a window facing onto vast shopping malls. The convenience of a built-in, 24-hour point-of-purchase in the home (not to mention in offices, cars, taxi cabs, airplanes, restaurants, and kiosks) will, over time, have dramatic effects on consumption patterns involving hundreds of billions of dollars."[113] From this point the story is predictable: an increased range of choices in an information-flooded market, accessible through an "interactive" interface with a universal medium, means broader customer choices, which, in turn, enhance the power of consumers to "exercise more control over their immediate environment."[114] In this view, networks deliver democracy to the extent that they encourage "a consumer-directed culture rather than producer-directed culture."[115] Under the auspices of network technology, at a certain point shopping and politics merge into an undifferentiated whole.

The commercial exploitation of network technology is just beginning. Reliable figures and projections regarding the amount of commercial activity currently mediated by networks are difficult to come by, though figures ranging from $100 billion to $500 billion in network-transacted business by the year 2000 are not uncommon.[116] It is certain that "commercial" (.com) represents the largest and fastest-growing domain category on the Internet, with more than 12 million commercial hosts registered by 1999.[117] In the abstract, networks are a near perfect medium for commercial transactions, particularly as more of the information involved in such exchanges is rendered in the form of bits. Previously, in order to purchase something, a customer would have to physically visit the site where the goods were being offered for sale, or at least the site where the transaction of sales took place.[118] Once there, he or she could peruse merchandise, solicit information from sales staff, and carry out payment by either the exchange of currency or the arrangement of credit. Because every aspect of this scenario is now reducible to the digital form of bits, the physical proximity of buyer and seller, and the physical exchange of currency, are unnecessary. With the sole exception of the transportation of the product itself (assuming the product is not reducible to bits, as is a newspaper, book, computer program, or financial service, in which case even physical delivery is eliminated), the entire process can be mediated by a computer network. Product information, pricing, the placing of an order, billing and payment can all be accomplished digitally using electronic data interchange (EDI) protocols, and bits can be moved in a great deal less time than it takes to bundle the kids up, get in the car, stop at the bank for cash, find parking at the mall, snag a salesperson, line up at the cashier, and wait for your change.

Considerable obstacles have as yet prevented network technology from reaching its full commercial potential as a medium of exchange and consumption – business-to-business electronic commerce is beginning to boom, while business-to-consumer transactions have been slower off the mark – but these are showing signs of rapid erosion. Indeed, the strategies for overcoming these difficulties illuminate much about the commercial character of this technology. The primary hurdles facing full commercial deployment and exploitation of network technology can be grouped into three categories: accessibility; the mode of transactions; and marketability. Each is considered in turn.

Accessibility

The problem of network accessibility for commercial use is essentially a technical one: generalized mediation of commercial transactions by computer networks requires the deep penetration of devices providing easy access to these networks into the everyday lives of a critical mass of consumers and, also, the construction of an infrastructure capable of carrying the massive volume of bit traffic such widespread commercial use would entail. This second problem – also known as the "bandwidth" issue – has animated much of the infrastructural effort being exerted by governments and private interests attempting to construct, or at least create the conditions for, an information "superhighway." Many emerging or potential commercial applications involve the transmission of massive amounts of bits, whether they are business transactions involving constant streams of data and complex operations, or consumer Web sites featuring sound, image, and video accompaniments to products for sale. Such sizable transfers of data take time: they are not *immediate* because there is still a *medium* to be traversed. Bits must travel *through* something, and that something is usually a wire with a limited capacity to carry bits. As the volume of bit-intensive traffic grows, the demands on existing transmission capacity escalate: a piece of information such as a high-resolution image is composed of a large number of bits, and the line-up to get those bits onto the wire that will carry them to a customer is growing, as more users with similarly voluminous transmissions enter the fray. Nevertheless, given that speed is one of the core attributes expected of this medium, commercial viability is contingent on the minimization of transmission time.

Consumers are unlikely to do their Christmas shopping at home if they have to wait five minutes for an excerpt from the recording they are considering to reach their home computer, and another five minutes for their order to be processed once they decide to purchase it. Digital video-on-demand – the Holy Grail of network commerce – has enormous bandwidth requirements that have, at least in part, prevented it from coming to fruition.

Attempts to make the "band" through which bits travel "wider" (hence "bandwidth") have taken a number of forms. Bits can travel through three kinds of wires. Listed in order from least to most capacious, they are: the twisted copper of regular local and residential telephone lines; the coaxial cable that carries broadcast television signals; and the fibre-optic lines of long-distance and telephone trunk lines. Bandwidth can be increased by combining the laying of more lines, the replacement of twisted copper with fibre-optic lines, and the routing of bit-intensive traffic to more capacious, and therefore more speedy, transmission lines. Another way of increasing transmission speed is through digital compression. Compression programs examine large chunks of data, recognize bit-patterns therein, and then substitute shorter codes for these bits before they are transmitted or stored. Once received, these data can be decompressed, when the original bit patterns replace the short codes that stood in for them during transmission and storage. Combined with various modulation and switching techniques, digital compression greatly enhances the capacity of existing transmission lines. Finally, the use of satellites capable of receiving and sending digital signals via microwave (i.e., "wireless" transmission) is an emerging and potentially expansive alternative to traffic-heavy terrestrial network media.[119] In one such system – the Teledesic program backed jointly by Microsoft's Bill Gates and Boeing – close to 300 low-earth-orbit (LEO) satellites will be launched to an altitude of 435 miles at an estimated cost of $9 billion. Expected to be operational by 2002, the system will enable widespread access, via twenty-five-centimetre antennae, to rapid synchronous and asynchronous transmission of digital data, voice, and video signals. Together with other, less ambitious LEO system proposals, the Teledesic program hopes to put more satellites into orbit by 2002 than has the entire world since the launch of *Sputnik*.[120]

The main barrier to simply increasing bandwidth in these ways is primarily economic, not technological. It is a question of who should bear the costs of high-bandwidth infrastructure construction (i.e., government, telephone companies, or cable companies), how and by whom those costs should be recovered, and who should set the standards of its use. In Canada, the Information Highway Advisory Council has determined that, "in this field, it is the marketplace that should determine the winners and losers ... Because the financing of the Information Highway should be left to the private sector, the firms and individuals who bear the risks of these investments should also reap the rewards."[121] The council is confident that the "private sector will, and should, drive the pace and scope of development. Balancing supply and demand is a self-reinforcing dynamic."[122] Nevertheless, the 1998 Federal Budget featured $55 million in public spending for the construction of CA*net 3, a high-bandwidth fibre-optic network designed for multimedia data transmission by Canada's high-tech and research industries.[123] Widespread commercial application of network technology requires access to substantial bandwidth, and governments appear prepared to enlist public resources to meet this demand.

The other barrier to commercial exploitation of computer networks involves the need to broaden access to devices that are both easy to use and capable of mediating complex consumer transactions. As pointed out in Chapter 3, by 1996 only about a third of Canadian households had personal computers, and only half of those were equipped with modems.[124] However, not all commercial activity mediated by networks necessarily requires a home computer. The commerce-enabling applications of networks include point-of-sale technologies such as those discussed above, and computer chips and interfaces are being designed into an increasing array of appliances that are not commonly described as computers. Indeed, the great bulk of the world's microprocessors are not found in personal computers, but are instead embedded in common household electronic appliances.[125] Many of these appliances can themselves be networked and used to transmit messages or information. For example, residences at the Stonehaven West model "wired" community in southern Ontario have included as part of their appliance package Northern Telecom's "VISTA 350 Interactive Telephone," which allows them to access an "electronic mall" featuring an assortment of "interactive shopping and information services."[126] Using their telephone – which is equipped with a small display screen – Stonehaven residents can pay bills, do their banking, view advertisements, compare prices, order prescriptions, make purchases, and even read news headlines without ever leaving the house or turning on their personal computer. Of course, because the telephone can receive, process, and transmit digital information, it *is* a computer of sorts, but the point is that it is a networked computer integrated into an everyday household appliance which turns that device into an instrument of commercial transaction.

Other efforts to build network connectivity into devices that are not full-blown computers include the development of terminal devices devoted strictly to network use (known as an "NCs," or "network computers"). NCs are built without hard-disk drives and so contain no internally stored programming. Instead, they are simply terminals that are networked to a central server that gives their users access to a limited range of applications. Though still at an early stage of development, NCs have as their primary attraction their ability to provide a relatively inexpensive and very user-friendly means of engaging in transactive network activities. Transactive network activities include such things as electronic mail, information retrieval, and the perusal of Web sites; they also include home shopping, banking, and purchasing.

Another attempt to generalize networks as means of consumption beyond those who regularly use personal computers takes the form of the drive to "converge" the technologies of digital networks, computing, and television into a single medium capable of delivering high-resolution audio and video content that is also "interactive." A number of obstacles – including marketing an affordable appliance and determining which configuration of telephone and cable companies will deliver the content – remain before interactive TV becomes part of the fabric of everyday life.[127] Nevertheless, indications are that the rush is on: in 1998, the Federal Communications Commission mandated that all 1,500 television stations in the United States must begin transmitting digital signals by 2003.[128] Accordingly, manufacturers have begun producing computer-television hybrid devices (known as "PCTVs"or "WebTVs") and major computer network interests such as America

Online and the Microsoft Network have begun streaming their offerings in the style of television "channels" and "shows," and have invested heavily in the production of attractive content. Microsoft alone has committed $1.5 billion to developing content and delivery for its Network channels.[129] In 1999, Microsoft purchased a stake in cable giant Rogers Communication in hopes of stimulating the launch of interactive television services by the year 2000.[130]

As I discuss below, tapping into formidable advertising revenues may be one commercial explanation for the drive to convergence. However, as the following vision for digitized video-on-demand articulated by Microsoft's Bill Gates suggests, the commercial allure of converging computers and television may, in fact, lie in its optimization of the transactive attributes of network technology: "In the future, companies may pay not only to have their products on-screen, but also to make them available for you to buy. You will have the option of inquiring about any image you see ... If you are watching the movie *Top Gun* and think Tom Cruise's aviator sunglasses look really cool, you'll be able to pause the movie and learn about the glasses or even buy them on the spot ... If a movie has a scene filmed in a resort hotel, you'll be able to find out where it's located, check room rates, and make reservations."[131] Networks, it would seem, form the indispensable technological infrastructure for the ultimate convergence – not only of computers and television but, crucially, of entertainment, consumption, and exchange.

The chief obstacle to complete commercial exploitation of computer networks is thus a distributive one (i.e., the penetration of network access and networked devices into the everyday lives of the broad mass of people) and so it is not surprising that the most intriguing strategy for surmounting it has been expressed in the language of a moral or political imperative. I refer here to the emerging consensus that "universal access" to computer networks is an indispensable social requirement of a civilized and just society and, therefore, that it is incumbent upon societies to endeavour collectively to provide such access. This position has been well expressed in a report by the RAND Corporation on the feasibility of universal access to electronic mail in the United States.[132] The report begins with the assumption that American society has reached the point where access to electronic mail, in so far as it provides for "much more egalitarian, deliberative and reflective dialogue among individuals and groups," has become an essential condition of full participation in public life.[133] Universal access to electronic mail, it is argued, will "reduce the feelings of alienation that many individuals in the United States feel and give them a new sense of 'community,' revitalize the involvement of the common citizen in the political process, etc., and in general strengthen the cohesion of US society."[134] Access is defined as "universal" when it is "available at modest individual effort and expense to (almost) everyone in the United States in a form that does not require highly specialized skills or, accessible in a manner analogous to the level, cost, and ease of use of telephone service or the US Postal Service."[135]

The only thing perceived as standing in the way of such a desirable civic outcome is the unfortunate fact that access to e-mail is not yet universal; the report presents evidence suggesting the major class distinction in the information society is a distinction between those who have network access and those who do not, a bifurcation that correlates significantly with levels of education and income and, to a lesser degree, race, geographic location, and age. Given the apparent centrality of electronic mail to the future of civic communities, the implications of this polarization are portrayed as potentially dire: "those who lack access to new communication technologies may be at risk of exclusion from the fabric of the nation's social and economic life ... sizable demographic subgroups who remain in the have-not segment may be deprived of the benefits associated with membership in the information society."[136] Such widespread exclusion from participation in the goods of citizenship is, of course, repugnant to a culture whose self-image is strongly democratic. However, because lack of connectivity is identified as the source of this exclusion – rather than economic and educational deprivation, or racism and age discrimination – it is *this* condition that animates popular concern and motivates action. Indeed, the authors of this report go so far as to predict that access to networks will be instrumental in overcoming broader social and economic inequalities. Consequently, the report recommends the US government undertake "public actions to encourage universal access," and seek to "extend network access to currently underserved populations" in the form of "widespread home access."[137]

The RAND report recommends that this be accomplished using the specific mechanism of consumer subsidies drawn from general revenue rather than by enforcing a strict regime of market price control or regulation on equipment and service providers. The state should "provide purchasing power directly to marginal consumers. A program of e-mail service vouchers would enable consumers to shop for terminal equipment, user training, and e-mail service in competitive markets."[138] Regulators, on the other hand, "should adopt a light-handed approach" until market demand develops to the point that "regulations governing interconnections may not be necessary."[139] Major statements by governments in both the United States and Canada – the final report of the Information Highway Advisory Council (IHAC) in Canada and the National Information Infrastructure (NII) *Agenda for Action* in the United States – have strongly advocated concerted efforts to achieve universal access, have suggested the possibility of targeted subsidies, and have made it clear that market restrictions and regulations should not be included in strategies pursuant to this goal.[140] In its response to the IHAC report, the Canadian government committed itself unambiguously to an aggressive universal-access strategy[141] – a commitment it honoured in its 1998 Budget in Canada with $200 million in funding for increased network access in schools and neighbourhood facilities.

The RAND report asserts that there is a "strong correlation between democracy and interconnectivity" and that this correlation is "substantially larger than that of any other traditional predictors of democracy."[142] It is largely on the basis of this very assumption – the assumption that network access has become an essential requirement of democratic citizenship – that governments have pressed claims for the commitment of public resources to secure universal access. Universal access to network technology has thus become a moral imperative: a democratic society simply *must* wire *all* its citizens if it wishes to remain democratic; to do otherwise is unconscionable in a society that believes itself to be democratic. This conviction is voiced clearly in the US government's NII *Agenda for Action*: "Because information means empowerment, the government has a *duty* to ensure that all Americans have access to the resources of the Information Age."[143] However, it is not at all clear that network access is a sufficient – or even a necessary – condition of democratic politics. It certainly is not the most important one. It would be difficult to dispute that network access ranks lower on the list of crucial requirements of a democratic and public-spirited civic life than, for example, widespread leisure and economic security. Phrased in language more germane to the discussion of network technology, we might say that connectivity does not, by itself, create the conditions for *inter*activity. Connectivity is enough, however, to facilitate *trans*activity. On its own, connectivity lends little to the pursuit of a democratic and robust public sphere, but it does come closer to satisfying the conditions necessary for simple exchange.

It thus becomes possible to suggest a compelling answer to the question of why governments and other influential actors have expended such rhetorical effort, and appear prepared to devote substantial resources, to secure universal access to a medium that has high hopes, but scant likelihood, of contributing independently to the democratic goals to which they present themselves as committed. The answer is that the motivations of the promoters of universal access are commercial rather than democratic, and universal access to the network medium will definitely encourage the realization of goals of this nature. Perhaps unwittingly, the RAND report makes this clear: "because e-mail is immediately popular with network users, it plays a crucial role in stimulating them to experiment with other features of an electronic environment. The value of e-mail's role as a catalyst to other, more advanced network use is significant ... [because] return on investments made by commercial businesses and other enterprises in network services also are likely to rely on use of more advanced, value added features ... By extending access to groups that otherwise are not able to make use of the technology, networks raise demands for a wide range of services."[144] In this light, it may be more plausible to argue that universal access is a strategy designed to optimize the commercial attributes of transactive network technology than to suggest it is a serious effort to encourage deep democratization. Universal access emerges as a total solution to the problem of penetration that has prevented the complete commercial exploitation of network technology. Nevertheless, it is understandable that promoters of universal access have chosen to conceal their commercial designs inside the Trojan horse of democratization, education, and an investment in social capital. Few people could be expected to support with enthusiasm a massive public subsidy – which ultimately constitutes a more or less direct transfer of public resources into the private hands of commercial equipment and service providers – to institutionalize a new system for accomplishing transactions of consumption and exchange. But democracy? community? education? helping poor folks by giving them computers? Who could oppose these worthy goals?

The Mode of Transaction

Commercial transactions involve the exchange of money, and if computer networks are to be successful as commercial media they must become effective as instruments of such exchanges. Given what has already been said about the technical properties of networked computers, it would appear they are, in fact, a perfect instrument for this purpose. Money is, after all, simply a numerical representation of a certain type of information and, as such, it can easily be rendered in the digital form of bits and exchanged via computer networks. Indeed, the exchange of a great deal of the world's money is already carried out in precisely this manner. As Ronald Deibert and others have pointed out, the internationalization of finance capital has proceeded apace with the development of networks and other digital media, creating a "massive, 24-hour marketplace" for financial transactions in stocks, bonds, securities, currencies, and the like, in which non-stop exchanges occur across borders, across time zones, and often with little direct human intervention.[145] It has been estimated that, in New York city alone, as much as $1.9 trillion *per day* is exchanged electronically via computer networks.[146] Certainly, at the level of high finance, the network transaction of money has become not just the norm but an essential condition of doing business.

Similarly, the everyday consumer transactions that characterize the economic lives of most people are increasingly being transformed from an exchange of bills, coins, or bank cheques between one person's purse and another's cash register into the exchange of bits between networked computers.[147] However, a generalized mode of simple monetary transaction remains to be established at the consumer level, and the complete commercial exploitation of computer networks awaits resolution of this problem.[148] Some forays into the world of network-mediated money transactions have already been made, with mixed results. For example, there are three different "card-mediated" methods of transacting money currently in use or being tested; all three rely on the encoding of digitized information about money on a magnetic strip affixed to a plastic card that can be "read" by a "card-swipe" reader. For example, credit-card transactions are now usually mediated by networks: information from the card's magnetic strip is read and relayed to the credit-card company's computer, which immediately authorizes the transaction and adds the purchase to the customer's balance. However, while this removes the need for retailers and credit-card companies to process peevish and slow paper transaction records, it offers little added benefit to the consumer. Better in this regard is the so-called debit card. By swiping his or her debit card at the register and verifying his or her identity with a personal identification number (PIN), a customer can deduct the amount of a purchase directly from his or her bank account, and the bank will forward that amount to the retailer. This method of payment has proved quite popular, but still suffers from some drawbacks. In the first place, the banks who administer debit networks typically charge consumers a service fee for each transaction made (perhaps to recover revenue lost from interest-accruing credit-card purchases), making this mode infeasible for multiple or small transactions. In the second – and this is a problem shared by credit cards – debit-card transactions lack the anonymity of cash purchases; because they are linked to detailed personal information via bank and credit accounts, transactions using these devices create a trail of data regarding an individual's consumer habits, which many liberal individuals with privacy concerns find offensive.

A third system of digital card payment known as the "stored-value" or "smart" card overcomes both these difficulties. Smart cards are imbued with a pre-paid amount of money that can then be spent in bits by the cardholder as purchase amounts are deducted from the card until it is empty (at which point it can usually be re-valued). The absence of a service charge for use (there is sometimes a one-time fee for the initial purchase of a card) removes the disincentive to multiple, small deductions. Additionally, because the value on the card is prepaid, there is no need to authenticate the identity of the user and, therefore, no connection to any personal information about the user via bank or credit accounts, thus securing anonymity. The bits that are read as dollars and cents from a user's card have already been paid for by the user with real money: she has either purchased a value-laden card from the retailer or vendor, or has inserted money into a machine that adds value to depleted cards. These cards are currently being used for repeated small transactions, such as those carried out at pay telephones, vending machines, and transit fare-boxes.

Trials are also under way to introduce more sophisticated smart cards for wider use across the retail sector. A consortium of international banking and credit-card interests are currently field-testing a system known as "Mondex."[149] It is not yet clear that the Mondex model will succeed, but its core features are evocative of the emerging character of digital transaction in general. Each Mondex card contains a chip with eight megabytes of memory "in which is embedded the protocol for receiving, storing, transacting and locking Mondex value in up to five different currencies."[150] Using an automated teller machine, or even a public or home telephone equipped with a read/write device, a person can draw funds from his or her bank account and store them on the Mondex card. The consumer can choose to "lock" these funds on the card using a personal code so that only he or she can release them, and then spend this money anywhere there is a device capable of reading the card. This system of use does not breach anonymity because the code simply interacts with the card, which is releasing money into another reader – no connection to accounts housing personal information is necessary after value has been added to the card. Money can be transferred from one card to another using any Mondex read/write device, including an electronic "wallet" designed specifically for this purpose: you can pay the baby-sitter, leave some money in the cookie jar, and give the kids some money for pizza simply by transferring bits from one card to another in a Mondex wallet. Alternatively, you can transfer the money to the pizza parlour's Mondex unit directly over the telephone. Mondex cards are also capable of handling multiple currencies at once, and were the first to enable electronic carriage of the "Euro" currency unit adopted by the European Union.[151]

With its ability to transmit money over the more or less open infrastructure of the public telephone system, Mondex is smarter than other stored-value cards, which typically rely on closed, proprietary systems.

· · ·

For example, consider the current stored-value card system for the operation of pay telephones. For the system to be profitable, the phone company must ensure that only cards to which *it* has added value in exchange for cash payment can be used to purchase telephone calls. If one could make phone calls using a card dedicated to photocopying, to which value has been added using a vending machine whose proceeds belong to a library, then the phone company would be receiving nothing for the service it provides, and the library would get paid for providing nothing. Thus, to be profitable *for whom they are supposed to be*, these systems must remain distinct, closed, and proprietary. The Mondex system overcomes this by integrating into a single system a large number of banks who underwrite the value of the bits and reimburse retailers and vendors regardless of where the value was originally added. The interest banks have in developing this technology is twofold: first, it may eliminate the great expense of processing paper-cash transactions altogether; second, it is projected that smart cards such as Mondex will soon be integrated with credit and debit cards into a single, multi-use instrument of digital payment and exchange. Computer networks are the infrastructure upon which these modes of transaction – each of which makes it easier to spend money – rely.

Cards such as the Mondex variant described above represent two key advances in the mode of transaction that are necessary for the optimization of the transactional utility of computer networks. The first is the removal of the need for a consumer to be in physical proximity to the point of sale, and the second is the use of the open architecture of fully public networks to carry out everyday commercial transactions. Reaching these two goals has also provided the impetus for attempts to develop protocols for small-scale monetary transactions that are wholly and truly "on-line" – that is, where money in the form of bits is exchanged between one computer networked with another via an open and publicly accessible network infrastructure. The same concerns that have dogged other modes of electronic transaction apply here – data security; veracity of the currency being exchanged; determining the authenticity of parties to an exchange while maintaining their anonymity; and the economy of small transactions – but they become more pronounced when the medium being used is open and public, as is, for instance, the Internet. With closed proprietary networks (i.e., inter-bank clearinghouse networks that coordinate overnight financial trans-actions between banks), proprietors exchange bits they can be confident stand for real money because access to the network has been limited in advance to those whom the proprietors *know* to be real entities with real money, and who observe shared conventions regarding its ex-change. All bets are off, however, when it comes to freewheeling and open exchanges of money over a public network. In this case, protocols of verification, authenticity, and privacy must be built into the process of exchange itself, *every time it is carried out*. The magnitude of such an undertaking has vexed sorely those trying to design a system for the routine and secure exchange, over an open medium, of massive quan-tities of monetarily valued bits that will rival the perfection of physical cash money or "hard" currency in this regard.

Various schemes involving complex verification algorithms, encryp-tion techniques, digital signatures, and voice-recognition devices are competing in the course of a concerted effort to cash in on the poten-tial for everyday commercial transactions via networked computers. It remains to be seen which technical combination, if any, will emerge to usher in a cashless society. One thing is certain: influential actors in both technological and financial fields appear confident that this day will come. Three major commercial interests in the area of networked computing – Microsoft, Netscape, and Intel – have announced plans to market computing devices equipped with read/write slots for stored-value, credit, and debit cards.[152]

However, the technological capacity for secure monetary exchanges via public networks will not be enough to convince people that their money or credit is safe on-line. For electronic commerce to really blos-som, everyday people must become habituated to trusting the medium by which those commercial exchanges are transacted. This is a particu-larly difficult condition to achieve, given that, in such transactions, people can neither see nor touch the products or services they are pur-chasing, nor the money they are using to do so. Trusting the medium of traditional retail cash and credit transactions is easy: you are there at the store; you see the goods or person whose services you are purchas-ing; you take money out of your pocket and exchange it physically for the item purchased; the retailer can see by the seal and watermarks on the currency that it is not counterfeit, or can authenticate your identity via inspection of well-established documents of identification; you are given a receipt that conforms to widely observed conventions regarding proof of purchase and ownership; and, finally, if something goes wrong, you know where to find the person responsible and how to get redress. Network transactions will have to attain this same level of acculturated trust if they hope to replace cash-and-carry as the dominant mode of commercial transaction.

The engagement of trustworthy actors and institutions in monetary exchange over open public networks will contribute greatly to the nurturing of this trust, which is why the emerging presence of banks – the most trustworthy and secure of all institutions – on-line is so significant. Nearly every major banking institution in Canada now makes personal banking, financial, insurance, and other services available to its customers either by telephone or by networked home computer.[153] Using a touch-tone telephone, customers entering a PIN can interface with a bank computer rigged with an interactive voice-response system and check account balances, transfer funds, check interest rates, and make bill payments. More complex transactions such as loan applications, mortgage renewals, and RRSP investments can be made by telephone through agents at a call centre. With a home computer networked to the Internet or other private service network, customers need merely to visit their bank's Web site, download the bank's personal finance-management-software portfolio (including sophisticated encryption software to ensure transactional security over open networks), and they are availed of a plethora of remote services. Beyond those services already listed as accessible by telephone, on-line bank customers can buy insurance, track stock and investment portfolios, buy and sell stocks, and solicit financial advice, all with a high degree of transactional security. In Canada, the federal Task Force on Financial Services reports that telephone banking has increased by 50 percent yearly since 1994, and on-line banking has increased at an annual rate of 10 percent.[154]

Private, proprietary networks are, of course, even more secure than the Internet, but, as I have been suggesting, banks appear to recognize the benefits of habituating people to trust the security of transactions conducted over an open and public network infrastructure. Appreciation of these benefits is signalled by the banks' willingness to underwrite any losses due to system breaches that may be suffered as network security is fine-tuned. What do the banks gain from offering these services? Some charge fees for on-line services and transactions, but others do not. Also, as mentioned above, the overhead costs of administering "branchless" on-line transactions are considerably lower – by some estimates 75 percent lower – than those accruing to in-person service delivery.[155] However, it may be that the primary benefit of the on-line presence of vanguard financial institutions such as banks will be realized in the contribution this steadying influence makes to normalizing on-line monetary transactions in the minds of ordinary consumers. If network transactions are secure enough for banks, then they are secure enough for everyday commerce, and what is good for commerce is good for banks. The sooner on-line commercial transactions recede into the unnoticeable fabric of common exchange and become as "natural" and habitual as leaving a pile of dollars on a table to cover the cheque in a restaurant, the better things will be for everyone whose livelihood depends on the movement of money from one set of hands (or one hard-drive, or one "smart" card) to another.

Marketability

The ability to market goods and services effectively is, along with widespread access to the means of transaction and a secure common mode of transaction, key to the successful exploitation of the commercial potential of network technology. Network technology, in so far as it combines delivery of content in textual, auditory, and visual formats with transactivity, would seem a perfect medium for the simultaneous advertisement and exchange of commodities – in other words, a perfect marketplace. The genius of television was confined to its facility as an advertising medium largely because of the inherent limits of any broadcast technology: while TV succeeded in gathering large audiences to view advertisements for products that the programming inserted between ads had fabricated needs for, it could not enable the transmission of content (i.e., transactions) in the other direction (i.e., from receiver to sender). As a multicast, transactive medium, computer networks seemingly overcome this limitation.

However, as a marketing technology, computer networks face a problem that television has already solved – namely, how to aggregate multiple individual users/viewers into a consistent, predictable audience of sufficient mass to make marketing efforts via the medium worthwhile for commercial interests. In this regard, what are often promoted as network technology's virtues – the ability of users to access the medium at various times in multiple ways, to navigate its content idiosyncratically, to produce as well as consume content, to choose among an exploding universe of diverse content sites and types – become barriers to commercial exploitation on a mass scale. Built into the medium are capacities and instrumentalities that encourage its user population to be fragmented and disaggregated, with their eyes and ears on different things at different times, and perhaps seldom in the same place twice in a row. In many respects, computer networks like the Internet and World Wide Web have characteristics that make them anathema to commercial marketing.

Ironically, the response of those hoping to capitalize on the commercial potential of network technology has been to make the medium behave more like television in order to salvage its marketing utility. Efforts by major network service providers to stream content into "channels" suitable for delivery via converged television-computing appliances have already been discussed, and these represent just one example of the trend to convert public networks from a resource that requires users to search for and "pull" desired information onto their screens into a medium through which packaged content can be "pushed" toward them. So-called push applications are being touted as a commercial boon for network technology. In the early days of the Internet and the Web, users looking for information would have to actively seek it out and "go" to the sites where that information was located. As resources, sites, and users proliferated, the demands of searching for information escalated (there was so much to choose from, wade through, so many ways to find it, and so many paths to reach it – the problem of information overload or "glut") and it became increasingly difficult to predict or ensure the exposure of significant numbers of people to given sites on an consistent basis. The characteristics of a proliferating, distributed network that sometimes made navigation and utilization of its resources daunting for users also compromised its utility as a device for commercial marketing.

One solution to both these problems was the development of so-called push servers. In basic terms, a push server is a program that conducts automated searches of designated network sites and resources on a scheduled basis, gathers the results of those searches, and delivers – or pushes – that content to a user's computer.[156] Initially, these servers were customized to suit the individual user's preferences. This approach quickly proved inconvenient – it still required individual users to develop and maintain their own search strategies – and it did not overcome the problem of idiosyncrasy thwarting commercial marketers. Consequently, fully customized push programs fell by the market's wayside, and were replaced by what might be called "semi-custom" push services, in which commercial providers configured push packages and users subscribed to those that interested them or fit their preferences, much as they would select a battery of specialty television channels from a cable-TV provider. Subscribers to the "PointCast" server, for example, received regularly scheduled deliveries of news, weather, and sports information on the Web, culled from sites that were searched automatically and consistently by the software.[157] The vogue of push service appears to be waning, but the model it established captured an enduring imperative in the development of Web software: commercial viability demands that users be relieved of the task of searching the network themselves *and* that significant and predictable numbers of users be delivered regular content on a consistent basis. As PointCast's on-line promotional tour described, "gone are the days of surfing the net for the news and information important to you. The Internet is brought to you ... Headlines move dynamically across your screen, the colors pop and all you have to do is keep your eyes open. Effortless. No surfing required."[158] Put simply, the push model is to gather content and deliver it in the form of channels and, in so doing, create audiences for Web sites. And, where there is an audience, there is utility and value in advertising.

PointCast, and many other "tuning" programs, can be downloaded free of charge from the Web. There is no charge because the investment in developing and distributing these programs is more than recouped by the advertising revenue they generate. When content from a limited range of regularly searched sites is delivered to a substantial and measurable number of network users, it becomes good business to advertise on those sites because you can be confident that your ads will be seen. Push services pushed computer networks away from their multicast origins back toward a more traditional broadcast paradigm, albeit with a slight twist. Television, with its relatively few channels and generalized content, was a perfect instrument for mediating the production of mass audiences for sale to manufacturers advertising their mass-produced consumer goods and services. When the multitudinous and often highly idiosyncratic content accessible through them is delivered using the push model, computer networks create audiences that are somewhat differentiated and fragmented, but still quite large and relatively homogeneous; with the extra intervention of pushers, networks emerge as the perfect instrument for mediating consumptive behaviour that complements the somewhat customized, reputedly flexible, short-cycled, semi-variegated products of the post-Fordist commercial economy.

Despite the non- (and often anti-) commercial character of its origins, the Internet has emerged as a promising vehicle for the advertisement of consumer goods and services. The Internet Advertising Bureau (IAB) reported that network ad revenue in the first half of 1997 exceeded US$343 million, a 322 percent increase over the same period a year before.[159] In terms of the products advertised, leading the way were consumer products (30 percent of total ad revenue), which outpaced both financial services (22 percent) and computer products (21 percent).[160] Most network commercial advertising takes place on the World Wide Web because of its graphical, audio, video, and hypertext facilities, and it generally takes three forms. Manufacturers, retailers, or service providers often maintain "destination sites" of their own where interested consumers can "visit" and browse or solicit product information (i.e., the Nike Web site). The high production values demanded of successful commercial advertising means such sites are typically expensive to develop and maintain. Further, destination sites typically capture only those consumers who are *already* interested in a company's product and offer little benefit in terms of enhanced exposure beyond the brand's existing loyalty or interest base. The marketing benefit of such sites is realized primarily in their provision of the sort of detailed information that is involved in the latter stages of a high consideration consumer purchase (i.e., an automobile).

A second form of network advertising takes place on "micro-sites." These are advertising sites that are not maintained by the manufacturer or retailer, but instead are "hosted" as an adjunct by a site whose non-commercial content is popular among the advertiser's target audience. Users of host sites are presented with a hypertext link that allows them to "click through" to the advertising pages of the micro-site appended to the main site. However, in line with the tendency of media audiences to circumvent commercial messages whenever possible, these micro-sites are vulnerable to being either ignored or "clicked past" rather than "clicked through." Given that this type of ad is usually paid for on a "per click-through" basis, they are not always a great source of revenue for the host site either.

By far the most effective, lucrative, and, consequently, popular form of network advertising comes in the form of what is known as "banners," which in 1997 accounted for roughly 80 percent of on-line ad placements.[161] Banners appear on screen at a host site not simply as a highlighted hypertext link, but rather as a graphic, and sometimes video, electronic billboard. Banners often include an image of the product, other still or moving graphic images, and accompanying slogans or text. They also often include hypertext links through which interested customers can visit the manufacturer's destination or micro-site, request samples or information, and even register in contests. However, as the findings of a recent IAB study indicate, the influence of click-throughs on overall ad effectiveness is negligible.[162] The study found that, while just a single exposure to an on-line banner ad could generate "dramatic" increases in advertisement awareness (i.e., recollection of having seen the same ad elsewhere or before), brand awareness, product-attribute communication, and intent to purchase, "nearly all of the impact measured was generated without a 'click-through' to the advertiser's site – proving the power of the ubiquitous banner."[163]

An example from the study proves illustrative in this regard. In one test, respondents were shown a series of three on-line banners that featured a blurred image of a Volvo automobile approaching the user from behind, viewed as if through a side-view mirror, accompanied by text reading: "So sleek. So swift ... Apparently it can travel through time ... Brace yourself. Click here." The study found that, *after a single exposure*, "the Volvo ad banner seems to increase users' belief that Volvo makes 'a good automobile' (an increase of 55%) and that Volvo 'offers something different than other brands of automobiles' (an increase of 57%). Because of the ad banner, those exposed are more likely to 'Have a higher opinion of Volvo than other automobiles' (an increase of 44%)."[164] The study's further findings regarding this example are illuminating: "Banner exposure itself was responsible for 96% of the brand enhancement, while a click-through only contributed 4%. While additional powerful messaging may wait on the other side of a banner at the advertiser's Web site, the analysis indicates that *the exposure itself carries nearly all the value* ... the real communications power is where the majority of the audiences can *see* the message."[165]

This illustration tells us much about the dynamic pervading advertising on computer networks and, coincidentally, something about the essence of the medium itself. Banner advertising is customarily paid for, not on a "click-through" basis, but, instead, at a predetermined rate per thousand "impressions." An "impression" occurs every time someone visits the site where the banner is on view (a number that can be tabulated exactly). High-traffic sites with large quantities of visits, or "hits" – such as starting points for Web-surfers, sites with popular content, sites where content is continually updated, sites that are trolled and delivered by push servers – charge high rates per thousand impressions, much as television shows with consistently or predictably large audiences can charge more for thirty seconds of advertising time than unpopular programs. Indeed, given the increasing appeal of push services and content streamed into pre-selected channels, and if the suggestions of the IAB study I have cited are even remotely persuasive, it is difficult to discern what differentiates computer networks, as a medium of commercial marketing, from broadcast television, the medium that networks are so often contrasted with. In general, computer networks are championed as the medium of people (i.e., consumers) who are not satisfied with being spoon-fed superficial, pre-packaged information, and who, instead, prefer to dig underneath the surface and avail themselves of the considerable resources accessible via the wired screen that allows them to get behind the impressions to the facts. Nevertheless, it would seem that click-throughs to more extensive, specialized, and detailed information are not what defines the appeal of this medium to commercial marketers. If mere exposure and impressions are what really matter in terms of consumer choices, then what works on computer networks will be what works on TV: catchy tunes, clever slogans, pretty images, pretty colours, and pretty girls. The only difference is that these things might work even *better* on computer networks than they do on television. As the IAB study beamed in identifying the peculiar advertising utility of the 'Net, "Web and print-based media have the advantage of active reader involvement and attention, being 12-18 inches away from their audience and requiring them to take action to consume the medium. The engaged state, which the Web encourages, seems to help provide higher attention to on-line advertising."[166] In the practices of advertising and marketing – as in all the other aspects of consumptive life in which it is implicated – what defines the commercial utility of network technology is its ability to facilitate transactive exchange. That some insist on labelling this facility "interactivity" serves only to distract us from this very basic observation.

Networks, Capitalism, and Democracy

The present discussion of the political economy of network technology began with a quotation (the epigraph to Chapter 4) in which it was suggested that network technology is instrumental to a massive shift in power away from the privileged class, which owns the means of production, and toward the working class, whose members sell their labour in order to secure resources for a decent existence. The implications of this statement are clear: despite the capitalist economic system that engendered it, this technology is revolutionary and democratic in its very essence. Indeed, it was shown that this view represents a formidable mainstream discourse that insists that the political and economic change being wrought by this technology is fundamental and irresistible. Such statements are difficult to reconcile with the picture emerging from these chapters of a hegemonic economic order simultaneously wielding and settling into the technology of its acceleration and ultimate completion. As Ronald Deibert has written, "if there is one clear 'winner' in the hypermedia environment, it is the collective interests of transnational capital."[167] This conclusion is, of course, incompatible with the assertion that networked computers are the technology of a revolution in the basic power arrangements of modern capitalist society. It is also, however, a nearly unavoidable one.

Rather than any revolution, the present and preceding chapter have shown that the proliferation of network technology represents an acceleration of the logic and effects of capitalism in the practices and relationships of production, work, consumption, and exchange. It does so because its peculiar properties – among them its ability to collapse control, information, and communications utilities into a single undifferentiated stream of bits – enable and encourage capitalization on the fertile environment for profiteering and accumulation created by transnational economic liberalization, privatization, and deregulation. Such acceleration has entailed changes that have been catalogued and analyzed in the foregoing pages, but includes none at the level of the fundamental power structure and relationships of capitalism itself. Contrary to popular images that place computer networks at the centre of an irresistible democratic restructuring of political and economic life, "the new world order of global communication, among the most profitable consequences of global capitalism, tends to reinforce the status quo."[168] In so far as they bolster the already formidable control of capital over the means of power, computer networks are an essentially conservative, not revolutionary, technology – conservative, that is, of the prevailing liberal and capitalist order. Indeed, as Robert McChesney has written, as this acceleration continues unabated, the likelihood of this technology being enlisted in any sort of profound systemic upheaval becomes increasingly remote: "Although the Internet clearly has opened up space for progressive and democratic communication, the notion that the Internet will permit humanity to leapfrog over capitalism and corporate communication seems dubious unless public policy forcefully restricts the present capitalist colonization of cyberspace."[169] As discussed in Chapter 4, if the postures and "reform" activity of the Canadian and US governments around this technology are any indication, it seems quite unlikely that such forceful restrictions will be forthcoming; on the contrary, the dominant thrust of policy in this regard has been to simply reduce friction by getting the state out of the way.

If the economics of network technology are distinctly capitalist, then what are its hopes for fostering or even contributing to a democratic political transformation? The answer to this question lies not so much in what one thinks about network technology as it does in how one defines democracy and appraises its relationship to capitalism. If one simply equates democracy and capitalism, as is the case in contemporary popular discourse, then the answer is so clear that the question is not even worth asking. If, on the other hand, one defines democracy as a political regime in which citizens enjoy an equal capacity to participate meaningfully in the decisions that affect them most closely as individuals in communities and if, along with David Noble, one sees capitalism as "a euphemism of sorts, a polite and dignified substitute for greed, extortion, coercion, domination, exploitation, plunder, war and murder,"[170] then the answer becomes more interesting, if problematic. Marxists and other socialists have, of course, always doubted the ability of capitalism, with its necessary and gross inequities of power, to provide the economic foundation for a genuine, robust, sustainable, democratic political order. To the extent that capitalism as an economic system and democracy as a political system have become synonymous in the present discursive climate, socialists have been effectively deprived of the language with which to make this argument.

However, one need not endorse Noble's passionate rhetoric in order to wonder whether it makes sense to regard as democratic a political system embedded in an economic order that, by its very nature, systematically deprives the vast majority of people of even a limited share of effective control over the means of power, and so also deprives them of control over the immediate social conditions in which they live. Such a system may not be evil, but it is certainly not democratic. By extension, we might wonder how a technology that aggravates rather than alleviates this situation can be called democratic. Specifically, is it reasonable to suggest that the large numbers of working people thrown into fundamental economic insecurity because network technology facilitates (and so demands) productive "flexibility" can be public-spirited democratic citizens? Is it sensible to believe that teleworkers whose working lives have been diminished by computer networks and degraded by omniscient surveillance will be able to develop and nurture the capacities of responsible citizenship? Is it likely that an environment in which the ownership of dominant communications media is concentrated to near-monopoly levels will be a good one for widespread, conscientious civic participation in common public decisions?

Far from overcoming the substantial impediments capitalism presents to democracy, network technology is instead being deployed in ways that augment them. Network technology enables the further entrenchment of those inequalities in control over the means of power that frustrate the equal ability of citizens to participate in the fundamental decisions that affect their everyday lives. As one writer has observed, "virtually all known theories of political democracy would suggest that such a concentration of media and communication in a handful of mostly unaccountable interests is little short of an unmitigated disaster."[171] People who are controlled *in* their working lives rather than being in control *of* them are effectively deprived of participating meaningfully in decisions in and about that sphere of their lives; people who are infantilized by the manner in which control is exercised over them in their working lives are unlikely to be capable of meaningful participation should the opportunity even arise. Conveniences of consumption and commercial exchange are not substitutes for civic participation, and they may even encourage the normalization of an economic complex that excludes it. Any technology – networks included – that contributes to the perfection rather than the relief of these conditions cannot reasonably be considered democratic.

Democracy may not be better than capitalism. My point here is simply that even a minimum of definitional specificity requires that we recognize they are not identical; more rigorous scrutiny suggests that they may even be at odds. If this is the case, and if, as these chapters have made clear, network technology is distinctly capitalist in character, then as a technology it is more likely to be democracy's enemy than its saviour. The popular discourse of enthusiasts has, admittedly, included consideration of the apparently burning issue of information overload as a potential democratic liability. However, the possibility that network technology and democracy may be opposed on a more fundamental level – the level of economy – has escaped them entirely. As Peter Golding has lamented, the result is that "somehow the dream of Jeffersonian democracy through optic fibers [has] been transposed into the increased chance of saving a twenty minute round trip to the video rental store."[172] Such is the price of detaching political hope from economic necessity.

a standing-reserve of bits

Some among the group of potters who fire there
speak of the kiln as though it were sentient.
I've watched Jack run his hand over its heated flanks,
a light stroke, the way a man might massage the swollen
belly of a pregnant woman. Effleurage, it's called.
The tender, encouraging motion of his hand is a sign that
one is not within the realm of a technology here,
but nearer cooperation with a mystery.

BARRY LOPEZ,
describing the relationship between an anagama kiln
and its keeper, in "Before the Temple of Fire,"
1998

CONFRONTATIONS WITH THE ESSENCE of a technology often occur in unlikely places, such as the washroom at the cinema. I had just watched
The Sweet Hereafter, a film made by Canadian director Atom Egoyan,
based on a novel of the same name by Russell Banks. The film tells the
moving story of a small town's struggle to deal with the tragedy of a
school bus accident that claims the lives of many of the town's children. In one of the film's most arresting scenes, the yellow school bus
misses a curve on the highway and is sent spinning across the surface
of a frozen lake. The bus slides to a halt and, for a brief moment, sits
silently upon the sheet of ice, as if waiting for Nature to decide whether
its passengers will live or die. Suddenly, the ice buckles beneath the
vehicle's weight, and the bus goes under. What if the overnight

temperature had been two degrees colder and the ice an inch thicker? Would the ice have borne the weight of those young lives if just one child had stayed home sick that morning, or if the gas tank was not full? Would the rate of survival have been different if the bus had rolled rather than spun? Among innumerable others, these questions and their unknowable answers weave the fabric of an overwhelming and wrenching mystery that formed the inescapable context in which the town and its people continued to live.

In the washroom, after the film, I overheard the following conversation between two men wearing jackets that identified them as workers in the film industry:

> *Bill:* Hey man! What'd you see?
> *John: The Sweet Hereafter.*
> *Bill:* No way. What'd you think?
> *John:* Pretty good.
> *Bill:* D'you like the bus scene?
> *John:* Yeah. It was good.
> *Bill:* It was all digital, eh? DGI.
> *John:* F#@*! No way! Looked real.

So, it wasn't a bus full of schoolchildren sinking slowly into a frozen lake. Not even close: it was a bunch of bits. The mystery, along with the film's response to it – that is, that some things just cannot be accounted for and that sometimes it is best not to try – began to evaporate. Perhaps this reaction was unfair. After all, whether its effects were produced digitally or otherwise, the whole thing was still a piece of artifice: the story was fictional; there was no bus full of children; the town did not really exist. These are things that I knew to be true whether these images were produced by a camera or a computer. Why was I so disappointed to learn that these images, which had moved me so terrifically and which had revealed such a beautiful truth, were produced using digital technology? Wasn't this cause to celebrate the technology's wonders?

There is, in all artistry, a delicate relationship between discovery and fabrication. All artistry involves fabrication, or "making." Good artistry, however, does so only up to a certain point: it makes that which is necessary to reveal something present, but perhaps not *immediately* evident, either in the human condition or in Nature, something that itself ultimately defies human creativity. Thus, genuine artistry negotiates between fabrication and discovery, making and revealing, and knows itself not to be creative. It necessarily follows that it is limited in both its ends and its means. The means used to mediate that which is present but not evident should not encourage artists to presume they can pass over the threshold that separates them from the power to create that which is not all present. Phrased differently: when it comes to artists, the less creative they are, the better.

In *The Sweet Hereafter,* the artist walked this line deftly, revealing with fabricated images a truth about the precarious nature of human existence that he could never have created. That this was accomplished digitally was still intensely unsettling. On one level, it was simply the shock that accompanies confrontation with a new technique of auditory or visual representation. Stained glass in Gothic cathedrals, the introduction of perspective in painting, moving pictures, the phonograph, and stereo recording have all been shocking to viewers and listeners accustomed to previous modes of representation. The truth cannot be fabricated, but it can be represented in oil, expressed in words, captured on film, and now, apparently, dangled from strings of os and 1s. As with previous techniques, we will soon become habituated to representations of the world that are composed of bits, and these will recede into the fabric of our everyday experience without interrogation.

On a deeper level, I think it was this prospect of an uninterrogated assimilation of digital representations of the world that was so disturbing. Developments in techniques of representation are typically welcomed because they promise to overcome the limitations imposed by previous instruments and media, and so enhance the creative potential of artists who use them. Innovations in artistic techniques tend to encourage creativity and, in so doing, entice artists to ignore the distinctions, limitations, and practices that define genuine artistry. The world as apprehended in and through binary digits is easy to manipulate and "re-create" in ways that, like the real world, are not clearly fabricated. In this, digital technology far outstrips previous artistic media. This is not to say these new techniques are without limitations of their own, or that they cannot be employed, as they were in the scene of the sinking bus, in aid of the activities of discovery and revelation characterizing genuine artistry. However, this is not the disposition they encourage, and herein lies the real source of my unease at learning of the bus full of os and 1s: digital technology represents another in a long line of technologies that, by allowing artists to transcend the limitations of previous media and instruments, encourage them to believe in their own essential creativity. In the digital age, as one popular commentator observes, "expression becomes art ... The real artists will be those who know how to imagine things that no one else does – the people brave enough to conceive the inconceivable."[1] In this sense, digital technology is part of a long trajectory of modern technologies that fool us into believing we are creative, rather than created, beings.

In modern times, it is said that we are all artists, enacting our designs upon nature both individually and collectively to construct identities and societies. Thus, my questions about film artists in the age of digital networks would seem to pertain also to the artists of identity and society. Toward what end do human beings who, in the modern age, are believed to be the artists of their very selves, use digital technology? Do they simply reveal what is present but not apparent, or do they presume to create that which was previously absent? And what effect does this technology have on the status of their artistry? What is lost and what is gained? The networked digital computer is often presented to contemporary individuals as the final technology of their ultimate self-creation, a view that is both suggestive of, and suggested by, prominent themes in postmodernist discourse. In this chapter, I argue that, in so far as they reduce the world – human beings included – to a standing-reserve of bits; networks culminate the distinctly modern technological condition described by Martin Heidegger: a condition characterized by rootlessness, calculation, and the denial of mystery. To illustrate, I examine the deployment and use of network technology in relation to contemporary practices of community and surveillance.

Networks and (Post)modern Identity

Ideas about the self-creating self are present in the foundational documents of modern Western political philosophy, and are a core attribute of modern discourse and social life. Whether they are Machiavelli's self-made men, Hobbes's instrumentally rational artificers, or even

Rousseau's savages acting on their *perfectibilité,* modern individuals believe themselves to be creative beings capable, at the very least, of making themselves into something more than what they were made. Indeed, the economic, social, and political trajectory of modernity can largely be read as an elaboration of Francis Bacon's suggestion that, by enlisting the techniques of science, humanity could relieve itself of its "estate," that is, its natural status or condition.[2] Liberalism, perhaps the greatest of modernity's ideologies, is premised on the belief that individuals are free to make of themselves what they will. As discussed in Chapter 1, various technologies have played a major role in this distinctly modern drama.

Among other things, it is this belief that human beings create their own identities, and also the conditions of their sociability, that puts the "modern" into "postmodern." Postmodernists basically retain their modern forebears' conviction that individual selves and the political communities in which they congregate are artificial constructions that are historically situated, conditioned, and contingent. What postmodernists reject is the assertion common to most modern philosophers that holds there are essential traits of human nature that ultimately limit, centre, or at least direct, the design and course of these constructions. For Hobbes it was appetitive, acquisitive, competitive calculation; for Rousseau it was compassion and a love of human existence; for Marx it was the capacity for freely self-determined labour. For postmodernists, these are all fictions, as are any "essentialist" accounts of natural human attributes that necessarily fix the origins, boundaries, or character of constructed human subjectivity. Indeed, the idea of subjectivity itself – coherent, purposive, intentioned, rational consciousness linked to action – is inimical to postmodern conceptions of the radical openness and indeterminacy of individual identity. Postmodernists reject the idea of a self constructed upon an endowed essence because they believe that essence is, itself, entirely constructed by language. This does not mean that individuals in society are *necessarily* free to build their identities as they fancy – individual identities are always contingent upon the historical, discursive, and institutional contexts in which they are situated, with certain materials rendered available or unavailable, according to the relationships of power prevailing in any given situation. It simply means that discourses about natural human essences are simply that: discourses grounded in power rather than truth, a means of endorsing one set of identity-traits rather than another. It is to this that Fredric Jameson refers when he delineates the "meaningless materiality of the body and nature and the meaning endowment of history and of the social."[3]

Postmodernists identify modernity as the historical period in which a few metadiscursive "grand narratives" colonized the field of individual identity by framing total accounts of what it meant to be a normal, productive, and healthy human being, in terms that legitimated prevailing or ascendant configurations of knowledge and power. Among these were narratives pertaining to the liberation of the rational actor, the collective emancipation of the working class, the achievement of prosperity through industry and material accumulation, and progress through science. Attending these narratives were discourses and institutions that provided materials for the construction of identity, and set the criteria for inclusion in, or exclusion from, both the mainstream of social life and one's essential humanity. Remaining within the range of identity positions sanctioned by the prevailing grand narrative meant that one was in touch with, or at least on the way to realizing, one's essential human nature. Non-conformity rendered one an alien, both from one's society and from one's true self. Thus, the conditions available to the modern subject were either harmony (or progress toward it) or alienation, depending on where one sat relative to the grand narratives forming the context in which one exercised one's subjectivity.

Postmodernists argue that *post*modernity is characterized by the collapse of these grand narratives as convincing accounts of the human condition, and so also as exclusive providers of material for the construction and orientation of identity. Jean-François Lyotard identifies an "incredulity towards metanarratives" as the definitive feature of the "postmodern condition."[4] Corresponding to this collapse of metanarratives is a disintegration of any essential "centre" around which identity can be wholly and coherently constituted, or from which it can be considered alienated. According to Jameson, "concepts such as anxiety and alienation (and the experiences to which they correspond ...) are no longer appropriate in the world of the postmodern ... This shift in the dynamics of cultural pathology can be characterized as one in which the alienation of the subject is displaced by the latter's fragmentation."[5]

Truly postmodern identities are described as, variously: fragmented, de-centred, partial, unstable, multiple, heterogeneous, incomplete, discontinuous, fluid, and highly differentiated. The postmodern identity is constructed, but it is constructed without a foundation. Once freed from the tyranny of grand narratives, the self is more free to construct itself than ever before. The postmodern self is an assemblage of identifiers culled from the myriad influences, relationships, positions, styles, and discourses available to individuals freed from the homogenizing demands of a compelling grand narrative. And, in the absence of a universal standard, or even an accepted narrative against which one can judge, no identifiers, nor any combination of them, can be posited as more essential, natural, or legitimate than any other.

Thus, the stakes of the identity game are lowered considerably. Decisions about what and how to "be" can be more playful than serious, more ironic than earnest, more superficial than deep, more accidental than intentional, more deceptive than honest, and more cosmetic than ethical without appreciable consequences. Accordingly, *collage*, palimpsest, *pastiche*, and play – the arts of children – are the *modus operandi* of postmodern identity assemblage, rather than thought, prudence, commitment, prayer, or deliberation.[6] Constituted as they are by networks of relationships and appropriations in which each is both influenced and influential, postmodern identities make up in intersubjectivity what they lack in autonomous subjectivity.

Setting aside the question of whether this is truly what people are like in late capitalist, liberal-democratic societies, it is certainly true that this vision of human identity – our consciousness of what we are as beings – resonates strongly with contemporary network technology. As Ronald Deibert writes, referring to the "fit" between postmodern views of the self and digital technology, "within the hypermedia environment, digitization and networked computing provide users with the ability to extract bits of data in different forms from disparate sources, and then paste them together into an assembled whole ... this transparent environment opens up and disperses personal information along decentered computer networks, much the same as postmodernists conceive of the self as a networked assemblage without a fixed center ... Identities on the 'net' – such as age, gender and occupation – are malleable because of the concealment that computer networks afford the user."[7] Deibert uses the contradictory words "transparency" and "concealment" to describe the complementarity of digital networks and postmodern discourse, but the second term captures this sympathy best. Users of networked computers find themselves confronted with increasingly opaque interfaces – despite being called "Windows," user-friendly programs hide more than they reveal about what is going on inside the computer – that appear as surface images to be manipulated rather than openings to be seen through. This, of course, resonates strongly with postmodern accounts of the present superficiality and "depthlessness," wherein "depth is replaced by surface, or by multiple surfaces."[8]

A noted proponent of the inherent link between network technology and postmodern selfhood, Mark Poster calls the computer – "with its dispersal of the subject in non-linear spatio-temporality, its immateriality, its disruption of stable identity" – "a factory of postmodern subjectivity, a machine for constituting non-identical subjects."[9] In the midst of a "second media age" characterized by "two-way, decentered communication," Poster observes, "the subject may be decentered or multiple or whatever."[10] Describing "the construction of the self in the Internet," Poster writes: "The shift to a decentralized network of communications makes senders receivers, producers consumers, rulers ruled ... subject constitution in the second media age occurs through the mechanism of interactivity."[11] He even suggests that, to the extent it subverts traditional mechanisms of domination, this technology contributes to a reconstitution of the world whereby "subject constitution becomes its designated goal and social end."[12]

Perhaps the clearest attempt to draw connections between networked computer technology and the postmodern *zeitgeist* is Sherry Turkle's book *Life on the Screen: Identity in the Age of the Internet*.[13] Turkle not only connects the rhetorics of postmodernism and information technology, but also attempts to illustrate their reciprocity with reference to specific practices. For this reason, and because she introduces themes I wish to address later in this chapter, her efforts merit attention here.

Turkle presents the age of network technology as characterized by a "nascent culture of simulation," wherein "the self is constructed and the rules of social interaction are built, not received," and in which we are "inventing ourselves as we go along."[14] Thus, as Turkle writes, "computer technology not only 'fulfills the postmodern aesthetic' ... Computers *embody* postmodern theory and bring it down to earth."[15] In particular, the computer has become involved with "ideas about unstable meanings and unknowable truths," and this engagement, in Turkle's estimation, has enabled our profound shift "from a culture of calculation to a culture of simulation."[16] Calculation still occurs within and between boxes containing circuitry, but, for Turkle, this is beside the point because the vast majority of people interact with computers on "surfaces" that occlude the calculation within. The confrontation between the "aesthetics" of calculation and simulation is a confrontation between, respectively, transparency and opacity. A culture of calculation demands transparency so that it might see through to the depths or origins of phenomena and figure out how and why they occur as they do. A culture of simulation is content with opacity because it is "nonchalant": it cares little about deeper meanings and (suspect) truths, preferring, instead, the freedom to appropriate and combine images and appearances as desire directs.[17] The culture of simulation is the culture of the postmodern world of digital networked media. As Turkle puts it, *"we have learned to take things at interface value. We are moving toward a culture of simulation in which people are increasingly comfortable with substituting representations of reality for the real."*[18]

The practices of self-construction and simulation peculiar to this aesthetic are defined by Turkle in contrast to those she identifies as central to the modern, calculative aesthetic. Thus, seduction replaces addiction in relationships, the caprice of play supplants the rigidity of rules in collective action, tinkering replaces thinking in the pursuit of ends, and, most important, *bricolage* replaces structure as the model for assembling one's self.[19] *Bricolage* is a French word deriving from *bricole*, a noun denoting a thing of apparently diminished importance. Thus, *bricoler* is to do odd jobs, a *bricoleur* is one who putters around or tinkers, and *bricolage* is the act of puttering. Drawing on Claude Lévi-Strauss's use of the word to describe non-Western associative science,[20] Turkle defines *bricolage* as follows: "problem-solvers who do not proceed from top down design but by arranging and re-arranging a set of well-known materials can be said to be practicing *bricolage*. They tend to try one thing, step back, reconsider, and try another."[21] For Turkle, this practice characterizes a change in the orientation of computer programming since the advent of networks; it also defines the way human beings interacting digitally via these networks construct their identities. To the extent that their "selves" are made up of bits mediated by networks, individuals can tinker freely, multiply, and reversibly with their identities; "self-fashion and self-create";[22] borrow, copy, and simulate at will; as they live their lives, not through bodies and souls, but, instead, through wholly constructed and ephemeral "virtual" beings. Computer networks are the technology that makes *bricolage*, as a practice of identity construction, possible. According to Turkle, network experiences "admit multiplicity and flexibility. They acknowledge the constructed nature of reality, self and other" and, through them, "we are encouraged to think of ourselves as fluid, emergent, decentralized, multiplicitous, flexible, and ever in process."[23]

Turkle marshalls evidence for these impressions from a number of different aspects of networked computing. For example, as evidence of a shift from a preference for figuring-things-out/calculation/depth/transparency to one for exploration/simulation/surface/opacity, Turkle cites the popular eclipse of operating systems that use a text/command based interface (i.e., DOS) by those that employ more user-friendly, icon-based, "point and click" interfaces (i.e., Windows, Macintosh). Early IBM personal computers using the DOS system required considerable expertise to operate, but made it easier for people with this expertise to "look under the hood" of the computer to observe, and even alter, how its programming worked. Icon-based systems such as the one employed by the Macintosh are easier for the uninitiated to use because commanding them simply involves pointing at an icon on the screen and clicking, but they are not as amenable to tinkering at the level of programming as DOS machines were. According to Turkle, "postmodern theorists have suggested that the search for depth and mechanism is futile, and that it is more realistic to explore the world of shifting surfaces than to embark on a search for origins and structure. Culturally, the Macintosh has served as a carrier object for such ideas."[24] She also points to the sorts of "distributed" and "emergent" artificial intelligences – self-organizing patterns derived from local interactions between decentralized actors and operations – enabled by networks, and suggests that these also represent manifestations of the "nondeterminism" and "widespread disaffection with instrumental reason" that mark postmodernity.[25]

Most illuminating, however, is Turkle's discussion of the various practices in which individuals more or less explicitly use network technology to construct and assume identities in the manner suggested by postmodernists. For example, Turkle examines the practice of role-playing in Multi-User Domains (MUDs). A MUD is a program stored in one computer that allows multiple users at other computers (that are networked to the MUD's location) to connect and "play" with each other in the domain established by the program. Once connected to a MUD, users are typically presented with a textual description of a simulated environment (i.e., a saloon), and a list of other users who are "present" (i.e., also connected to the program). Through the use of standardized commands, users can exchange messages with other users, "move" from one "room" in the domain to another, or alter the simulation itself. The play in MUDs, while almost always expressed in text, takes a number of forms: "conversations" with other users; individual or collective "quests" through the domain; ongoing construction of the domain's "architecture" and rules as one participates; and sometimes engagement with computer-generated allies, foes, and other entities.

The important thing about MUDs in regard to identity is that the people who participate in them are free to describe themselves in any way they wish. In a MUD, users adopt pseudonyms and become whoever and whatever they want to be – male, female, or hermaphrodite; a-, bi-, homo-, or heterosexual; rapist, architect, or priest; earthling or alien; ravishing or repulsive – and, because of their physical remoteness and the opacity of the medium, no one can prove them to be otherwise. It is here that Turkle locates the affinity between network technology and postmodern identity: "As players participate, they become authors not only of text but of themselves, constructing new selves through social interaction ... The anonymity of MUDs gives people the chance to express multiple and often unexplored aspects of the self, to play with their identity and to try out new ones. MUDs make possible the creation of an identity so fluid and multiple that it strains the limits of the notion ... MUDs imply difference, multiplicity, heterogeneity, and fragmentation ... When each player can create many characters and participate in many games, the self is not only decentered, but multiplied without limit."[26]

While MUDs represent a fairly thin slice of network use, Turkle asserts "they more generally characterize identity play in cyberspace."[27] Other network practices that are said to be emblematic of the self-creating, multiple, and fluid character of postmodern subjectivity include participation in discussion groups, bulletin-board postings, and engagement in real-time text-based conversations via Internet-relay chat. In each case, the anonymity afforded by a lack of physical proximity and the opacity of the network medium allow users to assume whatever attributes they wish, with little fear of being revealed for who or what they truly are. On-line, men can participate as females in discussion groups that are ostensibly restricted to women talking about their sexuality, skinheads can pose as Jews and chat with rabbis about Zionism, and children can respond to the bulletin-board postings of handgun lovers as if they were adults.

The multiplicity of identity enabled by networks gains coherence through *collage* or *pastiche,* practices discussed in postmodern writing that are, according to Turkle, brought "down to earth" in the form of World Wide Web pages connected by hypertext links: "On the Web, the idiom for constructing a 'home' identity is to assemble a 'home page' of virtual objects that correspond to one's interests. One constructs a home page by composing or 'pasting' on it words, images and sounds and by making connections between it and other sites on the Internet or the Web ... one's identity emerges from who one knows, one's associations and connections."[28] Here, the self is the product of the material a person includes in his or her Web "site," whether it is material he or she has produced, copied from elsewhere, or simply established a digital connection with. And because almost every hypertext link leads to a series of more hypertext links in endless proliferation, the networked self is dispersed, groundless, and boundless.

Turkle attributes the seductive "holding power" of network experiences like these to their ability "to help us think through postmodernism."[29] While this may be true, it is not so clear that postmodernism helps us think through network technology. Accounts such as the one provided by Turkle are useful to the extent they describe what many people do *with* network technology. They are less helpful if one's concern is also to discover and understand what network technologies do *to* us, a matter that is, perhaps, of far greater consequence. There are a number of reasons for this shortcoming. In the first place, accounts such as Turkle's concentrate on an idiosyncratic set of applications that are not representative of the range of ways in which most people encounter network technology. People may play with identity-creation in MUDs, in discussion groups, and in constructing their Web sites, but this is not what they are doing when they use on-line banks, conduct telework, or use a smart card to pay for hockey tickets. Postmodern analyses tend to underestimate the importance of what these, perhaps less conscious but, arguably, more pervasive network mediations mean for what we are as human beings. Second, the postmodern predilection for surface effects and its denial of depth effectively undermine serious consideration of what our embrace of network technology might be doing to us *underneath* the identity *collage* some of us sometimes use it to construct. I would suggest the deeper significance of networks vis-à-vis our essential selves is the same for both those who use the technology to construct their superficial and fleeting subjectivities, and those whose encounter with networks is less playful. It is to be discovered in the logic of digitization that networks enforce. However, such questions of depth and essence are uninteresting to postmodernists, and so we will have to turn elsewhere for the answers to them.

The Essence of Network Technology

The perspective on technology developed by Martin Heidegger suggests we should be less interested in the superficial effects of network technology (i.e., its concealment of boys pretending to be girls) and more interested in its essence. That is not to say Heidegger would have considered the various practices carried out by people using networked computers to be irrelevant. As outlined in Chapter 2, Heidegger understood the essence of technology to be located in its mediation between the ontic and the ontological – between the practices of existing beings and a thoughtful engagement with the Being of those beings. Technological practices, like all existential activities, are ontologically significant to the extent they express something at issue in terms of

Being. Thus, how human beings actually use network technology clearly matters in terms of understanding its essence.

However, despite their relationship, the ontic and the ontological are not identical. If they were, ontology would be redundant: there would be no need for questions about Being if Being was self-evident in each and every act of a human being. One could observe a potter and surmise that Being is constituted by the act of pottery, or a beggar and determine the essence of Being to be beggary, or a rapist and conclude Being is defined by wilful mastery. Such would be tantamount to asserting there is no such thing as Being, but rather just however one happens to exist at any given moment. This is, of course, exactly the position on human nature advanced by postmodernists, who present people's use of computer networks to "be" gender-indefinite, simulated, and multiply identified as proof of the non-existence of either essence or Being. Such accounts simply perpetuate the condition whereby considerations of Being are excluded from behavioural examinations of being, a condition Heidegger identifies with the spiritual illness of the modern age.

"Language," Heidegger has written, "is the house of Being."[30] As a house, language is the dwelling of Being, but not, as some postmodernists have surmised, its *creator*. Through language we articulate our discovery of the world and we manifest Being, but language is constitutive of neither. Like any ontic practice, speech acts certainly matter in terms of Being, but that does not mean our essence is wholly constituted by whatever we happen to say we are at any given moment. A boy who says "I am a girl" to his fellows in a MUD is telling us something about his essence, but that something he is telling us is not that he is essentially a girl. Thus, while there is always something of the essence of Being contained in any particular speech act, the essence of Being is not simply what that act *declares* it to be. As Heidegger writes, "man does not decide whether and how beings appear, whether and how God and the gods or history and nature come forward into the lighting of being, come to presence and depart."[31] The essence of language is that it is "the house of the truth of Being," not the creative speaker of this truth.[32] Like any container, the house of language conceals what it contains: its appearance is a good indicator of the Being inside, but the two are not identical. To get at the essence of Being housed by language, we must penetrate its surface.

As with language, so with technology. To comprehend the essence of a technology, Heidegger suggests, it is necessary to gather those attributes of human practice in which Being is at issue into an ontology – a thoughtful questioning about the nature of Being – and specify the role of technology in the condition accounted. This necessarily involves looking behind the immediate appearances of ontic activities employing the technology in question, in order to discover that which unites them despite their apparent variety. Gender-swapping, the maintenance of multiple identities, and simulated personae all certainly tell us something about the variety of ways computer networks enable people to exist or "be" in the world, as do the myriad other human activities mediated by network technology. They also tell us something about Being. To discover what this is, Heidegger might advise us to advance beyond the superficial and cosmetic variety of the technology's effects – that is, to stop "gaping at the technological"[33] and concentrate, instead, on what it is that unites or gathers them at the level of ontology.

To do this we must clarify what networks cause, not simply as tools, but, more comprehensively, as a penetrative and pervasive technology. As *causa efficiens,* networks are tools or instruments involved in initiating a wide range of effects, many of which have been described in the preceding section and in previous chapters. None of these, however, captures independently or wholly the considerable breadth and depth of that which is caused by network technology in general. Networks are the *causa efficiens* of on-line banking, teleworking, global currency transfers, MUDing, and electronic mail. Simply listing these effects produced by network tools does not provide a compelling account of the essence of network technology. As a technology, the essence of networks consists in how they gather the material, formal, and teleological causes of the effects they produce and, in so doing, "bring forward into appearance" (*apophainesthai*) the very world in which these instruments and effects are situated.[34] As with all technologies, the essence of network technology is to be discovered in its mode of revealing (*alētheia*) the world in which human beings live. According to Heidegger, the mode of revealing that defines the essence of modern technology is not a bringing-forth (*poiēis*) of that which is rooted in Nature, but, rather, an enframing (*Gestell*) that challenges Nature to be a standing-reserve (*Bestand*) of exploitable resources. At the level of Being, this mode of revealing places human beings at odds with their rooted and meditative essence: the world as provided by Nature is transformed from a ground to be rooted in, into a cache from which resources can be demanded and extracted; contemplation is replaced by calculation oriented to the exploitation and accumulation of these resources.

What is the mode of revealing that defines the essence of network technology? Modern industrial technology produced a number of effects: automobiles; the factory proletariat; suburbs; and hydro-electric dams, to name but a few. All were emblematic of the view that the Earth comprised a standing-reserve of physical and human resources ripe for exploitation and consumption. Similarly, from amid the plurality of outcomes effected and practices mediated by networks, there emerges a certain unity. Below the surface of every effect and application of network technology is a gathering of binary digits that stand in as representative of some aspect of human existence, experience, or practice. Everything that networked computers do – whether it involves work or play; production or consumption; communication, information, or control – they do with bits. Numerous practices and outcomes can be listed among the *effects* of network technology. What unites them all, and so suggests the technology's *essence,* is the inescapable fact that they all must be reducible to the form of bits.

That which is irreducible to the form of bits cannot be mediated by network technology. Nevertheless, this technology continues to insinuate itself into an increasing number of everyday human activities, to the point that much of social, political, and economic life as it is currently conducted would grind to a halt in the event that networks suddenly disappeared. A world in which computer networks are the dominant and indispensable technology is, necessarily, a world of bits. The question is whether networks are a technology that simply *brings forth* the world as such or, conversely, a technology that *sets upon* the world and *demands* that it be such. It is possible, I suppose, that the world is, in fact, nothing more than a collection of electromagnetic impulses capable of being represented by strings of positives and negatives, presences and absences, os and 1s, and that networked computers are the technology of *poiēsis* that has arrived to reveal our true nature to us finally and decisively. In this case, the uploading of the entire world, complete with all its beings, forces, energies, and objects, simply awaits the construction of a large-enough memory chip and a fast-enough processor. This view of network technology does have its exponents. Notable among them is Nicholas Negroponte who, in his book *Being Digital*, designates digitality as "a force of nature."[35] Kevin Kelly adopts a similar tone when he asserts that "the realm of the *born* – all that is nature – and the realm of the *made* – all that is humanly constructed – are becoming one," and paints a picture of "the whole world networked into a human/machine mind."[36] Asked whether digital technology is inherently anti- or pro-Nature, network enthusiast Douglas Rushkoff responds, "Technology *is* nature."[37]

It is perhaps more plausible to suggest that, as network technology becomes more pervasive as a medium of human practices, its dominant presence simply demands that more of human activity and the world in which it takes place be reduced to the form of bits. Network technology reveals the world, and the human beings that inhabit it, as a cache of bits, not by bringing forth our essentially digital nature, but, rather, by setting itself upon that world and those beings, and demanding that they be converted into a form that it can manipulate and mediate. Having reached a point of critical saturation as an information, communication, and control utility, network technology enforces this demand automatically, as failure to submit to the regime of bits forces people and the practices in which they engage onto the margins of social, political, and economic life. Here I am not simply referring to those who hold out against the faddish pressures of e-mail and cellular phones. What I am talking about is the difficulty of imagining a functioning person in, for example, Canada, whose life activities have not been digitized to a considerable extent. Whether it is the digitization of their medical records, the digitization of their relationship with the state (tax records, pensions, employment insurance, citizenship information), the digitization of their consumer activity (credit/debit/smart card use, check-out scanners, database marketing), the digitization of their working life (use of computers and other digital devices at work, employment records, telework, surveillance) or the digitization of entertainment (audio CDs, Web-surfing, digital video), contemporary citizens must confront and embrace widespread digitality in the practices of being if they are to retain even remote contact with the social, economic, and political mainstream. Much as industrial technology enframes Nature and challenges the Earth to act as a standing-reserve of physical resources – "a gigantic gas station"[38] – network technology sets upon the world and demands its service as a standing-reserve of bits, a gigantic database. This is the mode of revealing in which the essence of network technology is located.

Network Technology versus Rootedness

What is at issue for Being when the significant practices of being in the world are reduced to a standing-reserve of bits? In other words, what are the ontological consequences of the digitization of the ontic? What can we gather about the essential Being of beings whose lives are mediated by this technology? Heidegger felt that one of the most unfortunate consequences of modern technology was its denial of the essential rootedness (*autochthony*) of Being by enframing Nature as a resource to be exploited rather than a place of dwelling to be spared. As Heidegger puts it, "the basic character of dwelling is to spare, to preserve ... Mortals dwell in that they save the earth ... Saving the earth does not master the earth and does not subjugate it, which is merely one step from boundless spoliation."[39] As was made clear in our discussion of the political economy of networks, this technology represents not so much a departure from industrial capitalist production regimes as an augmentation of them, and so it extends rather than retracts the modern tendency to enframe Nature as a standing-reserve of physical and human resources. In a sense, the standing-reserve of bits *is* the standing-reserve of physical and human resources at one remove of abstraction. Networks as a control utility simply make this enframing, entailing as it does the denial of rootedness, easier to accomplish. Indeed, perhaps the definitive feature of the perfecting capitalism described in the preceding chapters, in which networks play an indispensable role, is its lack of grounding, its fluidity, its uprootedness. One need only recall accounts of the ease with which billions of dollars can be "moved" from one "place" to another via networks, or the development of "virtual" corporations that "exist" nowhere but in the network connecting a matrix of computers, to appreciate the uprooted-ness this technology enables and encourages.

Similarly, a denial of rootedness figures prominently in the prac-tices described above as comprising distinctly postmodern applications of network technology. There is not much in these on-line practices that would suggest digital beings are more rooted than their modern analog counterparts. The qualities of dispersal, decentredness, multi-plicity, disembodiment, and ephemerality attributed to the network-mediated self are all anathema to rootedness. They may be liberating in a certain sense, but liberty and rootedness are far from synonymous. The complex relationship between liberty and rootedness has been ex-plored most deeply by Simone Weil, who lists liberty and freedom of opinion among the "needs of the soul" but stresses that these needs are inherently limited by corresponding needs for obedience and truth, and by the need "to be rooted," which she lists as "perhaps the most important and least recognized need of the human soul."[40] The net-worked selves described by Turkle and other postmodernists are grounded by neither biology, geography, history, physicality, or hon-esty. In cyberpunk science fiction, from which much postmodern writ-ing about networks draws its lexicon, the human body is often referred to as "meat," a contemptible resource that selves consisting of dis-embodied bits would rather do without, and to which no particular status beyond its limited use-value should be attached.[41] There is, of course, nothing radically human about meat. We gather steaks under the disembodied name of beef because we do not wish to recognize their natural roots in the steers in the field. People whose bodies are designated meat are denied their human roots, and networks are the technology that gathers beings in this uprooted form of bits.

A standing-reserve of bits also eludes the requirements of place – Weil lists "place" as first among the "natural" sources of rootedness[42] – that characterize genuinely rooted practices, beings and Being, a condition that even pertains to those whose encounters with network technology are less explicitly intended to extricate themselves from entanglement in their roots. Not every person who uses network technology does so in an attempt to escape, conceal, or multiply their identity. Some are just looking for advice on their gardens, but this does not negate the possibility that the network medium is itself an essentially uprooting technology. Put crudely, the practice of information gathering via the World Wide Web does not root someone in the same way that withdrawing a book from the local public library does. Both are mediating technologies that connect users to sources of information that are remote to their immediate experience. However, the library is rooted, by virtue of its spatial fixity and proximity to the place where those using it work and live, in a way the Web cannot be. By using their local library, people root themselves in the place where that technology resides. The Web exists everywhere and nowhere, and by using it people are rooted everywhere and nowhere, which is to say they are not rooted at all. Thus, as a technology, libraries gather beings and their practices into an account of the rootedness of Being in a way the Web cannot.

Similarly, once the bulk of people's practices, activities, and attributes are converted into the form of bits adding to the standing-reserve, they tend to resist grounding in extra-digital forms. Bits mediated by networks defy roots because their very nature is to move. Rootedness is a comparatively static condition – roots grow, but seldom move – while networks and bits are essentially dynamic and flexible. One hesitates to speak nostalgically of banks, but a good way to illustrate this is to consider the practical consequences of the colonization of banking by network technology described in Chapter 5. As people and their accounts have become digitized, and as the movement of their funds is carried out exclusively by computer networks, we find ourselves confronted with the prospect of the "branchless" bank. The corner bank branch is no church, but its physicality is symbolic of the institution's roots in the community of corporeal beings whose pasts and futures are often invested in it. However, the digitization of accounts and the proliferation of networks through which these can be accessed and managed has rendered the corner branch inefficient to the point of obsolescence. As formerly "in-person" banking services assume highly mobile digital forms that can be delivered dynamically, non-digital, rooted technologies such as the corner branch (not to mention the human teller) cease to make sense. In this way, digital network technology collects its users and summarily uproots them.

Simone Weil has written that "a human being has roots by virtue of his real, active and natural participation in the life of a community which preserves in living shape certain particular treasures of the past and certain particular expectations of the future."[43] Heidegger also decries the threat posed to rootedness by mass, electronic communications media: "All that with which modern techniques of communication stimulate, assail, and drive man – all that is already much closer to man today than his fields around his farmstead, closer than the sky over the earth, closer than the change from night to day, closer than the conventions and customs of his village, than the tradition of his native world."[44] Heidegger points out that electronic communications media accomplish this uprooting despite the fact those exposed to them remain physically situated in their homes, and suggests that these latter are in fact more "homeless" than those who actually leave their native place: "Hourly and daily they are chained to radio and television. Week after week the movies carry them off into uncommon, but often merely common, realms of the imagination, and give the illusion of a world that is no world."[45]

Proponents of network technology as a communications medium often promote it as a solution to this sort of alienation rather than a contributor to it, insisting that networks enable participation and so nurture precisely the sort of rootedness in community described by Weil as a human need. Clearly, these claims conflict with my account of a technology that, in enframing the world as a standing-reserve of bits, perpetuates the uprootedness identified by Heidegger as an essential attribute of the modern technological condition. If network technology enhances people's ability to participate in community life, how can its essence be uprooting?

To answer this question, one must distinguish between two distinct ways in which networked computer technology mediates community participation or, perhaps more specifically, between two distinct *types of community* in which network technology is involved. The first type are exclusively on-line communities, in which the network *is* the community. Here I am referring to extra-geographical, non-localized aggregations of individuals whose interaction is carried out solely across computer networks. These are the communities that form among otherwise dispersed patrons of bulletin-board services, members of electronic mailing lists, players in MUDs, and, most significantly, participants in network-mediated discussion groups. Despite being separated by continents and oceans, participants in an ongoing discussion group about, for example, elder care, develop relationships that are said to make of them a community that is stronger and more vital than most suburban subdivisions. The most famous of these "virtual communities" is known as the WELL, or Whole Earth 'Lectronic Link, a network of multiple discussion lists that, though based in southern California, attracts participants from all over the world to debate, exchange ideas, commiserate, and engage in small-talk across a broad range of subject areas. Stories about intense relationships, communal support in times of crisis, and collective celebration on the WELL have become the stuff of network legend.[46]

There is no doubt that physically separated people whose intercourse is limited to the exchange of bits over a network can develop intense relationships, or that groups of people enjoying these relationships can share common priorities, or that these relationships and priorities can endure and develop over time to encompass a history and tradition all their own. In short, virtual communities can and do exhibit many characteristics that any serious definition of community would have to admit as genuine.[47] The question, however, is whether these communities encourage or discourage rootedness. People certainly become attached to the on-line communities in which they participate, but attachment and rootedness are not identical. At the very least, they establish qualitatively different types of obligation. Attachment is a function of interest: people decide when and where to attach themselves, and they can detach themselves without vital consequences when they lose interest. The obligation in a relationship based on attachment – such as membership in a network community – lasts only as long as the parties' interests are satisfied, and so resembles that established by a contract. Roots, on the other hand, run deeper than mere interests, create obligations that often conflict with interests, and bind more strictly than contracts. People who are interested in doing so attach themselves to communities; people who are rooted *belong* to communities. It is for this reason that attachment can only stand in for, but never adequately replace, rootedness.

Participants in on-line communities often cite the declining presence and vitality of traditional public spaces, and contend that networks constitute an alternative to physical sites of community that have atrophied. Many draw on Ray Oldenburg's idea of a "third place" beyond home (our "first" place) and work (our "second" place) where social beings can interact, primarily through conversation, on a relatively equal and informal basis, and suggest that computer networks constitute this necessary site of community socialization in a world where coffee shops, park benches, and beauty parlours have ceased being sustainable "third places."[48] Despite the fact on-line communities exist only in the machines and fibre-optic cables connecting network users, their members are often eager to assert that "it feels like a real place in there."[49]

Though they might *feel* like it, the fact remains that computer networks are not real places, and while their virtuality might present certain benefits for community formation, these same attributes compromise the rootedness of those communities once they are established. For example, it is often suggested the opacity of networks undermines the effect of physically apparent social cues pertaining to rank, status, or other attributes, and so admits a broader spectrum of people into interactions from which they might have been excluded in a physical setting. In an on-line community, no one knows that you are poorly dressed, working-class, dark-skinned, obese, disfigured, female, balding, or confined to a wheelchair unless you choose to make these facts known.[50] Network opacity grants individuals access to relationships from which they might have been excluded due to discriminatory social conventions, had the communities in which these encounters occur been situated in the face-to-face places of the real physical world.

Despite the obvious virtue of overcoming superficial and unjust bases of discrimination, the reality remains that individuals who access participation in this way do so by concealing a significant aspect of who they really are – a concealment that typifies the uprooted identities of on-line beings discussed above. Uprooted individuals can participate in communities, but those communities will be communities of the uprooted. It is not my task here to determine which of community or rootedness is preferable. However, in this regard we might consider which is healthier: to confront that which is true and cannot be changed, admit our differences, and purge our communities of discrimination based on those superficial differences that are not germane to its constitution, maintenance, or projects; or, to embrace a medium that encourages us to ignore our natural attributes, conceal our differences, and pretend we all conform to the very idealized standards and conventions of identity and appearance that are pathological, unjust, and discriminatory in real, off-line communities. As Julian Stallabrass writes, "the extreme mutability and multiplication of identity possible in cyberspace collides with the desire to build communities based upon honest communication with people of diverse backgrounds and interests. Role-playing, and the potential for dishonesty that goes with it, militates against community."[51] On-line, individuals uprooted from their natural selves have the luxury of social perfection, the generalization of which in virtual communities cannot help but make their socially imperfect natural attributes even more conspicuous in their off-line communities, and their real lives more stigmatized and marginal.

A second set of claims regarding the virtue of on-line communities revolves around the ability of individuals to maintain attachments to them even when their geographical locale changes or is unstable.[52] In this view, the transience of contemporary life, resulting in large measure from the flexibility demands of neoliberal economies, is offset by the ubiquity and universal connectivity of networks. People can migrate, yet continue participating in their on-line communities of interest simply by logging-on to the network. As one observer has described, in on-line communities mediated by network technology, the nomads of flexible capitalism "had a place where their hearts could remain as the companies they worked for shuffled their bodies around America. They could put down roots that could not be ripped out by the forces of economic history. They had a collective stake. They had a community."[53] Because of networks, people accustomed to "conversing" with a particular group of interlocutors every evening about ice hockey, for example, need not relinquish that community simply because unemployment forces them to move to a new city.

This argument does well to illustrate what is at stake in distinguishing between community and rootedness. Clearly, networks are instrumental in forming and maintaining communities of interest able to thrive in an environment of escalating uprootedness. Of course, it is also true these same computer networks are instrumental in generating the very condition of uprootedness in which these on-line communities are situated. Individual identities are not the only things uprooted by networks. As described in the previous two chapters, digital networks are also the key technology undergirding the dislocation, instability,

and transience that are endemic features of flexible capitalist economies. Thus, while networks may represent a "third place" where on-
line communities might form and subsist, they are also, probably more
significantly, the technology that uproots us fundamentally from the
"first" and "second" places of home and work. Networks make it easier
for community and mobility to co-exist. Rootedness and mobility, however, are necessarily opposed; that which is rooted is immobile. In this
respect, on-line communities are more a symptom of a general condition of uprootedness than a cure for it. As is typical, the symptom and
the disease share a common medium in network technology, but the
real cure is external to it.

It is conceivable that uprooted, network-mediated, on-line communities contribute to diminishing the vitality of rooted off-line communities rather than to enhancing it. In a recent study of on-line
communities in Calgary, Alberta, it was found that membership in
network associations had "corrosive effects" on civility: "Respondents
who were most engaged online tended to be relatively disengaged with
(and distrusting of) the 'real' community. It appears that these online
associations could be damaging to civil society as the more time an
individual spends engaged in virtual communities the less time that
individual has to be engaged in the real community."[54] Faced with a
choice between the effort required to set down roots in a succession of
new home and work situations with entrenched conventional social
prejudices, and a medium that enables a person to portray himself as
whoever he must be in order to be embraced by a community that is
"there" regardless of his unstable home and work locations, many would
choose the latter option. Accordingly, "virtual communities are accelerating the ways in which people operate at the centers of partial, personal communities."[55] As more people devote their finite social energies
to personal and partial on-line communities at the expense of those
off-line, their attention and care for where they live and work is likely
to wane. On-line communities – "a technologically supported continuation of a long-term shift to communities organized by interest rather
than by shared neighbourhoods or kinship groups"[56] – may aggravate
the condition of uprootedness to which they respond.

That being said, a second type of community participation mediated
by networks occurs where the technology is employed as an instrument

of communication and information distribution for localized communities that already exist off-line. Here the relationship between the network and community is reversed: these are "community networks" rather than "network communities." Community networks have arisen out of a concerted movement to make network communication and information resources available to local communities so that participatory life in them might be enhanced.[57] Douglas Schuler, a noted student and advocate of community networks, describes them as follows: "These community networks, some with user populations in the tens of thousands, are intended to advance social goals such as building community awareness, encouraging involvement in local decision-making, or developing economic opportunities in disadvantaged communities. They are intended to provide 'one-stop shopping' using community-oriented discussions, question-and-answer forums, electronic access to government employees and information, access to social services, e-mail, and in many cases, Internet access."[58] So-called Freenets – of which there are more than 150 in the United States, and 75 in Canada, including the national Capital Freenet in Ottawa – are perhaps the best-known examples of the community-network phenomenon.[59] Community networks such as these are typically founded and run by volunteers with public subsidies, and provide various levels of cost-free network access to users, either from home by modem or from public terminals located in community facilities. Some more business-oriented community networks are funded and maintained by private corporations or consortiums wishing to piggyback commercial vitality on civic rejuvenation.[60]

In any case, the promotion of community networks is a fairly explicit attempt to redress what is widely perceived as a deficiency of civic resources and a decline of "social capital" in contemporary North American political cultures.[61] Networks, it is argued, are a key technology in the effort to resuscitate civic communities culturally, economically, educationally, ecologically, and politically, a process requiring concerted citizen-led, participatory, and cooperative action supported by governments and the private sector.[62] In these "new" communities, "conversation is the main activity."[63] Computer networks, conveniently, have "immense potential for increasing participation in community affairs."[64] Networks and communities thus need each other in order to reach their respective horizons. Though they prefer "geographic" over "virtual" communities, civic networking advocates such as Schuler assert that "community networks offer a new type of 'public space,'" and hope for a "marriage of community and technology" in which "some of the tension between 'community' and 'technology' can be removed and technology can be made to better serve human needs."[65]

There is simply no doubt that widely accessible computer networks can provide communities with significant communication and information resources, and in so doing can enhance participation and conscientious citizenship.[66] There are, however, limitations to this utility. As most proponents of civic networks affirm, communication is a necessary but not sufficient condition for the formation and maintenance of community. There are a number of ways individuals can and must participate as members of a community, and communication is only one of them. Others might include work, child-rearing, worship, and collective action against adversity, not all of which can be reduced to communication. Nevertheless, to the extent that it does enhance the communicative aspects of community, network technology is a potentially valuable resource. The question in the context of this discussion, however, is whether the limited role network technology *might* play in community enhancement overrides the generally uprooting quality of its essence. Just as communities are not built on conversation alone, communication does not exhaust the ways in which network technology penetrates our communities, or our lives outside them. Community networks are but one, relatively narrow, application of this technology. If we are divining for network technology's essence among the various ways it affects our manner of being, then the community-enhancing potential of civic networks must be considered in light of the myriad other uprootings conspicuous in a world enframed as a standing-reserve of bits. For example, the point of attachment networks sometimes provide, even to geographical communities, must be measured against the manner in which this technology imposes transience in the realms of work and identity. Once again, the genuine hope vested in community networks may simply be symptomatic of a malaise in relation to which network technology stands as a cause, rather than a cure.

Networks and Calculation

For Heidegger, the primary danger contained in the enframing essence of modern technology was its tendency to privilege and institutionalize calculative modes of thought that threatened our meditative nature. As an instrumental mode of thought oriented to effectiveness, efficiency, and accumulation, calculation is a mode of thinking well suited to a world enframed as standing-reserve. Calculative thinking is at the core of the purely technical relations that characterize the modern technological condition. Conversely, meditation, a contemplative mode of thinking, is useless in a world enframed as standing-reserve. The question here is whether network technology, which enframes the world as a standing-reserve of bits, represents a departure from, or a continuation of, the holding sway of calculative thinking that Heidegger identified as essential in the modern technologies that preceded networked computers.

There is considerable debate on this question, with the line of disagreement corresponding generally to whether it is computers or their connection in networks that is emphasized. As described in Chapter 3, computers have a memory, can be programmed, and perform operations on abstractions (numbers, words, images) represented as bits – qualities that distinguish computers from more simple calculating devices. Nevertheless, everything a computer does involves calculation, even when that calculation is manifested as memory, programmed operation, or representation. Whatever else they are being enlisted to do, in order to succeed computers must add, subtract, combine, separate, and invert binary pairs according to Boolean logic. This *sine qua non* has animated many critical assessments of the social and epistemological impact of computer technology.

Writing before the widespread proliferation of networks, Joseph Weizenbaum placed the computer at the zenith of a trajectory in which human judgment has been progressively eclipsed by calculation.[67] In this regard the computer has "merely reinforced and amplified those antecedent pressures that have driven man to an ever more highly rationalistic view of his society and an ever more mechanistic image of himself."[68] Weizenbaum concludes that, due to its instrumentalist biases and its mechanization of language and reason in the form of calculation, the computer effectively excludes philosophical thought, ethical action, and, ultimately, wisdom itself.[69] David Bolter reaches similar conclusions in his discussion of Western culture in the age of "Turing's Man."[70] For Bolter, the computer represents "the triumph of logic" and is "the embodiment of the world as the logician would like it to be."[71] Contemporary – especially postmodern – commentators often claim that networks have precipitated the dawning of an age of open-ended uncertainty. Disagreeing, Bolter thinks the calculative operating demands of computers trump any vision of their inherent flexibility: "Uncertainty in a computer language will not produce poetry, express emotion, add color, or do anything at which natural language excels. It will simply produce an error ... the logical nature of circuits and storage registers makes ambiguity impossible."[72] The result is that Turing's man develops a "concern with functions, paths, and goals [which] overrides an interest in any deeper kind of understanding ... Turing's man analyzes not primarily to understand but to act ... For Turing's man, knowledge is a process, a skill. A man or computer knows something only if he or it can produce the right answer when asked the right question."[73]

Those who emphasize the *networking* of computers as the most salient feature of the current technological condition have tended to reject the characterization of the technology as essentially calculative and instrumental. As Sherry Turkle writes, "the lessons of computing today have little to do with calculation and rules; instead they concern simulation, navigation and interaction. The very image of the computer as a giant calculator has become quaint and dated. Of course, there is still 'calculation' going on within the computer, but it is no longer the important or interesting level to think about or interact with."[74] For Turkle, networks extend computers viewed as "emergent" systems rather than calculative machines, wherein "intelligence does not follow from programmed rules but emerges from the associations and connections of objects within a system."[75] In this conception, once powerful computers are connected in a network, the "thinking" they become capable of exceeds algorithmic data-processing to encompass the spontaneous, self-organizing patterns into which otherwise distinct intelligences naturally coalesce. Through networking, the computer defies the romantic caricature of a cold, rigid, mechanical calculating device and, instead, its operations resemble more natural, organic, biological processes such as the flocking of geese, the schooling of fish, the swarming of insects, and the development of synaptic pathways in the human brain.[76] Crucially, the intelligence/thinking deriving from networks is said to lack the features that distinguish calculation and instrumental reason: seriality, linear causality, centralized control, impermeable boundaries, hierarchy, rigidity of rules, and predictability. Instead, networks substitute parallelism, mutual influence, decentredness, expansiveness, adaptability, polyarchy, and contingency.[77]

From this perspective, the essence of computer networks is connection – the connection of humans to humans *and* computers to computers – not calculation.[78] That is to say, its proponents emphasize the communication, and not the information or control, utility of networks. As explained by Derrick de Kerckhove, "the Internet gives us access to a live, quasi-organic environment of millions of human intelligences perpetually at work on anything and everything with potential relevance to anyone and everybody. It is a new cognitive condition I call 'webness.'"[79] This "webness," according to de Kerckhove, "is a condition for the accelerated growth of human intellectual production."[80] That may be true. What is not so clear, however, is whether this accelerated intellectual production is as far removed from the calculative essence of computers as claims about the connective essence of networks suggest it is. Some insist network connectivity makes an anachronism of fears surrounding the calculative and instrumental rationality of stand-alone computers. It is assumed that connection and calculation are somehow mutually exclusive, and scant consideration is given to the possibility that connectivity simply extends, rather than mitigates, the calculative essence of computers. Does the calculative essence of computers – and previous modern technologies – persist despite their connection, or does it disappear? Does connectedness entail a new form of intersubjectivity based on the playful and creative interaction of the Web, or does it require submission to the calculative mode of information exchange characteristic of computer mediation more generally?

The answers to these questions lie in the role networked computers play in the various practices of being wherein Being is at issue. I have suggested these practices can be grouped under categories corresponding to the information, communication, and control utilities of networks.

I have also suggested that, of these, it is the third category – the control utility – in which the distinctiveness of computer networks as a technology is captured most clearly, due to the propensity of communication and information to collapse into control under the auspices of digital media. For this reason, I propose the essence of computer networks should be evident, if anywhere, in a consideration of the practice that most clearly defines its utility as a technology of control – namely, the practice of surveillance.

"Surveillance" combines the French words *sur* and *veiller,* the latter deriving from the Latin *vigilare,* meaning "to keep watch." There being no English verb "to surveil," "surveillance" rather denotes a condition of supervision, observation, or invigilation. As David Lyon argues, a society in which an increasing number of everyday practices are mediated by networked computers is, regardless of whatever else it might be, a "surveillance society": "Surveillance [today] concerns the mundane, ordinary, taken-for-granted world of getting money from a bank machine, making a phone call, applying for sickness benefits, driving a car, using a credit card ... picking up books from the library, or crossing a border on trips abroad ... Computers record our transactions, check against other known details, ensure that we and not others are billed or paid, store bits of our biographies, or assess our financial, legal or national standing ... Computers and their associated communications systems now mediate all these kinds of relationships; *to participate in modern society is to be under electronic surveillance.*"[81] Lyon identifies surveillance as "the single most controversial and potentially alarming social issue prompted by the massive expansion of computer power in human affairs."[82] If the volume of scholarly, governmental, and popular attention being paid to computerized surveillance and the related topics of privacy and encryption is any indicator, this is an understatement.[83] Of course, neither the practice of surveillance, nor concern about it and its relationship to privacy, is entirely new. However, it is true that, in the modern world, which features both highly complex bureaucratic economic and political organization, and a heightened sense of the status of private individuals, surveillance and attendant concerns about its impact on privacy have escalated.[84] Theorists and chroniclers of modernity ranging from Weber to Orwell to Foucault have stressed the crucial administrative and disciplinary functions served by surveillance in modern societies.[85]

Network technology has served to extend the surveillance capacities of modern societies and institutions more broadly, and also to entrench these practices more deeply. As discussed previously, large, complex organizations and systems have formidable control requirements. These include the need to collect and maintain comprehensive records of the attributes and behaviour of actors involved in the organization or system. Surveillance, then, is a practice inherent in control. Specifically, in order to remain functional (i.e., under *control*), large-scale systems must develop routinized and reliable techniques for the identification, isolation, and correction of deviance. As James Rule points out, "surveillance entails a means of knowing when rules are being obeyed, when they are broken, and most importantly, who is responsible for which ... also indispensable is the ability to locate and identify those responsible for misdeeds of some kind."[86] These control requirements pertain to any functioning system or organization.

Clearly, effective surveillance relies heavily on information and its communication collapsed into a regime of control, and so it is not surprising that networked computers have emerged as a surveillance technology par excellence. The technology's ability to gather, store, retrieve, and process massive amounts of digital information from numerous sources, and to communicate that information between multiple systems fulfils the requirements of effective surveillance completely. As more of the "raw material of human experience"[87] is mediated digitally and stored in databases – that is, as the standing-reserve of bits grows – the comprehensiveness of networks as a surveillance technology expands. Oscar Gandy provides a thorough accounting of the various ways in which people contribute to the construction and accumulation of "machine-readable, network-linked data files" about themselves.[88] Gandy's list includes information submitted to various agencies and bodies regarding personal identification and qualification (driver's licences, passports), finances (bank records, tax returns), insurance (auto, health), social services (unemployment benefits, pensions), utility services (telephone, cable), real estate (titles, liens), entertainment and leisure (travel, subscriptions), consumer activities (credit-card and debit-card purchases), employment (histories, applications), education (records, rankings), legal activity (court records). As the range of everyday practices mediated by networked devices capable of registering and transacting digital information increases, the range of human activity not subject to surveillance of one sort or another decreases.

The keeping of records in regard to such activities did not originate with the computer. What networked computers have done is to make the gathering, storage, and retrieval of such records, in the digital form of bits, far easier to accomplish. Once rendered in the form of readily manipulated bits, human behaviour is easily measured and analyzed, as required for the establishment and maintenance of control. In order to be meaningful for surveillance purposes in large systems, not only must bits be gathered in great volumes, they also must be processed. They must be identified, categorized, collated, classified, used to construct profiles, linked with other sets of bits, matched with and compared against related files, and subjected to algorithms designed to form generalizations and generate predictions. In other words, *they must be submitted to calculation,* a task for which computers are uniquely suited. Networking extends the reach of the computer's calculative contribution to surveillance, both geographically and across the organizational/ system boundaries. When the bulk of socially, politically, and economically significant human behaviour is reduced to the universal language of bits, and when a medium for the easy exchange of these bits is similarly universal (at least for those with an interest in its control utility), then the gathering of bits from across the globe and the sharing of them between organizations becomes as easy as swiping a card through a reader at the cash-register. So long as human beings continue to act in a networked world where the bulk of their socially significant actions is represented in the form of bits, the standing-reserve of bits will be bottomless, and networks also ensure that this reserve can be readily drawn upon, and its contents subjected to calculation, for a variety of ends, with increasing ease.

Lyon has categorized the distribution of surveillance in the contemporary context into four primary domains: government administration, policing and security, the workplace, and the consumer marketplace.[89] I have discussed workplace surveillance in Chapter 5, and Lyon provides numerous examples of computerized surveillance operating in the other three domains, while emphasizing that the definitive characteristic of modern networked surveillance techniques is that they enable surveillance *across* these somewhat artificial boundaries (for Lyon, this transgression of organizational and systemic boundaries is perhaps the major distinguishing attribute of network-mediated surveillance).[90] Others have also compiled highly illuminating accounts of the operation of networks as a surveillance technology, from a variety of perspectives.[91] Here, I wish to briefly explore one particular manifestation of consumer surveillance as exemplary of the essence of network technology.

There are two reasons for this focus. The first is that it is here, in the extension of surveillance to the sphere of the consumer marketplace, that the contribution of networks as a surveillance technology is most profound. As the efficiency of coercive techniques of social control has receded, techniques of consumption management have emerged as their successor. As Lyon writes, "consumption has become the all-absorbing, morally guiding and socially integrating feature of contemporary life in the affluent societies. Social order is maintained through stimulating and channeling consumption, which is where consumer surveillance comes in."[92] Consumption is the primary behaviour through which modern individuals exhibit conformity with the prevailing social order, and express their acceptance or rejection of its rules. Similarly, it is through the correction of dissonance in the sphere of consumption (e.g., "niche" marketing; "green" products) that the imperatives of social order and the behaviour of deviant individuals are reconciled. As consumption emerges as a primary system of socialization, normalization, and control, network technology is stepping in to meet the need for a comprehensive surveillance technology in this area. Together, consumption and computer networks present a "massive intensification" of surveillance throughout modern societies in general.[93] The second reason I wish to focus on this particular example from the realm of consumptive surveillance is that I have encountered no other application of network technology that exemplifies more clearly the point I am trying to make here via Heidegger – namely, that networks enframe the world as a standing-reserve of bits and, in so doing, perpetuate modern conditions of uprootedness and calculative thinking. In a sense, the name given to this practice – data mining – speaks for itself.

As defined recently by Ontario's Privacy Commissioner, "data mining is a set of automated techniques used to extract buried or previously unknown pieces of information from large databases. Successful data mining makes it possible to unearth patterns and relationships, and then use this 'new' information to make proactive knowledge-driven business decisions."[94] Data mining differs from other techniques of processing digital information in that it is designed to reveal relationships that are previously unknown to the user. Unlike simply programming a computer to categorize and compare, for example, a retail chain's weekly sales by store and region, data mining excavates information, in the form of patterns and relationships, that the user may have never considered looking for.

The first step in effective data mining is the consolidation of a vast amount of digitized information in a centralized "data warehouse." The sources of such data are many, and can be both internal and external to the organization maintaining the warehouse. Internally, organizations collect and store digitized data culled from network-mediated transactions with their customers or clients. These transactions can take a number of forms – purchases, applications, use of services, and so on – depending on the organization, and can have occurred at any number of disparate sites. For example, the vast amount of data generated by digitally mediated transactions across the outlets of a large retail chain can be collected and stored automatically in a centralized database. Even if these transactions are completed in ways that allow the customer to remain anonymous (i.e., because he or she paid in cash), voluminous data about the items purchased, in what combination, at what hour, in what quantities, in which geographic areas, and numerous other variables, are generated and collected. If the customer completes the transaction in a manner that identifies and links him or her with the purchase personally (i.e., because he or she paid with a credit or smart card, and used a discount, loyalty, or "Air Miles" card), the amount of data generated increases exponentially. Data from external sources can also contribute to an organization's warehouse. Because bits are a universal language, data collected by one organization for one purpose can be bought or otherwise procured, added to the warehouse of another, and combined or cross-referenced with the data there. For example, the credit-card company MasterCard International processes millions of cardholder transactions daily, and has announced plans to sell the data gleaned from those transactions to its business partners who offer credit-card services. The size of contemporary data warehouses is literally monstrous, with some enterprise-wide systems reaching volumes up to twenty-four terabytes.[95]

Once gathered, data in a warehouse are then "scrubbed" to remove repetition and other noise (for instance, if the purchaser in a single transaction is identified by both his or her credit card and "loyalty" card, the double-registration of his or her name and address is redundant and can be purged). The warehouse full of clean data is then processed by a series of mining algorithms – an algorithm is a set of rules for calculation – that seek to discover previously unseen relationships between the data.[96] Data-mining algorithms search for trends, patterns, associations, sequences, clusters, classifications, and generalizations, and can also generate forecasts and predictions. Without being directed specifically to analyze a particular relationship, a data-mining algorithm might reveal, for example, that people who maintain large, interest-bearing balances on the credit cards also tend to regularly purchase a certain type and quantity of goods at the same time every month. Results of data-mining investigations are also highly reliable, due to the extremely large sample size they draw upon, and the vast array of variables they are able to process. For anyone seeking to exert control over a large and complex system, data mining is a technique that offers considerable advantages.

Data mining is a surveillance technique applicable to any enterprise whose system information is collected in the form of bits mediated by networked computers, including those in the fields of public administration, security, service provision, and workplace management. However, it appears likely that data mining is a tool particularly suited to surveillance and systems control in the commercial marketplace. By the end of 1997, it was estimated that 80 percent of the world's largest 2,000 companies were engaging in data-warehousing and -mining strategies.[97] Estimates place the amount spent worldwide on data mining in 1997 at $16 billion, and suggest that the quantity of data stored in global warehouses doubles roughly every eighteen months.[98] By 1998, over 56 percent of Canadian households (7.2 million) "belonged" to the Loyalty Management Group's Air Miles program, which rewarded consumers with credit toward air travel for shopping at participating retailers. In exchange, sponsoring retailers – which include banks, petroleum companies, supermarkets, and the Liquor Control Board of Ontario – get access to the cache of data ready to be mined in the program's warehouse.[99]

Examples of the use of this technique abound. The Wal-Mart chain of retail stores warehouses data from point-of-sale transactions from its 2,900-plus stores in six countries, and grants warehouse access to its more than 3,500 suppliers.[100] In the United States, the chain has used patterns excavated by data mining to determine that diaper buyers shopping on Thursdays buy more additional items than those who shop on other days, and that men who buy beer rarely buy anything else. The company also claims to be able to reconstruct the path each customer takes through the store, based on the digital record of his or her purchases.[101] As a result, Wal-Mart reduced the price of diapers on Thursdays, and rearranged the layout of their stores to force men who are looking for beer to walk past other items popular among their demographic along the way.[102] Telephone companies use data mining to search for patterns in calling practices, customer loyalty, and defections to competitors in order to craft more attractive offerings, and also to identify anomalies that might signal fraudulent use.[103] Numerous companies employ data-mining techniques to enhance their direct, database marketing appeals. Unforeseen relationships between consumers from a particular demographic or geographic location and a particular type of product or service, between the purchase patterns and sequences of one group of products or services and another, or between innumerable other unpredictable variables can assist marketers in tailoring their appeals directly to consumers in identified groups.[104] In the United States, over thirty consulting firms now offer data-mining services for political campaigns.[105] In Canada, political parties have begun using this technique to craft and deliver election platforms.[106]

In a sense, data "mining" is a misnomer. Unlike real miners, data miners do not have to know exactly what it is they are looking for, or even if there is anything there to be found – they are more like "prospectors" – a crucial difference that accounts for the genius of the technique. Misnamed or not, data mining has been established as one of the definitive surveillance techniques of the networked world. It is a practice made possible by the particular technical attributes of network technology: the digital representation and mediation of an ever-increasing range of human activity and practices; the prevalence of networked devices through which the conversion of these practices into bits can take place, and across which these bits can be transacted and exchanged; the growing and seemingly boundless capacity for replicating and storing these bits in a form that is easily retrievable and does not degenerate; and the staggering but increasing speeds at which these bits can be subject to multiple and complex processing. In other words, data mining is a defining technique in a world enframed as a standing-reserve of bits, a world whose essence is defined by its apprehension through calculation.

The privacy concerns raised by this situation are substantial. It is interesting to note, however, that attempts to address these concerns typically affirm rather than challenge the enframing of the world and its inhabitants as a standing-reserve of bits. Recent privacy legislation proposed in Canada – discussed in Chapter 5, above, in relation to workplace surveillance – does not even contemplate that the digital collection, use, and exchange of personal information is something that perhaps should be fundamentally limited, or perhaps even discontinued altogether. Instead, consistent with its purpose as "an Act to promote electronic commerce," the Personal Information Protection and Electronic Documents Act (Bill C-6) simply establishes the conditions under which the standing-reserve of bits can continue its phenomenal growth.[107] For example, the legislation does not stipulate that firms cannot electronically gather information on their customers or employees and sell that information to other firms – it simply requires that they state their purpose and obtain consent (sometimes express, sometimes merely implied) before doing so. In a speech supporting the legislation, Canadian Industry minister John Manley said, "As our competitors around the globe scramble to put in place the frameworks that will create the consumer confidence to make electronic commerce a practical reality, the privacy protection in Bill C-6 will put Canada at the forefront."[108] In other words, the legislation is explicitly designed to encourage the ongoing accumulation and trade of digitized personal and other information – the continuing growth of the standing-reserve of bits.

As David Lyon has pointed out, neither surveillance nor its enhancement by network technology is unambiguously evil. Surveillance activities by the state, for example, have tended to expand in tandem with the extension of democratic political rights, and the development of economic welfare and health programs, as means of exerting control over these large social systems and the bureaucracies they entail.[109] Sweden, often heralded as the model of progressive, social-democratic, welfare statism, is also "arguably the technologically most advanced surveillance society in the world."[110] Surveillance systems make sure citizens get to vote, sick people get medical care, and poor people get income assistance. As Lyon puts it, "the advantages of modern state-run surveillance systems should not be sneered at."[111]

However, my point here is not to specify whether network technology's contribution to surveillance is itself good or bad – this is a complex determination: caller-identification technology allows women to screen out abusive callers; it also allows abusive husbands to establish the whereabouts of spouses who are trying to escape them – but rather simply to suggest that surveillance applications reveal network technology's calculative essence in a world enframed as a standing-reserve of bits. Networks function so well as a surveillance technology because they collect so many of the practices involved in being human in contemporary society in the form of bits, and then facilitate the subjection of this information to calculation and its attendant processes: identification, classification, measurement, comparison, combination, and prediction. The completion of network technology's enframing of the world as a standing-reserve of bits becomes clear when even those genuinely critical of unchecked networked surveillance feel compelled to characterize digitized personal information as a "resource" that must be "managed" like any other.[112] As a former Canadian Privacy Commissioner has written, "the invasive, indiscriminate use of the computer in gathering, storing, and comparing personal data *for purposes either benign or malign* reduces individuals to commodities, subjugates human values to mere efficiency."[113]

By this estimation, networks fall squarely within the essence of modern technology described by Heidegger. Networks enframe the world as a standing-reserve of bits because they demand that human practices be converted into bits in order to be mediated and included in the institutional life of society. As networks proliferate to the point of becoming the primary medium of social existence, those practices that are not, or cannot be, represented as bits simply cannot be an important part of that existence, and so they are excluded from the collective reckoning of social, political, and economic goods. In his Foucaultian account of the "superpanoptic" effects of network-mediated surveillance, Mark Poster describes an "additional self" consisting wholly of data in the form of binary digits.[114] In a networked world enframed as a standing-reserve of bits, this digital image takes the place of the actual person it represents in all socially significant registrations and determinations: it is one's "additional self" that is taxed, insured, and granted or denied a mortgage. Moreover, this data image is taken as the authoritative representation of what that person does, and what that person is all about, when it comes to the calculations performed by powerful actors in the administrative systems – whether public or private – in which that data image circulates. Though Poster would reject such essentialist talk, his highly descriptive notion of the "additional self" evokes the two attributes of computer networks that I have been presenting, via Heidegger, as the essence of a technology that enframes the world as a standing-reserve of bits: uprootedness and a privileging of calculative thinking.

The Saving Power?

In Heidegger's conception, modern technology's enframing of the world as a standing-reserve of bits produced an ontological condition at odds with the rooted and meditative essence of Being. Heidegger also felt it was precisely in this mistaken enframing that technology revealed the possibility of escaping this condition and entering "a more original revealing ... a more primal truth."[115] In the foregoing, I have tried to show that, in so far as it enframes the world as a standing-reserve of bits, network technology bears *in particular* the same uprooting and calculative biases Heidegger identified as essential to modern technology in general. The corollary is, of course, that networks also pose the same dangers to Being as their technological predecessors. All this would seem to indicate the latent "saving power" Heidegger identified in modern technology has not, in fact, been made manifest at a societal level. To conclude this chapter, I would like to consider whether there exists a Heideggerian "saving power" dormant in network technology, and whether there are grounds for hoping it can be awakened.

A technology can be said to harbour the power to save us from the condition it imposes to the extent that it stands as a vivid and appreciable manifestation in the world of that very condition. In this sense, computer networks are the house in which the uprooted and calculating essence of contemporary technological Being dwells and, while the walls of that house are not transparent, its very manner of concealment reveals an ontology, an account of the essence of Being it contains. Salvation is possible when thinking beings recognize in this ontology an account that conflicts with their genuinely rooted and meditative essence, establish limitations on the holding sway of this account, and seek out a relationship to their true essence that is more adequate than the one mediated by technology. Phrased more simply, technologies harbour a saving power because in considering them we are confronted with tangible proof of how far we have strayed from our genuine essence. As Heidegger has pointed out, accomplishing this recognition requires "catching sight of what comes to presence in technology, instead of merely gaping at the technological."[116] That is, it requires consideration and understanding of the condition enframed by technology, rather than a reckoning of its effects that simply serves to reinforce this enframing by mimicking its logic.

I have argued that the essence of network technology derives from its enframing of the world as a standing-reserve of bits, an enframing biased in favour of rootless and calculative practices, and therefore an enframing that gives a mistaken account of Being as essentially uprooted and calculating. The question is whether this technological enframing has been met with substantial challenges, on ontological grounds, at a societal level. Have we contemporary beings caught sight of what comes to presence in the massive deployment of network technology, or have we merely gaped at its effects and conceded that its shortcomings can be addressed only technically? To be sure, the onset of network technology has not been without its critics, both scholarly and popular.[117] Some of these critiques have even entailed consideration of the appropriateness of this technology for the basic nature of human beings and the world in which they live. Most, however, concentrate on what are understandably perceived as existing or potential ill *effects* of networks, and steer clear of ontological issues.

At a societal level, popular discourse about network technologies has certainly not been dominated by concerns over their relationship to the genuine essence of Being. Predictably, the concerns animating popular discourse in this area are expressed as a sensitivity to technical "problems" amenable to technical "solutions" that fall squarely within the ambit of the technology itself. Thus, worries about network privacy and security motivate the drive to refine encryption techniques. The gap between digital "haves" and "have-nots" should be addressed by universal access to networked devices. Sluggish delivery of digital material over networks due to excessive traffic cries out for faster chips and an expansion of bandwidth. The availability of pornographic and other objectionable material via computer networks necessitates the development of techniques for identifying and blocking access to proscribed sites. Anxiety about information overload stimulates the perfection of "intelligent agents" capable of automatically sifting through the haze of data smog. The inconveniences of hard- and software incompatibility prompts development of universal technical standards and cross-platform programming languages. The integrity of digital property requires the encoding of identification and use-tracking codes. The daunting elements of technical complexity drive improvements in user-friendly interfaces and the incorporation of "participatory design" models.

When societal consideration of a new technology is limited to identifying technical problems and technical solutions, the general condition in which technology holds sway is reinforced rather than challenged. This, by and large, has been the case with network technology. The essence of networks as a technology that enframes the world as a standing-reserve of bits, and Being as uprooted and calculative, has been neither interrogated nor confronted with an alternative account at any significant social or political level. Canada's most important public, collective examination of what is at stake in the development of network technology, the Information Highway Advisory Council, framed its investigations within fifteen "public policy issues." Not one of these came even close to prompting questions Heidegger would have us ask about how the technology enframes the world, about what account it provides of the nature of Being human, about whether what comes to presence in this technology is in harmony with our essence, or about whether the technology should be embraced, rejected, or modified on these grounds. Instead, the first question summarily sealed the entire exercise well within the boundaries of the technological: "How fast should the advanced network infrastructure be built?"[118] "How fast?" is a calculative, rather than a meditative question. A society thinking about the essence of technology would ask whether the network ought to be built at all. Asking "how fast?" is simply an indication of surrender.

Were we prepared to exhibit a comportment toward network technology more appropriate to the essence of Being it would be based on releasement, or letting go, rather than unbridled embrace. Letting go, in the Heideggerian context, does not mean letting go of the reins and allowing network technology to proceed as it will, because to do so is to abdicate responsibility for maintaining an authentic relationship to the true essence of Being. This type of letting go is actually an embrace and, in surrendering Being to technology, is impious. The impiety is evident in statements such as the following, by Kevin Kelly, an influential pundit who draws spiritual guidance from computer games, which he refers to as "god games": "To be a god, at least to be a creative one, one must relinquish control and embrace uncertainty. Absolute control is absolutely boring ... The great irony of god games is that letting go is the only way to win."[119] For Heidegger, letting go meant releasing ourselves from the holding sway of the technological, not giving in to it. It meant establishing a critical distance at which the intimacy between human beings and the essence of Being is protected from corruption by an undue attachment to the enframing essence of technology. It meant affirming that the essence of Being calls for limits on the use of technology. It meant recognizing precisely that, despite the promises of technology, human beings are not gods.

As Heidegger wrote, "man is not the lord of beings. Man is the shepherd of Being. Man loses nothing in this 'less'; rather, he gains in that he attains the truth of Being. He gains the essential poverty of the shepherd, whose dignity consists in being called by Being itself into the preservation of Being's truth."[120] This is why "only a god can save us" from technology's assault on Being – because only an encounter with a god will remind us that we are not gods; that we have not achieved divinity through our technology; that, instead, we have relinquished our essentially rooted and meditative humanity. Sadly, any such god remains trapped behind the ramparts of the network revolution, and the enframing of the world as a standing-reserve of bits proceeds at breakneck pace. Our relationship to network technology in its various manifestations is becoming more comprehensive and intimate, rather than more circumscribed and detached. It is becoming increasingly difficult for us to simply "let go" of the technical devices mediating our digitized existence, and to separate them from our "inner and real core."[121]

As our distance from network technology diminishes, so, too, does our "openness to the mystery," the second aspect of maintaining a relationship to technology that Heidegger identified as comporting with the essence of Being. A world enframed as a standing-reserve of bits is a world of os and 1s. It is a world we can create and re-create, a world at our command, a world without mystery. Techniques and devices can be used poetically to bring forth the true into the beautiful, to illuminate the presence of the gods, and to set out the dialogue between the human and the divine, in ways that yield to, and protect, the truth of our rooted and meditative essence. Such is the manner of wood-fired kilns built into hillsides, whose mystery we can feel but never presume to fully understand, command, or have created. Computer networks, on the other hand, enframe the world as a standing-reserve of bits and so share the anti-poetic essence of modern technology in general. The accounts given above of the uprooting and calculative practices enabled and encouraged by this technology confirm this. There are exceptions, but there is little consolation to be derived from them. Egoyan's digital bus on a frozen lake was a beautiful representation of the truth of life's enduring mystery, but it also presented before our eyes, in a manner scarcely even visible, the world as it becomes under the sway of network technology: a standing-reserve of bits. Openness to the mystery requires an appreciation of that which cannot ever be reduced to, or represented as, bits, and a recognition that therein lies the true essence of Being. This is an openness to which network technology appears opposed in its very essence.

government, politics, and democracy: network technology as stand-in

· · · · ·

THE "CYBER" in "cyberspace" – a euphemism frequently used to name the networked world – is taken from "cybernetics," the science of systems control. The words "government" and "cybernetics" both derive from the Greek *kubernētēs,* for "steersman." However, despite their common etymology, "cybernetics" and "government" are not synonymous. "Government" pertains to the exercise of rule in a political community that, in contemporary states, typically combines the practices of making and enforcing (or choosing not to make or enforce) laws. Accordingly, "government" is necessarily conscious, active, purposive, and authoritative. It is also *essentially political* and *distinctly human:* government subjects human affairs to political determinations that only human beings are capable of undertaking. Cybernetics, on the other hand, refers to a type of control that is mechanical and unconscious, automatic and instrumental, inhuman and apolitical. There are cybernetic systems in the human body, but they differ little in kind from those found in lesser creatures, and resemble strongly those directing the operations of a variety of machines. Unlike government, therefore, cybernetics is not distinctly human; it does not distinguish human beings from either beasts or inventions.

Perhaps this explains the alleged antipathy between cyberspace and government. As we will see presently, a prominent feature of the discourse accompanying the proliferation of computer networks is a conviction that cyberspace – the world gathered by this technology – is immune to human government and defies sovereign political control. Borrowing the Heideggerian language of the preceding chapter, we might say that contemporary discourse takes as given the necessity of

· · ·
236

surrender to network technology's enframing essence, gapes at the technological, and concedes the impossibility of its limitation: we cannot let go of this technology and so we must let go of our sovereignty over it. This chapter investigates this dramatic concession. Drawing on the theoretical concerns elaborated in Chapter 2, I want to clarify what is at stake in adopting this belief uncritically. Specifically, I argue that, while there is nothing in the *technicality* of digital media as instruments that exempts them from sovereign governance, their *technological* character – and the technological character of the world they simultaneously fashion and are situated in – effectively precludes the exercise of genuine politics. This, of course, has serious ramifications for the technology's democratic potential. Finally, I consider how it is that a technology involved in the collapse of politics can also be consistently heralded as democratic. The answer, I conclude, lies in the peculiar ability of network instruments to "stand-in" for things they really are not, including democratic government.

Networks and the Sovereignty of Politics

For Aristotle, the useful arts and practical sciences (*technai*) were, along with practical wisdom (*phronēsis*), pure science (*epistēmē*), intelligence (*nous*), and wisdom (*sophia*), means by which human souls expressed the truth. In this view, *technai* and their products are neither necessary nor ends in themselves: the existence of artefacts is contingent upon the volition of their producers, and they are means to ends that are external to them. To be genuine and healthy, arts must proceed rationally as means to good ends. However, the goodness of ends cannot be established by the means themselves or, in this case, by the technique or technology being employed. The worthiness of any given end can be determined only by human beings – whose nature it is to deliberate upon good and bad, right and wrong, just and unjust – invoking *politikē*, the science whose object is the highest, intrinsically worthy good. In attending to the highest end, in relation to which the goodness of all other ends is determined, politics is the master science, deemed by Aristotle to be not only comprehensive, but also sovereign. Consequently, all genuine *technai* (i.e., techniques that are rational means to good ends) are subject to the sovereignty of political deliberation about the goodness of their ends and their integrity as means to these ends.

· · ·
237

Means, by nature, are limited by the ends they serve. Because the science of ends is a political one, technologies such as computer networks are means that must be limitable by, and subject to, sovereign political determinations, which achieve their ultimate expression in laws made by governments.

Is this the case with networks and the world they help fashion? The question here is not so much whether the ends of this technology are good, but whether, *as means,* networks are even amenable to such consideration and the limits it might entail. Is it possible to impose and enforce limits on the networked world, or has this technology rendered political sovereignty moot, and, if the latter, what does this mean for the democratic prospects of network technology? Considering these questions involves two related claims: first, that computer networks are technically impervious to any sort of sovereign authority whatsoever; and, second, that, to the extent networks *are* governable, the sovereign nation-state is not capable of doing the job. I consider each in turn.

An Ungovernable Instrument?

In response to concerns raised at the 1996 World Economic Forum about the threat network technology poses to state security, Nicholas Negroponte, director of MIT's influential Media Lab, had this to say about cyberspace: "You can't control it, it's uncontrollable. If someone tells you that you can, they are probably smoking pot."[1] John Perry Barlow agrees, adding that governments have "neither the right nor the means" to legislate practices and activities mediated by networked computers.[2] The prevalent view that "the Internet is ungovernable" is central to popular imaginings of a lawless electronic frontier, where any order that might exist is spontaneous, voluntary, and provisional.[3] In cyberspace, no rules are preferable even to good rules, because rules of any kind presuppose conditions of centrality, containment, and transparency that are inimical to the core technical properties of digital computer networks: distributed access and transmission, endless proliferation, and opaque mediation. After all, packet-switched, digital communication across distributed networks is designed to seek alternative routes to its destination when confronted with obstructions, errors, or other constraints. Thus, the network environment is "a fundamentally uncontrollable region."[4] "In the Network Era," writes Kevin Kelly, "openness wins, central control is lost."[5]

It is certainly true that digital networks pose challenges to those seeking to subject the practices they mediate to centralized authority, and that these challenges issue from the core technical properties of the medium, including its opacity and distributed connectivity.[6] I explore the character of these challenges below, and suggest how attempts to meet them help establish that networks, as instruments, are not, in fact, *necessarily* immune to politics. Before doing this, however, it is interesting to consider what might account for the persistence of this belief.

The insistence that networks cannot be governed is a variant of the disposition Langdon Winner has identified as modern society's somnambulist resignation to "autonomous technology": "a belief that somehow technology has gotten out of control, and follows its own course, independent of human direction."[7] The present version of this belief is curious – while the prospect of autonomous technology has customarily been regarded with dread because of its negative implications for human freedom, those who declare that network technology is autonomous do so because they sense in this autonomy the possibility of expanded, rather than contracted, liberty.

As Winner points out, autonomous technology is technology presented as not only independent of human judgment, control, and political intervention, but also as the replacement of these. Under the regime of technology perceived as autonomous, "*technē* has at last become *politēia*."[8] If networks as a medium are impervious to human political control, then the activities they mediate are insulated from whatever limitations the exercise of that control might involve. Such a situation has obvious appeal to those – such as pornographers, terrorists, capitalists, and criminals – whose activities are typically constrained by state or other authorities in one way or another, and who are always looking for means of concealment and evasion. Thus, capitalists such as Walter Wriston write approvingly of the contribution networks make to the "twilight of sovereignty."[9] It is instructive that Wriston manages, without apparent irony, to invoke the spirit of popular democracy in the course of asserting that network media are, and should remain, beyond the control of democratic governments: "No matter what political leaders do or say, the screens will continue to light up, traders will trade, and currency values will continue to be set, not by sovereign governments, but by global plebiscite."[10] Here, democracy is reduced to a free market, participation to a decision to buy low and sell high, and sovereign government to a distant echo. This is cause for celebration rather than alarm because it is a certain kind of liberty, rather than substantial democracy, that is being aimed for. As Wriston puts it, computer networks are "beyond the power of any sovereign government" and "that is good news for those who believe in freedom."[11] Clearly, a great deal is at stake in the matter of whether these instruments will be subject to, or exempt from, the rule of law and binding public deliberation about good ends. My point here is simply that arguments that insist networks are technically exempt from legal authority cannot be separated from the ideological belief that such exemption is politically desirable as well.

A good way to illustrate this dynamic is to consider the issue of digital cryptography in the United States, which is something of a crucible in terms of the relationship between networks and sovereignty. In 1991, an American computer scientist named Phil Zimmerman wrote a piece of "public-key encryption" software called "Pretty Good Privacy" (PGP) that allows people to use the substantial computing power on their desktops to encrypt e-mail messages using nearly unbreakable codes. In a public key system, a user actually has two keys – one public and one private. Public keys are used to encrypt messages that only the corresponding private key can decipher. So, for example, a person can encrypt an e-mail message with the public key of the recipient, and the message could be read only when deciphered by that recipient's private key. The system is nearly unbreakable because the processors in desktop computers are now powerful enough to encrypt messages using key lengths that would take exponentially greater – and basically unavailable – computing power to break.[12] In a 1997 brute-force decryption experiment, it took 78,000 networked computers 96 days to crack a message encrypted with the 56-bit Data Encryption Standard used for international monetary transfers in the banking industry. The PGP encryption software employs keys of this bit-length. It would take the same number of computers 67 years to crack a 64-bit encryption algorithm and 128-bit key lengths will be indecipherable for the foreseeable future.[13]

Fearing increased state invasions of privacy in the midst of the Gulf War, and acting "for the good of our democracy,"[14] Zimmerman made PGP available for free over the Internet. At the time, software containing strong encryption algorithms was classified as a munition under US International Traffic in Arms Regulations (ITAR), making it illegal for export without government approval. However, despite export prohibitions, the software quickly spread worldwide to become a de facto international standard for e-mail encryption. In what has been described as "a Promethean gesture," Zimmerman "stole encryption from the government and gave it to the people."[15] For his efforts, Zimmerman became the subject of a three-year criminal investigation – closed without indictment or explanation in 1996 – by the US Customs Service. The story of PGP is often recounted as an example of the network medium's technical indifference to governments attempting to assert their sovereign authority. The US government's inability to halt the illegal spread of encryption software such as PGP is cited as confirmation of the on-line world's immunity to law and its enforcement.

However, a second case involving the distribution of cryptographic software suggests the exemption from sovereign authority afforded to networks may be more political than technical in nature. In 1995, a mathematician named Daniel Bernstein brought a suit against the US government when the State Department required him to seek a permit to export the source code for an encryption program he had developed called "Snuffle."[16] In a series of rulings, a US District Court determined that encryption source code – like pornography – is a form of speech, and so is entitled to protection under the First Amendment to the US Constitution. The court found the government's export prohibitions to be a prior restraint on free expression, and so enjoined the state from enforcing them. Thus, in this case, it was not the technical peculiarity of networks that prevented the application of a particular law to a practice mediated by them, but, rather, the enforcement of a prior political commitment. There is nothing about digital networks that prevented Mr. Bernstein from being identified, held responsible for his actions, and compelled to cease and desist. Instead, it was merely the observance of a long-standing commitment in the United States to place a certain conception of liberty ahead of order on the scale of political virtues. If cyberspace were indeed immune to sovereign authority, then the American state, as represented by its courts, would be as unable to protect freedom there as it is to violate it. The availability of one or the other option suggests political decision making is the issue here, not technical necessity.

In this regard, it is interesting to note the primary sponsor of Bernstein's legal challenge is the Electronic Frontier Foundation (EFF). The EFF is a self-described "civil libertarian" organization that, as part of an "industry-led alliance" called Americans for Computer Privacy, is dedicated to "preserving the right of all Americans to use any encryption product or technique they wish," and to opposing "any government attempts to regulate the domestic use of encryption."[17] The interests of hyper-liberal civil-libertarians and entrepreneurial capitalists merge somewhat on this issue. The former seek protection against government intrusion into private electronic communications, and the latter are eager to compete in markets for strong encryption programs, markets that are growing in response to escalating levels of on-line commercial and financial transaction. Both require a relaxation of controls on cryptographic tools in order to realize their goals. On the other hand, governments recognize that vacating the field of digital cryptography in an increasingly network-mediated world will seriously compromise their ability to enforce laws wherever that enforcement requires scrutiny of documents, transactions, or other communications and information. This difficulty would apply equally to the rooting out of child pornographers, the interception of drug traffickers, the infiltration of seditious conspiracies, and the identification of tax evaders.

This clash of interests has led to a flurry of regulatory proposals, debate, legislation, and lobbying, especially in the United States, where the Clinton administration has been keen to craft an accommodation among the demands of privacy, commerce, and state security. In 1993, the administration announced its plans for the Clipper chip, a piece of hardware to be installed in telephones and other networked digital devices that would allow state authorities to tap into personal voice and data communications (the chip for data communications was to be called "Capstone") when authorized by legal warrant.[18] The Clipper plan was, for various political and commercial reasons, "laughed out of Washington."[19] In response, the administration proposed a relaxation of prohibitions on encryption software contingent upon the development of a public-key management infrastructure, in which private decryption keys would be held by escrow agents trusted mutually by users and law enforcement agencies both to protect the general confidentiality of keys and to release them upon certification of lawful authority.[20] This plan was criticized strongly by privacy activists, and prompted a barrage of legislative activity in the US Congress, both supporting and opposing the executive initiative.[21] In introducing one piece of cryptography legislation aimed at liberalizing prohibitions, the sponsoring senator pronounced, "The guiding principle for this bill can be summed up in one sentence: Encryption is good for American business and good business for Americans."[22] Indeed, widely used, secure encryption techniques may be the key to inspiring the consumer confidence necessary to convert networks into full-blown media of commercial transaction, and international markets for encryption products are potentially lucrative. It is this very pot of gold that has enticed major players in the computer industry to consider cooperating with the state's key-escrow plans, despite the dire warnings of civil-libertarians, for whom privacy is an end rather than a means. In 1998, the US Commerce Department announced that a coalition of ten companies could begin exporting strong encryption software, equipped to allow restricted access to law enforcement agencies, to an approved list of allied countries under relaxed rules.[23]

Still, an effective key-recovery system would require international as well as domestic compliance, and it is not yet clear that such a consensus is forthcoming. In 1997, the European Commission rejected US key-escrow plans as an undue infringement of data privacy.[24] Canada, though historically more committed to peace, order, and good government than to the exuberant protection of free expression, currently maintains fairly liberal regulations concerning encryption software. As a signatory to the thirty-three-nation Wassenaar Arrangement, Canada prohibits the export of encryption hardware and customized (i.e., for military use) software to non-allied nations, but places no restrictions on mass-market and public-domain software. Since 1996 Canada has allowed the export of 56-bit key-length encryption software, has not required permits for exporting customized encryption technology of any strength to the United States, and imposes no constraints on either the domestic use or the import of any strength of

cryptographic products.[25] The government of Canada is also committed to implementing and developing a recoverable public-key infrastructure for all internal and external government on-line communications,
a system it eventually hopes will interface with other private- and public-
sector key infrastructures.[26]

The complex political wrangling surrounding the issue of cryptography – involving, as it does, the limits imposed by constitutions, the
competition of interests in the legislative arena, the reckoning of economic implications, and the pursuit of international cooperation – suggests the exercise of the state's sovereign legal authority is anything
but irrelevant to the future of digital networks. If networks were truly
immune to sovereign authority and the political deliberations directing it, then it simply would not matter what the state had to say about
encryption software. This is clearly not the case. Societies could, for
example, enjoin their governments to prohibit entirely the use, sale,
and distribution of digital encryption software, just as they could ask
them to make it illegal to distribute pornography. Networks might make
it difficult to enforce such laws, but not impossible – indeed, network
technologies provide law enforcement authorities with as many investigative resources as challenges.[27] Thus, the considerations presently
preventing or dissuading governments from doing so are political,
rather than technical, in nature: institutionalized or cultural injunc-
tions to protect certain rights, and a reluctance to hamstring potentially explosive commercial markets. As in the Bernstein cryptography
case, the US Communications Decency Act, which sought to prohibit
the digital distribution of pornography to minors, was rendered ineffective by the US Supreme Court when it was deemed offensive to
constitutional speech protections, not by any feature of networks as
instruments.[28] On the other hand, the government of Singapore, whose
attention to public order outweighs its concern for free speech, has
seen fit to enact and enforce broad prohibitions against pornographic
and other prohibited material distributed via computer networks.
Singapore holds both content and access providers – deemed equivalent to broadcasters – responsible for material they make available for
public consumption that "undermines public morals, political stability
or religious harmony."[29]

In other words, if the sovereign authority of state governments is under threat, this threat is being posed by the imperatives of hyper-liberalism and hyper-capitalism, rather than the technicalities of hyper-media. Nevertheless, a discourse maintaining that networks are technically immune to enforceable laws persists. It is ironic, but instructive, that some of the most vigorous opposition to laws that would curb commercial or civil liberty on networks issues from those who maintain the medium is essentially ungovernable. This was certainly true in the case of the ill-fated Communications Decency Act, which was subjected to a ferocious campaign waged by network activists who argued the legislation was simultaneously unconscionable and unenforceable.[30] We might expect that confidence in the medium's technological insulation from sovereign authority would inspire indifference to law-making efforts deemed necessarily futile. It has not. This may be less of a logical contradiction than a considered political strategy: in the discourse animating the development of computer networks, the relationship between the claim that on-line activities are immune to governance *in fact,* and the ideological conviction that they *ought* to be so, is one of collusion rather than confusion. Stated crudely, if you keep telling people that networks cannot be subjected to legal authority, pretty soon they will begin to believe it. And once they believe it, the demand to subject computer networks and the activities they mediate to political judgment and public authority recedes.

The relative absence of government (i.e., its relegation to the minimal role of protecting private property and enforcing contracts in an otherwise unfettered market) has always been good for capitalists. The question here, though, is whether networks can or will be directed to, or limited by, good ends that involve more than simply efficiency and profitability – that is, whether networks can or will be subjected to political deliberation. As suggested above, despite real (but not insurmountable) difficulties, network media can be legislated upon, policed, and governed. What remains to be seen is whether they will be governed publicly and politically, or simply managed privately. The stakes here are considerable: the possibility of public, political governance is a necessary and minimum condition for realizing the democratic hopes pinned to network technology. Crucially, the decision about whether networks will be submitted to *politikē,* or whether they will be depoliticized, is itself political, and its outcome may have more to do with the current discursive climate of liberalism and capitalism than anything inherent in the network medium itself. Spurious arguments about the instrument's essential immunity to public authority are merely an invitation to private management pursuant to private ends.

Networks and Globalization

A second, less extreme approach to defining the relationship between networks and governance affirms the medium is amenable to enforceable rules, norms, and codes, but holds that sovereign, law-making, national governments are not suited to the task of establishing and executing them. This inadequacy is only partially attributed to what has been characterized as an overall waning of the capacity of nation-states to exercise sovereignty under the conditions of a globalized political economy. Phenomena such as capital mobility, liberal trade and investment agreements, the growth of transnational corporations and global social movements, and the increasing influence of transnational political and financial organizations – all of which, as outlined in Chapter 4, rely heavily on network technology – are said to have compromised the power of the nation-state in the late twentieth century to exercise sovereign control over, among other things, computer networks. As the authors of a recent report on the future of Canadian foreign policy put it, "globalization is erasing time and space, making borders porous, and encouraging continental integration. In the process, national sovereignty is being reshaped, and the power of national governments to control events, reduced."[31]

However, aside from (or in concert with) these general limitations, it may also be the case that certain attributes of digital networks make it difficult for nation-states to effectively subject them to sovereign government. Chief among these is the extra-territorial character of network media. In terms relevant to governance, a proliferating, distributed network whose extent is defined by multiplying and shifting terminal access points is effectively *extra*-territorial because it is *multi*-territorial. Networks cannot be isolated within a distinct territory because they span multiple territories. To the extent that sovereignty, in its classic definition, involves a monopoly on the coercive enforcement of laws *within* a given territory, networks present considerable challenges, especially in liberal polities where the legitimacy of that monopoly is ostensibly contingent on the consent of citizenries whose membership is also territorially bounded.[32]

Whose laws regarding, for example, obscene material, confidentiality, or intellectual property should apply to networks straddling physical, jurisdictional boundaries between states that set out conflicting standards of legality? Should a piece of miscreant literature distributed via a computer network be policed according to the laws of the country where the material originates, or those of the location where it is consumed? Often, the effects of network-mediated activity are realized in jurisdictions that are remote to the origin of that activity, which may itself be removed from the actual whereabouts of the active party, and are often initiated by the affected party – such as when a teenager in Arizona "visits" a Web site on a server in British Columbia that is maintained by a party in France. Which of American, Canadian, or French laws should apply in this case, and by whom should they be enforced?

Even when questions such as these are answered, there remain considerable difficulties in identifying violators and enforcing obedience. Identification and the assignment of responsibility, key requirements of any law-enforcement regime, can be hard to establish on-line. The anonymity afforded by the medium's opacity, the use of sophisticated encryption and re-mailing techniques that obscure the content and origin of messages, and the ease with which network connections can be hidden, broken, relocated, and reconfigured, compromise efforts at identification. This not only makes it difficult to identify and locate people engaging in unlawful activity, but also can thwart efforts to establish that activity as illegal in the first place. It may be illegal to distribute sexually explicit material to children, but if pornographers or distributors cannot reliably determine that their customer is a child, then how can they be held responsible for doing so without severely curtailing their liberty to engage in activity that is otherwise legal?

Furthermore, to what extent should network access-providers be held responsible for the content made available through their services? With previous communications media, the distinction between carriage and content was easy to establish: telephone services *carried* messages between private parties that the phone company could not reasonably be held responsible for; television stations broadcast *content* for which they could be held responsible and which could be subject to regulation. Multicast computer networks obscure this distinction between carriage and content, as users and service providers can be carriers and content generators simultaneously. Typically, the imperatives of free expression have dictated that common carriers be free of restrictions on what they mediate, while content providers licensed to use limited, public-airwave resources could be legitimately subjected to somewhat more regulation. It is not clear which standard should apply to computer networks that expand exponentially, and across which communication resembles private conversation as much as it does public exhibition. Thus, the management of obscenity purveyed over networks has proven vexing for states whose concerns about public morality are often offset by their commitment to free speech, with outcomes typically decided according to which of these two concerns is of greater importance in a given national context. So, for example, Singapore regulates network communication more stringently than does Germany, whose prohibitions on offensive material are greater still than those of the United States.[33]

The uncertain status of digital property also contributes to the challenge networks pose to sovereign authority. Property has always been a key site where relationships of sovereignty and subjection are defined and exercised. What establishes a property in a stream of bits whose materiality is marginal at best, and how can such properties be protected when networks make accessing, copying, altering, and distributing bits so cheap and easy? Copyright protection generally accrues to any original expression fixed in a tangible medium. What does this mean for expressions that are barely fixed in a not-so-tangible medium where ease of duplication, embellishment, and circulation render the status of originality dubious at best? These and other questions have led to the popular perception that traditional means of protecting intellectual property, such as copyright laws, have been rendered ineffective by computer networks, and that the only option in this regard is the widespread use of digital encryption techniques that will limit access to protected materials to authorized parties.[34] Encryption, as discussed above, presents its own problems to sovereign governments, for example, when encrypted data pertain to commercial and financial transactions that, by virtue of their encoding, are removed from the scrutiny of revenue agents. If the technicalities of networks suggest encryption as a substitute for copyright and other forms of property protection, then what is to become of the central expression of a state's sovereignty, its ability to tax? Networks already pose considerable problems in this regard – in which jurisdiction and at what rate should taxes be assessed for on-line transactions conducted internationally via networks?[35] At a 1998 meeting, members of the Organization of Economic Cooperation and Development (OECD) agreed that taxes should be levied at the point of consumption, but implementing such a regime may prove more difficult than agreeing upon it, especially given the spread of encryption technologies, not to mention the OECD's parallel commitment to market-driven self-regulation in this sector.[36]

These questions are certainly vexing. It has been argued that any attempt to apply distinctive national laws to the transnational activity mediated by networks would necessarily fail to meet liberal standards of legitimacy because the laws would not be based on the consent of the territorially dispersed "citizenries" over whom they were being enforced. "Governments," according to this view, "cannot credibly claim a right to regulate the Net based on supposed local harms caused by activities that originate outside their borders and that travel electronically to many different nations. One nation's legal institutions should not, therefore, monopolize rule making for the entire net."[37] The borders of a community in cyberspace are established by access to it – by connectivity and passwords – not the extent of its physical territory. If a person can hook up and log on, then he or she is part of whatever network community he or she is attached to, regardless of his or her actual location. To meet the liberal test of legitimacy then, any rules enforced in these communities must derive from the consent of all those situated within a network-mediated community's "password boundary." Thus, when it comes to networks, liberalism demands that "established territorial authorities ... defer to the self-regulatory efforts of cyberspace participants," and that they be flexible enough to accommodate "multiple rule sets."[38] Moreover, it is argued, attempts by territorially defined authorities (i.e., the governments of nation-states) to restrict on-line activity pursuant to local objectives should be resisted by on-line, non-territorial rule-making authorities. As David Johnson and David Post write, invoking the liberal right to revolt against a government to which one has not consented, "for the Net to realize its full promise, on-line rule-making authorities *must not respect* the claims of territorial sovereigns to restrict on-line communications when they are unrelated to vital governmental interests."[39]

Thus, it appears that, despite the considerable difficulties governments or similar authorities might face in crafting instruments capable of enforcing law on the networks, the decision about whether, how, and by whom this should be carried out is a political, rather than strictly technical, determination. Significantly, the argument in favour of principled restrictions on, and resistance to, the extension of national authority to the on-line world affirms that networks and the activities they mediate are susceptible to binding rules: "Cyberspace is anything but anarchic; its rules are becoming more robust everyday ... Fundamental activities of law-making – accommodating conflicting claims, defining property rights, establishing rules to guide conduct, enforcing those rules, and resolving disputes – remain very much alive in the newly defined intangible territory of cyberspace."[40] To emphasize, one might say the operative question is not whether network-mediated activity *can* be subject to the limitations of laws and rules, but rather whether it *should be* and, if so, *by whom.* These are distinctly political, and not especially technical, questions.

As discussed above, those who promote the belief that digital networks are technically immune to legal authority of any kind express a normative preference rather than a fact, and the same appears to be true of those who advance the more modest claim that it is particularly the sovereignty of national governments that has been rendered impotent by the technicalities of networks. This is not to underestimate the challenges, briefly catalogued above, facing territorial nation-states vis-à-vis digital networks. It is, rather, to point out that contingent normative preferences for a particular relationship between sovereign political government and network media often masquerade as technical imperatives and, by virtue of this disguise, discourage rather than invite deliberation upon that relationship. Put bluntly, if the nation-state is unable to exercise sovereign authority over network media as a matter of technical necessity, then there is not much to talk about. If, on the other hand, the question is really whether nation-states *should* do so, then an entire frontier of potential political contestation is opened.

However, it may also be the case that this frontier closes as soon as it opens, but for reasons that might be described as more technological than technical. That is to say, more than any technical attributes of digital networks, there may be something in the world these instruments gather *as a technology* that precludes the exercise of sovereign, democratic government over them. Phrased differently, computer networks as instruments can definitely be governed politically, but can they be governed as a technology in a world that is technological?

Network Technology and the Universal, Homogeneous State

Writing in 1965, George Grant observed that, "by its very nature the capitalist system makes of national boundaries only matters of political formality."[41] This is true in two senses. In the first place, capitalism is about the free accumulation and movement of capital and profits, and borders between nations are a barrier to movement that need to be minimized in a world whose economy is capitalist. Second, a commitment to such an economy, over time, renders political distinctions between nations unwelcome, and the maintenance of any apparatus for sustaining those distinctions unnecessary. Political distinctions become unnecessary because it is unambiguously clear which type of politics is most appropriate for a society whose economic base is capitalist, and that is liberal politics. Modern liberalism, as Grant defines it, is "a set of beliefs which proceed from the central assumption that man's essence is his freedom and therefore what chiefly concerns man in this life is to shape the world as we want it."[42] Elsewhere, he makes the connection of this politics to capitalist economics more explicit when he defines liberalism as a belief in "the freedom of the individual to use his property as he wishes, and for limited government which must keep out of the marketplace."[43] The fitness of liberalism for capitalism represents the overwhelming consensus of the modern age, a consensus that is gathered more generally under the sign of technology.

Taken together, capitalism, liberalism, and technology form a trinity of sorts, outside or beyond which there exist no political options capable of being persuasive in the modern (and, even more so the

· · ·

postmodern) world. The modern capitalist, liberal, technological state is thus a "universal and homogeneous state [that] is the pinnacle of political striving."[44] Specifically excluded from this state is any conception of good that might circumscribe the conduct of human agents. As Grant writes, "it is the very signature of modern man to deny reality to any conception of good that imposes limits on human freedom. To modern political theory, man's essence is his freedom. Nothing must stand in the way of our absolute freedom to create the world as we want it. There must be no conceptions of good that put limitations on human action. This definition of man as freedom constitutes the heart of the age of progress."[45] Liberalism, with its (spurious) profession of neutrality as to ends, cannot tolerate a good that flatly prohibits certain activities or ends by deeming them unambiguously harmful. Capitalism cannot tolerate serious limitations on profit making and private accumulation and remain capitalism. Technology cannot realize its essence if its setting upon the earth and human beings is constrained by the limitations imposed by a transcendent good. Beings committed to the belief that their essential humanity is expressed in their ability to make themselves and their world cannot be obligated to a good emanating from some other conception of their essence. Nor can the societies into which they congregate. And to propose they can and should be so constrained – that is, to propose a conception of good other than human freedom – is to place oneself outside the universal homogeneous state and to subject oneself to a solitude that is inevitably terminal.

This makes the maintenance of local particularisms – that is, indigenous deliberation on local expressions of good ends and the effecting of these ends via the exercise of political sovereignty – impossible under the sway of technological modernity. "Modern civilization," writes Grant, "makes all local cultures anachronistic. Where modern science has achieved its mastery, there is no place for local cultures."[46] Grant warns that prospects in this regard are particularly dire for Canada, situated as it is in such close proximity to the United States, "a society that is the heart of modernity ... the spearhead of progress" – a situation which means that "Canada, a local culture, must disappear."[47] The danger is not that Sable Island – its sands exposed to Atlantic winds by the grazing of wild horses sheltered from predators – might disappear into the sea. The danger is that Canada's destiny as a modern nation is to recede into a sea of capitalist, technological liberalism in which it will be indistinguishable from others in that sea, its politics reduced to choosing among the latest fashions in management science.[48]

It is probably advisable to specify here the precise nature of the homogeneity to which I am referring. For Grant, the modern state is a universal and homogeneous one because all societies under its sway maintain a commitment to the ends of liberalism, capitalist accumulation, and technological development. This is not to say that there will be no variation in the means chosen to reach these ends. Indeed, a recent, comprehensive comparison of policy in the United States and Canada in the era of globalization discovered evidence of both policy convergence and ongoing divergence between the two countries.[49] Divergence was most clearly identified in the area of social policy, wherein the Canadian state has struggled to maintain a traditional commitment to redistributive social programs that is stronger than that of the United States. Even within areas of marked convergence, such as macroeconomic and industrial policy, where the imperatives of market liberalization have captured policy in both countries, the authors of this study found "significant degrees of freedom remaining for national choices."[50]

These policy differences should not be underestimated, particularly because they are precisely the sorts of choices that often have a dramatic effect on the lives of people living in complex modern states. However, they also should not be overestimated. This study itself finds more evidence and momentum for policy harmonization with the global force and interests of the United States than it does for widespread policy distinction in Canada. More important, the degrees of freedom countries like Canada might enjoy in terms of choosing desirable means are contained within a broader consensus regarding the ends of capitalist economics, liberal politics, and technological progress. Canada may choose to temper its capitalist economy with a more vigorous welfare state and more protective labour laws than those adopted by the United States, but it is unlikely that Canada could choose not to endorse the imperatives of liberalism, growth, and progress, and instead pursue a good outside this horizon. Indeed such a choice is basically inconceivable, and it is this condition, not complete policy harmonization, that the "universal homogeneous state" names.

This appears to be the position in which Canada finds itself with respect to network technology. Indeed, digital computer networks, the technology of the moment in a technological world, bring the condition revealed by Grant into high relief. Canada is not alone within the liberal, capitalist, progressive horizon – the United States and most countries of Europe are also wired to the universal homogeneous state – but its government's approach to the development of network technology is particularly expressive of what it means to be so situated. The Canadian government's embrace of network technology has been total and unreserved. In Chapter 4, I discussed the considerable financial investment Canada has made in the construction and maintenance of an information "highway." This continuing investment has transpired in the midst of a commitment to this technology that is perhaps best described as spiritual. In 1996, the federal government issued a plan in response to the recommendations of its Information Highway Advisory Council (IHAC). In it, the government begins by testifying that "the Net is indeed an opportunity to glimpse into the future," and pledges "a government-wide effort by more than 30 federal departments and agencies ... intended to ensure that the enormous enabling power of Canada's Information Highway can be harnessed to create jobs and open up new realms of economic possibility."[51] The plan is presented as a freely determined and pre-emptive act by a sovereign government, securing, with forethought, its sovereignty into the future: "an action plan designed to ensure that Canadians have the Information Highway they need and want – not something imposed upon them."[52]

This is impressive, but perhaps vain. One wonders to what degree Canada's governors genuinely believed they had a choice in this matter, a choice to act otherwise – a choice, for instance, to forgo developing this technology altogether. If there is one undeniable truth about this technology, it is that networks are a minimum infrastructural requirement for "countries that hope to participate in global commerce,"[53] and that "those who reject it will condemn their countries to second class status or worse."[54] The Canadian government clearly found motivation in these pressures. Its appreciation of "the challenge and the urgency" was expressed as follows: "If we fall behind our major trading partners in building our Information Highway ... the social costs in terms of lost job opportunities will be enormous. Our national cultural dialogue will languish and our governments will be less able to keep up with the rapidly changing realities of the electronic age."[55] Herein lies the intractable dilemma of the modern age: in order to retain independence, nations must embrace technological advance, but in so doing they lose all hope of genuine independence. As Grant has written, "those who want to maintain separateness also want the advantages of the age of progress. These two ends are not compatible, for the pursuit of one negates the pursuit of the other. Nationalism can only be asserted successfully by an identification with technological advance; but technological advance entails the disappearance of those indigenous differences that give substance to nationalism."[56] In this light, a massive commitment to build network infrastructure is not a choice that asserts political sovereignty; it is a confession of its impossibility.

Canada has long experienced the agony of trying to maintain a modicum of cultural integrity in a mass-media environment captured decisively by the country's southern neighbour. Traditionally, Canadian media policy has attempted to balance the needs of industry and nation, negotiating between the often conflicting demands of cultural protectionism and commercial development. It has done so by adopting a bifurcated approach that posits a separation between broadcast and telecommunications media.[57] The Broadcasting Act stipulates that "the Canadian broadcasting system should encourage the development of Canadian expression by providing a range of programming that reflects Canadian attitudes, opinions, ideas, values and artistic creativity, by displaying Canadian talent in entertainment programming and by offering information and analysis concerning Canada and other countries from a Canadian point of view."[58] Toward this end the act includes provisions for licensing, Canadian content, and ownership requirements, a relatively high degree of regulation by the Canadian Radio-Television and Telecommunications Commission (CRTC), and mandates the publicly owned Canadian Broadcasting Corporation. Justification for such interventionism derives from the scarce and public nature of the broadcast spectrum, and from the fact that the primary activity of broadcasters is the distribution of content, an activity with significant cultural implications. In terms of broadcasting, policy objectives have thus often been weighted in favour of cultural protectionism.

Telecommunications media, on the other hand, are engaged in common carriage rather than content provision. That is, telecommunications industries – whether telephone or cable or satellite companies – do not broadcast content from a central source; rather, they carry content between numerous independent, often private, senders and receivers. The Telecommunications Act states a number of objectives, among them the strengthening and enriching of Canada's economic and cultural fabric and the promotion of Canadian ownership (the act stipulates that no more than 20 percent of any telecommunications enterprise in Canada can be owned by non-Canadians). However, the act also is intended to promote the international competitiveness of Canadian telecommunications firms, enhance efficiency, and foster greater reliance on market forces.[59] Accordingly, economic and industrial considerations – and not cultural protectionism – have been the driving force behind telecommunications policy.

It is difficult to measure the effectiveness of this arrangement in terms of facilitating Canada's ability to exercise its sovereign authority to protect its culture in a continental media environment. On the one hand, it is undeniable that content and ownership regulations in the broadcasting sphere have resulted in far higher exposure to Canadian content than would have been likely in their absence. However, if cultural protection in the Canadian context means enforcing a condition wherein the bulk of cultural material to which Canadians are exposed is their own rather than, say, American, then it is fair to say that this approach has failed. It is instructive to note that this failure has become increasingly apparent as, since the 1980s, Canadian communications policy has shifted from being animated by the culturally protectionist spirit of the broadcast paradigm to the economically progressivist spirit of the telecommunications paradigm. As Marc Raboy writes, "the rationale for public support of content development shifted from nation-building to industrial growth ... Cultural nationalism continued to be the predominant theme of policy discourse, but actual policy programs focused on beefing up the industry side of the cultural industries equation ... In fact, while the policy debate focused on developing an appropriate interface between technology and culture, technological development and cultural production in Canada had been harnessed to create new concentrations of corporate wealth and power oblivious to the social and cultural objectives encoded in legislation and regulatory policy."[60] Accordingly, "foreign (largely American) content continued to flow through the veins of the Canadian communications system."[61] Despite efforts to encourage domestic ownership and content, no amount of state protectionism has been able to make Canadian cultural artefacts prominent among the offerings of continental mass media. Notwithstanding the recent success enjoyed by Canadian-born female singers offering generic pop fare, the result has been an overwhelming overrepresentation of American cultural products in the popular media landscape. And as Mary Vipond points out, Americanization has not simply been imposed on Canadians against their wishes, but rather is something many Canadians actually value. She quotes John Meisel, a former chair of the CRTC: "Canadians regard their right to watch American TV programming with the same passion as Americans regard their right to bear arms."[62]

The question is whether Canadian cultural sovereignty is likely to fare any better in a media environment structured around network technology. Two technical factors are relevant to this consideration. The first is that transposition of culture into bits rather than atoms increases the permeability of physical borders between cultures. Bits can travel across borders with as great or greater ease than books, magazines, films, and audio- or videotapes. The assertion that "governments cannot stop electronic communications coming across their borders, even if they want to do so" may be an overstatement, but it does point to a very real challenge.[63] Governments could restrict transborder data flow by disabling networks or by controlling access to them strictly, but once network infrastructure is in place, and access to it widespread, the difficulties of doing so escalate considerably. A second technical factor that should be taken into account is that networks tend to obviate the distinction between broadcasting and telecommunications, and reduce the differences between content and carriage. It is under this rubric – the phenomenon of so-called convergence – that the pretences of cultural protection and the imperatives of economic and technological development meet headlong.

While networks can be simultaneously broadcast (or, more accurately, multicast) and telecommunications media, our approach to them cannot be both culturally protectionist and economically progressive;

the demands of these two positions are too much at odds. The market interventions required to protect Canadian culture run afoul of the freedom required for capitalists to accumulate their profits and for technology to develop as it must. The result, in terms of Canadian policy toward the development of network technology, is schizophrenic.

On the one hand, we find a great rhetorical commitment to the safeguarding of Canadian culture in the network-mediated universe. The federal government's action plan declares: "The Information Highway must also provide us with a new and more powerful means of enriching and invigorating the ongoing cultural dialogue that defines our national identity, our shared values and the common social purpose that provides the foundation for democratic institutions. It must, in short, deliver Canadian cultural content that reflects our linguistic duality and cultural diversity."[64] The strategy the government has outlined for reaching this goal is based on enhancing the profile of Canadian content in the network media market, not insulating Canadians from the predominantly non-Canadian content already dominating that market. The IHAC's second final report expressed the rationale for this approach clearly: the only plausible defence against "the deluge of U.S. software, video games and multimedia products into the Canadian market" is to stimulate the production of "diverse, high-quality Canadian alternatives."[65] Accordingly, the Canadian government proposes to protect Canadian culture with an industrial strategy, that is, by "expanding opportunities for economic growth and job creation; employing a range of measures to support the production, distribution and promotion at home and abroad of Canadian cultural content ...; fostering an ongoing national cultural dialogue within Canada; and promoting the dissemination of the government's public information holdings."[66] Canadians supporting the production of indigenous cultural artefacts via their government is certainly a noble pursuit, but it has never been enough to secure the integrity and distinction of their culture in the face of mass media that are continental. It is unlikely to suffice in the context of a medium that is universal. Nowhere in the government's strategy is there evidence of a willingness to apply the more radical instrument of content restriction to network media. The importance of the CRTC and the Broadcasting Act are affirmed, as is the spectre of vulnerability to a flood of non-

Canadian cultural products.[67] However, an endorsement of the regulation of content as a possible response to the new reality is nowhere to be found.

Perhaps this is because the technical resistance of network media to sovereign intervention of this kind would render such a commitment hollow. More likely it is because the core thrust of policy in this area is driven by a commitment to a liberalism that cannot abide interventions pursuant to the public good that place limits on industrial and technological development. This commitment has definitely animated the Canadian government's initial approach to network technology. In 1995, the first IHAC final report began with the assumption that "in the new information economy, success will be determined by the marketplace, not the government," and recommended the Canadian state limit its "primary role" to "set[ting] the ground rules and act[ing] as a model user to inspire Canadians."[68] Stressing that "there is an overriding sense of urgency to move ahead with competition and the development of the Information Highway," IHAC advised the federal government to establish a competitive environment for network development by relaxing regulation and liberalizing foreign-ownership requirements, supporting research and development, facilitating standards development, and providing universal access. While the report recommends that "Canadian cultural policy must be reaffirmed and strengthened in relation to the new information infrastructure" and that "government policies should continue to recognize and implement measures that give priority to the services of Canadian programming undertakings," it stops short of endorsing radical, or even strong, content restrictions.[69]

It is this liberal, market-driven approach the government adopted in its 1996 action plan concerning network technology, the primary focus of which is to create a "competitive, consumer-driven policy and regulatory environment ... conducive to innovation and investment by Canadian industry in new services on the Information Highway."[70] And this approach is affirmed in IHAC's second final report, which reiterates the belief that, "in an age of headlong technological transformation, it is critical that market forces determine what technology is appropriate" and expresses strong "reservations about governmental efforts to engage in formal regulation of the Internet at this time."[71] Of course, no other position makes sense in a dynamic technological world, wherein the main priority for governments is "the need to ensure growth, expansion and innovation in the network infrastructure," or in a capitalist economy wherein "strengthening the emerging role of the Internet as a platform for electronic commerce should be the central economic strategy" of government.[72] Here, government is reduced to all that it can be in a state that is universally and homogeneously liberal, capitalist, and technological: sentimental about the remnants of its people's culture; a policeman (but not lawmaker) for the market; a gas jockey fuelling the engines of technological expansion. I should perhaps add educator, a role traditionally conceived as central to the exercise of good government. We learn the most about societies from the manner in which they educate their youth, and the substance of that education. The Canadian government has been particularly proud of its Schoolnet initiative, a program whose primary purpose is to provide students and teachers with "exciting electronic services that will develop and stimulate the skills needed in the knowledge society."[73] It is instructive that, in its publicity for this program, the government gives prominent place to its endorsement by Bill Gates, founder and president of the Microsoft Corporation, an American who is indisputably the world's most formidable capitalist and technologist.

The ethos guiding the Canadian approach to network technology – an ethos animated by liberal notions of freedom, capitalist visions of prosperity, and an abiding faith that technological progress is central to both – is typical of those adopted by North American, European, and some Asian governments, with minor national variations. In the United States, a trio of laws – the High Performance Computing and Communications Act (1994), the Telecommunications Act (1996), and the Next Generation Internet Research Act (1998) – affirmed the pattern of massive public investment in network infrastructure accompanied by market deregulation. Official discourse adorning these legislative commitments to liberal, capitalist approaches to technological progress expressed well the credo of the universal homogeneous state. Regarding the 1996 deregulation, the US President's Office declared that "this legislation will lead all Americans into a more prosperous future by preparing our economy for the 21st century and opening wide the door to the Information Age."[74] The investment of hundreds of millions of public dollars in 1998 was beyond political interrogation because there is apparently nothing this technology cannot do: computers, networks, and software "drive economic growth, generate new knowledge, create new jobs, build new industries, ensure our national security, protect the environment, and improve the health and quality of our people."[75] It is a mark of the universal homogeneous state that technology is deified. And when it is so elevated there is simply no political ground upon which citizens could reasonably argue for limits to its progress. As American president Bill Clinton explained, "our nation's economic future and the welfare of our citizens depend on continued advances in the information technologies."[76] If this is true, how can the imperatives of this technology be responsibly questioned, let alone refused?

The European Community has embraced similar articles of faith. Its Action Plan for developing network technology affirms, somewhat paradoxically, that "Europe's way" to the Information Society must begin by conceding that "the process [of digitization] cannot be stopped," and accepts "the need for an acceleration of the liberalization process and the achievement and preservation of universal service and Internet Market principles of free movement."[77] It is not clear, given these convictions, what distinguishes "Europe's way" from America's (or Canada's) at the level of core commitments, nor even how Europe could *imagine* a distinctive "way" for itself relative to a phenomenon that is indifferent to its determinations. The same might be said of the positions advanced by many Asian governments. Japan's government is certain that development of a "global information and communications infrastructure" is "indispensable" to re-vitalizing that country's economy, and is therefore anxious to "play a leading role in liberalizing the info-communications market."[78] The Japanese government's vision for the country's future further confides: "we have placed our faith in the market mechanism and made it clear the administrative stance ought to be market-oriented rather than protective and nurturing." It may simply be correct that countries will be unable to prosper economically in the foreseeable future without progressive development of their network infrastructures, and that such development requires parallel commitment to a liberal politics that complements the unfettered operation of capitalist markets. It may even be that such choices are preferable to the relative material hardship that would probably result from a rejection of this "reality." None of this, however, detracts from the universality or homogeneity of these various commitments; that they appear reasonably imperative simply makes them easier to accept.

The question of whether digital networks are likely to mediate a democratic public life must be situated within the context of the political possibilities of the universal homogeneous state, the state established by this technology in triumvirate with liberal ideology and capitalist economics. Some, who cannot see beyond the horizons of this state – that is, who cannot imagine or remember a good it does not encompass – certainly believe that democracy names the politics of the networked world. This doctrine comprises a major element of the faith of our times. The spirit of this faith and its politics is expressed in the following statement by cyberspace frontiersman and capitalist Walter Wriston: "Modern communications technologies ... are creating a global market that takes constant referenda on what in many ways is beginning to look like a global culture ... All of a sudden, everyone has access to everything ... Tens of millions of Chinese and Indians, Frenchmen and Malays are watching *Dallas* and the *Honeymooners* which, in their way, may be more subversive of sovereign authority than CNN. The people plugged into this global conversation are voting – for Madonna and Benetton and Pepsi and Prince – but also for democracy, free expression, free markets, and free movement of people and money."[79] Networks, then, carry the cultural virus that finally annihilates all particularism, and are the perfect media for the universal homogeneous state. Information and the medium that delivers it is, in this vision, the great equalizer. However, when equality – democracy's oldest friend – is reduced to homogeneity, it becomes democracy's greatest enemy. A nation cannot be self-governing when political options derived from a local culture that is its own have been obliterated. And a nation that cannot be self-governing cannot be a democracy, regardless of how many activist groups use networks to press their cause, or how rational, undistorted, and free of domination the conversations that networks mediate might be.

This is to say nothing of the status of a politics in which options that would fundamentally constrain one or another of liberty, profit making, or technological progress are effectively unavailable. George Grant never elaborated a comprehensive theory of democracy. One might expect that, as a Platonist, he would be suspicious of democracy as an irrational form of rule under which the immoderate appetites of numerical majorities, expressed as self-interested opinions, govern public life. Indeed, if liberalism is the perfect ideology for a capitalist and technological society because it rejects the authority of transcendent constraints on human agency, then democracy, which asserts that only human beings can and should decide which laws they will observe as limits on their activity, is probably the perfect form of government for that society. Nevertheless, Grant was concerned to point out that democracy was not the same thing as American, capitalist, technological liberalism. Referring to the ideology of Canadian prime minister William Lyon Mackenzie King, Grant wrote that one would have to be "sufficiently held by liberal theory to believe that the United States [is] a democracy," clearly indicating that the two were far from identical.[80]

What is the difference between capitalist, technological liberalism and democracy? In a genuine democracy, there are no options that are completely unavailable to "the people" as possible choices for the direction of public life. That is to say, in a democracy, it must at least be possible for the people to decide they will observe the constraints certain virtues might impose on human freedom, material accumulation, or technological advance. In other words, genuine democracies must at least have recourse to a good that exists beyond the horizon of liberty, profit, and progress, a good that might impose significant limits on the pursuit of these ends. No such recourse is available in a liberal, capitalist, and technological society, in which a human being's essence is believed to be his or her freedom, and the fulfilment of that essence is achieved through unlimited acquisition and endless progress. To the extent that the universal homogeneous state is a liberal, capitalist, and technological one that must, by virtue of its ends and what it truly is, deny outright the imperatives of the good – in which "the people" as a collectivity cannot effectively choose virtue over liberty, wealth, and progress – it cannot be a democracy. This is not to suggest that democracies *must,* or are even *likely* to, make this choice consistently. It is simply to require they at least have the *option* to do so. The universal homogeneous state of modern, capitalist, technological liberalism denies people this opportunity, and so cannot accommodate genuine democracy. In so far as it contributes substantially to the entrenchment of this state, network technology is an instrument of democracy's continued impossibility in the modern world.

Network Technology as Stand-in

The fact remains that network technology is widely believed to be the medium through which a democratic revolution is being, or will be, enacted. At the very least, the consensus suggests, computer networks enhance existing democratic practices and institutions measurably. In the foregoing chapters, I have argued that this belief cannot be sustained on economic, ontological, or political grounds. Nevertheless, it persists. In this concluding section, I wish to consider briefly what might account for this persistence.

In the first place, it should be acknowledged that computer networks are not completely without democratic benefits. Networks are used by many citizens to distribute and access political information, and to communicate with each other as well as with elected representatives. Used in this way, networks can certainly make a contribution to improving civic life, especially at the local level.[81] Such applications no doubt account for a substantial portion of the democratic hope popularly invested in this technology, although recent research indicates that network technologies tend to reinforce existing patterns of democratic behaviour rather than mobilizing new actors and practices.[82] Whatever the case, dialogic conditions do not exhaust the requirements for a democratic life, and dialogic applications represent only partially the manner in which this technology confronts us. As I have argued, network technology fails to live up to its democratic image precisely because its many non-dialogic applications are un- or anti-democratic, and these eclipse, and even undermine, the democratic potential of applications such as network-mediated civic discussions. The question is, why have many of us become seduced by a democratic potential that is at best marginal in relation to the predominantly anti-democratic attributes of this technology? Is it simply that, as the children of Prometheus, we have been blinded by hope?

As outlined in Chapter 2, Plato insisted a distinction be maintained between those genuine arts (*technai*) through which human beings pursue "the highest welfare of body and soul" and the simple knacks or unreflective routines (*empeiriai*) that stand in as poor substitutes for these arts, and by which people flatter their appetites with various plea-sures.[83] Habituation to good practices is essential for most people who wish to live a good life, as is the satisfaction of appetites for things necessary to sustain such a life. Knacks that stand in for genuine arts do not, however, accomplish either of these goals. Instead, *empeiriai* either habituate people to practices that are less than fully good, or gratify essential appetites without fully satisfying them. Diminished ends and gratification are often enough for weakened souls. When human beings lack the intelligence to identify truly good ends, or the courage to uphold what they correctly believe to be good, and moderate their appetites accordingly, they become vulnerable to the lure of *empeiriai* that masquerade as, and stand in for, genuine *technai*.

Plato's distinction illuminates much of what is at issue in the present discussion. Human beings have exhibited an abiding appetite for the unambiguous goods of community with their fellows and good self-government, and for the potential good of democracy (I realize I am on shaky ground here. Plato, the great critic of democracy, might rather have suggested that it is a stand-in for good self-government. When I specify democracy as a *potential* good here, I am referring to it in the sense of the preceding section, wherein democracy was presented as at least being *open* to goodness). However, for reasons that are far beyond the reach of this investigation, the human virtues of wisdom and courage have atrophied in the modern era, a condition that has left human beings vulnerable to the appeal of stand-in technologies that flatter but do not satisfy their appetites for a good life, and that, through their use, habituate people to live and accept a diminished existence. As a new century begins, network technology has emerged as perhaps the ultimate *empeiria*, the ultimate stand-in, capable not only of satisfying the baser human appetites for material wealth and mastery, but also of gratifying certain nobler human appetites with-out actually satisfying them.

At the present moment, networks and bits appear ready and able to stand in for almost anything. They stand in for trade unions and col-lective bargaining, as employers now find it more convenient to make offers directly to employees by electronic mail rather than through their representatives.[84] They stand in for the apparatus of the welfare state, as private companies in the United States who administer income assistance – some of which are also in the business of selling lottery tickets through automated kiosks – find efficiencies in distributing benefits through computerized terminals.[85] They stand in for education and teachers, as learning institutions throughout North America embrace the virtual classroom and library.[86] They stand in for medical care by facilitating the "timely electronic provision of health services and medical expertise in remote areas" so that doctors might be relieved of the burdens of actually living in such areas.[87] Indeed, network technology and the information it mediates are ready to stand in for knowledge and wisdom themselves – Canada's IHAC informs its citizens that, once the information highway becomes the pervasive infrastructure of their society, "knowledge will become increasingly available to everyone, allowing us all to make wiser decisions in all aspects of our lives."[88] It used to be that the journey out of the cave into the light of knowledge was long and painful; now, with network technology standing in for educational arts whose demands are more arduous, wisdom is at our fingertips, and it won't hurt a bit. We just have to develop a knack for it.

Most significant in the present context is the ability and propensity of digital networks to stand in for the genuine arts of government and democracy. Indeed, as Lawrence Grossman observes, inadvertently, in his book *The Electronic Republic,* this is a substitution to which we have already begun to become habituated: "Cable shopping channels have installed high speed, large capacity computerized systems to process millions of viewers' telephone credit card orders. The same or similar technology can be recruited to tabulate votes, process polls, and count the results of initiatives and referenda, dialled in from anywhere."[89] That the techniques of network-mediated politics could never really replace the genuine art of self-governance is signalled in Grossman's addendum to the above, in which he indicates the medium's essential indifference to the true ends of that art: "The question is not whether the transformation to instant public feedback through electronics is good or bad, or politically desirable or undesirable. Like a force of nature, it is simply the way our political system is heading."[90] Genuine government is always about deliberating upon what is good and what is bad, and distinguishing what is politically desirable from what is not. To the degree it is indifferent to such considerations, network technology can never be an adequate substitute for the *technē* of government, and will never fully satisfy the appetite human beings have for governing themselves well.

The possibility of politics in the age of networks thus derives from the consequences of the digital ontology discussed in Chapter 6. People whose lives are mediated by a technology that uproots and subjects them to the empire of calculation are particularly vulnerable to the superficial appeal of the politics described here as characteristic of the universal homogeneous state – a politics wherein thoughtless submission to perceived technological imperatives is identified with freedom and prosperity. In the digital age, the tragic consequence of alienation from our rooted and meditative essence is participation in a public and collective life that elides the genuine essence of politics itself. A more authentic politics would comprise, to use the Heideggerian language of the preceding chapter, a more original revealing. The positive content of this revealing is – if one respects the essential particularity of the rootedness necessary to accomplish it – impossible to specify in advance. For this reason, I have been reluctant to pronounce upon what the content of a genuine politics in the age of network technology might be (at most I would venture that it must be thoughtful and deliberative), opting instead to ruminate on the very possibility of such a politics. For better or worse, this question seems to revolve around the possibility of democracy in the era of digital networks.

At its best, democracy is a political art, a subsidiary of the art of government that involves citizens deliberating as equals upon good ends and the best means for achieving them. In this manifestation, it issues from the noble appetite people have for governing themselves well. At its worst, democracy is a degenerate technique for the public registration of privately formed and self-interested opinions, and is hostile to genuine politics and government. In this manifestation, democracy flatters the noble appetite for self-government, but satisfies only the baser material appetites for wealth, gain, and power.

If computer networks are to be involved in democracy at all, they are likely to be instruments of democracy at its worst, rather than at its best. This has little to do with the technical properties of digital networks. As George Grant said, "the computer does not impose on us the ways it must be used."[91] However, we cannot abstract the networked computer as an instrument from the technological condition in which is situated. Networks as instruments distribute information and facilitate communication, but these do not encompass the resources necessary for the practice of the genuine arts of self-government, politics, and democracy. Indeed, much of the present investigation can be read as a meditation upon the economic, ontological, and political conditions necessary for democratic self-government, the failure of the modern technological world to meet those conditions, and the likelihood that networks, as a technology, will perpetuate rather than alleviate that failure.

For the majority of modern citizens, public-relations spectacles combine with the periodic registration of private opinions derived from self-interest and propaganda to stand in for democratic self-government. In a time of weakened spirit, wherein wisdom and courage are conspicuous by their absence from public life, these surrogates are able to flatter our collective appetite for a more genuine politics. They are, however, ultimately unable to satisfy that appetite, a fact that may account for the residual cynicism, alienation, and disaffection afflicting many of the so-called advanced democracies. This malaise provides fertile soil for claims that network technology can satisfy more substantially our yearning for an authentic democratic politics. For reasons discussed at length in the foregoing pages, I submit that network technology is but the latest in a succession of stand-ins for the real thing. Under the sway of these stand-ins, we have become habituated to practising a diminished politics that bears a name it does not deserve. The regime of network technology offers scant hope for the shattering of this ignoble delusion.

notes

• • • • •

Chapter 1: Prometheus Wired

1 John Perry Barlow, as quoted in Evan Solomon, "Unlikely Messiah," *Shift* 4, 1 (1995): 31.

2 Ted White, MP, Reform Party of Canada. Personal correspondence, 22 November 1995.

3 Barlow, as quoted in Solomon, "Unlikely Messiah," 31. Emphasis added.

4 The tragedy of Prometheus has been recounted by the likes of Ovid, Hesiod, Aeschylus, Plato, Byron, Goethe, Longfellow, and both Shelleys. For a brief summary of the development of the myth, see Olga Raggio, "The Myth of Prometheus: Its Survival and Metamorphoses up to the Eighteenth Century," *Journal of the Warbourg and Courtauld Institutes* 21, 1 (1958): 44-62.

5 Hesiod, "Works and Days," in *Hesiod: The Poems and Fragments*, trans. A.W. Mair (Oxford: Clarendon Press, 1908), lines 42-50.

6 Ibid., lines 97-8.

7 Aeschylus, *Prometheus Bound*, trans. Herbert W. Smyth (London: Heinemann, 1922), lines 250-2.

8 Francis Bacon, "*De Sapientia Veterum* or, Wisdom of the Ancients," in *The Philosophical Works of Francis Bacon*, trans. James Spedding, ed. John M. Robertson (London: George Routledge and Sons, 1905), 821-58; Karl Marx, "The Difference between Democritean and Epicurean Philosophies of Nature," in *The Marx-Engels Reader*, 2nd ed., ed. Robert C. Tucker (New York: Norton, 1978), 9; Mary Shelley, *Frankenstein, or, The Modern Prometheus* (London: Oxford University Press, 1969); Friedrich Nietzsche, *The Gay Science*, trans. Walter Kaufmann (New York: Random House, 1974), 240-1.

9 David S. Landes, *The Unbound Prometheus: Technological Change and Industrial Development in Western Europe from 1750 to the Present* (Cambridge: Cambridge University Press, 1969).

10 On the origins of mechanical time, see David S. Landes, *Revolution in Time: Clocks and the Making of the Modern World* (Cambridge, MA: Belknap, 1983).

11 Ibid., 53-94.

12 See Lewis Mumford, *Technics and Civilization* (New York: Harcourt, Brace and World, 1963), 198. See also Landes, *Revolution in Time*, 286.

13 Landes, *The Unbound Prometheus*, 41.

14 This is the argument advanced by Werner Sombart in *Der Moderne Kapitalismus*, 4 vols. (Munich: Duncker and Humbolt, 1928).

15 Landes, *The Unbound Prometheus*, 276.

16 Shelley, *Frankenstein*, 52-3.

17 Ibid., 51.

18 Ibid., 85-96.

19 Elizabeth L. Eisenstein, *The Printing Press As an Agent of Change: Communications and Cultural Transformations in Early Modern Europe* (Cambridge: Cambridge University Press, 1979). For the relationship between printing and nationalism, see Benedict Anderson, *Imagined Communities: Reflections on the Origin and Spread of Nationalism* (London: Verso, 1991).

20 See Marshall McLuhan, *The Gutenberg Galaxy: The Making of Typographical Man* (Toronto: University of Toronto Press, 1962), and Harold A. Innis, *The Bias of Communication* (Toronto: University of Toronto Press, 1991).

21 Jean-Jacques Rousseau, "A Discourse on the Moral Effects of the Arts and Sciences," in *The Social Contract and Discourses*, trans. G.D.H. Cole (London: Everyman, 1988), 26.

22 The works of Noam Chomsky, Edward Herman, and Michael Parenti explore this theme in great detail. See, for example, Edward S. Herman and Noam Chomsky, *Manufacturing Consent: The Political Economy of the Mass Media* (New York: Pantheon, 1988), or Michael Parenti, *Inventing Reality: The Politics of the Mass Media* (New York: St. Martin's Press, 1986).

23 All of these techniques had been previously used in print media. For an excellent account of the history of modern advertising, see William Leiss, Sut Jhally, and Stephen Kline, *Social Communication in Advertising: Persons, Products and Images of Well-Being* (Scarborough, ON: Nelson, 1990).

24 Ibid. See also Dallas W. Smythe, *Dependency Road: Communications, Capitalism, Consciousness and Canada* (Norwood, NJ: Ablex, 1981).

25 *Bit* is the short form of *binary digit*, and refers to the electromagnetic pulses, rendered as 0s and 1s, that form the language of the modern computer. The nature of bits is discussed in detail in Chapter 3.

26 Mark Poster, *The Mode of Information: Poststructuralism and Social Context* (Chicago: University of Chicago Press, 1990), 18.

27 Ibid. See also Mark Poster, *The Second Media Age* (Cambridge, MA: Polity, 1995).

28 George P. Landow, *Hypertext: The Convergence of Contemporary Critical Theory and Technology* (Baltimore: Johns Hopkins University Press, 1992).

29 Ronald J. Deibert, *Parchment, Printing and Hypermedia: Communication in World Order Transformation* (New York: Columbia University Press, 1997), 177-201.

30 Ibid., 214.

31 For an example in which semiotics and critical theory are combined to good investigative effect, see Timothy W. Luke, *Screens of Power: Ideology, Domination and Resistance in Informational Society* (Chicago: University of Illinois Press, 1989).

32 Jean Baudrillard, *The Ecstasy of Communication*, trans. Bernard Schultze and Caroline Schultze (Montreal: Semiotext(e), 1988), 22.

33 Poster, *The Mode of Information*, 20.

34 Landow, *Hypertext*, 3.

35 Baudrillard, *The Ecstasy of Communication*, 99. Emphasis added.

36 Ronald Beiner, *Political Judgment* (London: Methuen, 1983), 109.

37 Ibid., 144.

38 Fredric Jameson, *Postmodernism, or, The Cultural Logic of Late Capitalism* (Durham, NC: Duke University Press, 1991), 48-9.

39 Chris Hables Gray, *Postmodern War: The New Politics of Conflict* (London: Guilford, 1997), 22.

40 Poster, *The Second Media Age*, 55.

41 Daniel Burstein and David Klein, *Road Warriors: Dreams and Nightmares along the Information Superhighway* (New York: Dutton, 1995), 254.

42 Alvin Toffler and Heidi Toffler, *Creating a New Civilization: The Politics of the Third Wave* (Atlanta, GA: Tower, 1995), 21.

43 Walter B. Wriston, *The Twilight of Sovereignty: How the Information Revolution Is Transforming Our World* (New York: Macmillan, 1992). Mr. Wriston was CEO of Citibank/Citicorp from 1967 to 1984.

44 George Gilder, *Life after Television: The Coming Transformation of Media and American Life* (New York: Norton, 1994).

45 Executive Office of the President, Washington, DC, *The National Information Infrastructure: Agenda for Action*, reprinted in *The Information Revolution*, ed. Donald Altschiller (New York: H.W. Wilson, 1995), 9-39.

46 David Johnston, Deborah Johnston, and Sunny Handa, *Getting Canada On-line: Understanding the Information Highway* (Toronto: Stoddart, 1995), vii.

47 Notable exceptions would include Clifford Stoll, *Silicon Snake Oil: Second Thoughts on the Information Highway* (New York: Doubleday, 1995), and David Shenk, *Data Smog: Surviving the Information Glut* (New York: HarperCollins, 1998). Both of these tend to be more nostalgic than critical. For better expressions of dissent about the nature and depth of change promised by digitization and networks, see Heather Menzies, *Whose Brave New World? The Information Highway and the New Economy* (Toronto: Between the Lines, 1996); William Leiss, "The Myth of the Information Society," in *Cultural Politics in Contemporary America*, ed. Ian Angus and

Sut Jhally (New York: Routledge, 1989), 282-98; and Michael Traber, *The Myth of the Information Revolution: Social and Ethical Implications of Communication Technology* (London: Sage, 1986).

48 Wriston, *The Twilight of Sovereignty*, 2. Emphasis added.

49 Gilder, *Life after Television*, 61.

50 Burstein and Klein, *Road Warriors*, 104-5. Emphasis added.

51 See James A. Monroe, *The Democratic Wish: Popular Participation and the Limits of American Government* (New York: Basic Books, 1992).

52 Lawrence K. Grossman, *The Electronic Republic: Reshaping Democracy in the Information Age* (New York: Viking, 1995). Mr. Grossman is the former president of the Corporation for Public Broadcasting (PBS) in the United States, and a former head of NBC television news.

53 Ibid., 33-4.

54 Ibid., 4, 120.

55 Ibid., 148.

56 Wriston, *The Twilight of Sovereignty*, 153.

57 Every major Canadian political party now maintains a Web site: www.blocquebecois.org, www.liberal.ca, www.pcparty.ca, www.fed.ndp.ca, www.reform.ca. Most of these sites list candidates'/MPs' electronic-mail addresses. For the Canadian government's gateway to official documents on-line, as well as to provincial government sites, see www.canada.gc.ca.

58 In Canada, the Reform Party has been particularly active in this area. See Darin David Barney, "Push-Button Populism: The Reform Party and the Real World of Teledemocracy," *Canadian Journal of Communication* 21, 3 (Summer 1996): 381-413. For commentary on the American experience in these areas, see Graeme Browning, *Electronic Democracy: Using the Internet to Influence American Politics* (Wilton, CT: On-line Press, 1996); Christa Daryl Slaton, *Televote: Expanding Citizen Participation in the Quantum Age* (New York: Praeger, 1992); and F. Christopher Arterton, *Teledemocracy: Can Technology Protect Democracy?* (Newbury Park, CA: Sage, 1987).

59 As quoted in Barney, "Push-Button Populism," 381.

60 Howard Rheingold, *The Virtual Community: Homesteading on the Electronic Frontier* (Reading, MA: Addison-Wesley, 1993), 14. For more on the WELL, see Katie Hafner, "The Epic Saga of the Well," *WIRED* 5, 5 (May 1997): 98-142.

61 For an introductory review of democratic political theory, see David Held, *Models of Democracy*, 2nd ed. (Cambridge, MA: Polity, 1996).

62 Ronald Beiner, "Introduction," in *Democratic Theory and Technological Society*, ed. Richard B. Day, Ronald Beiner, and Joseph Masciulli (Armonk, NY: M.E. Sharpe, 1988), xi. This antagonism is also discussed elsewhere in the same volume. See Ronald Beiner, "Ethics and Technology: Hans Jonas' Theory of Responsibility," 336-54, and H.D. Forbes, "Dahl, Democracy and Technology," 227-50.

Chapter 2: On Technology

1 Martin Heidegger, "Letter on Humanism," in *Martin Heidegger: Basic Writings,* ed. David Farrell Krell (New York: Harper and Row, 1977), 220.
2 Edward Andrew, *Closing the Iron Cage: The Scientific Management of Work and Leisure* (Montreal: Black Rose, 1981), 17.
3 Plato, *Gorgias,* trans. W. Hamilton (Harmondsworth: Penguin, 1960), 450-62.
4 Ibid., 462c.
5 Ibid., 465a.
6 Ibid., 465b.
7 Ibid., 465d.
8 Ibid.
9 Ibid., 465e.
10 Ibid., 465a.
11 Ibid.
12 Ibid.
13 Aristotle, *Nichomachean Ethics,* trans. Martin Ostwald (Indianapolis: Liberal Arts Press, 1962), Book VI (4), 1140a1.
14 Ibid., 1140a11-15.
15 Webster F. Hood, "The Aristotelian versus the Heideggerian Approach to the Problem of Technology," in *Philosophy and Technology: Readings in the Philosophical Problem of Technology,* ed. Carl Mitcham and Robert Mackey (New York: Free Press, 1972), 347-9.
16 Aristotle, *Nichomachean Ethics,* Book I (2), 1094a18-1094b6.
17 Ibid., Book VI (3), 1139b15. Emphasis added.
18 Ibid., Book VI (4), 1140a10.
19 Ibid., Book I (2), 1094a26-7.
20 For descriptions of Marx in these terms, see Alvin H. Hansen, "The Technological Interpretation of History," *Quarterly Journal of Economics* 36 (November 1921): 72-83; Robert Heilbroner, "Do Machines Make History?" *Technology and Culture* 8, 3 (July 1967): 335-45; and William M. Shaw, "'The Handmill Gives You the Feudal Lord': Marx's Technological Determinism," *History and Theory* 18 (1979): 155-76.
21 See Langdon Winner, *Autonomous Technology: Technics-out-of-Control as a Theme in Political Thought* (Cambridge, MA: MIT Press, 1977), 80-5; Donald Mackenzie, "Marx and the Machine," *Technology and Culture* 25, 3 (July 1984): 473-502; and Nathan Rosenberg, "Marx as a Student of Technology," *Monthly Review* 28, 3 (Summer 1976): 56-77.
22 Karl Marx, *The Poverty of Philosophy,* trans. Harry Quelch (Moscow: Progress, 1984), 102.
23 Ibid.
24 Ibid.
25 Karl Marx, "Preface: A Contribution to the Critique of Political Economy," in *The Marx-Engels Reader,* 2nd ed., ed. Robert C. Tucker (New York: Norton, 1978), 4.

26 For the pitfalls of equating the forces of production and technology, see Mackenzie, "Marx and the Machine," 476-8.

27 Karl Marx and Friedrich Engels, *The German Ideology* (Moscow: Progress, 1976), 37.

28 Karl Marx, *Capital*, vol. 1, trans. Samuel Moore and Edward Aveling (Moscow: Progress, 1978), 173.

29 Ibid.

30 Ibid., 352, note 2.

31 Ibid., 351.

32 Karl Marx, "The *Grundrisse*," in *The Marx-Engels Reader*, ed. Tucker, 283.

33 Marx, *Capital*, vol. 1, 353-5.

34 Marx, "The *Grundrisse*," 279-83.

35 Ibid., 280.

36 For Marx's discussion of how machines contributed to profit-making, see Marx, *Capital*, vol. 1, 365-94.

37 Ibid., 394, 363.

38 Marx, *The Poverty of Philosophy*, 130.

39 Marx, *Capital*, vol. 1, 455.

40 Karl Marx and Friedrich Engels, *Manifesto of the Communist Party*, trans. Samuel Moore (Moscow: Progress, 1986), 41.

41 For Marx's general views on alienation, see Karl Marx, "Economic and Philosophic Manuscripts of 1844," in *The Marx-Engels Reader*, ed. Tucker, 70-81.

42 Marx, *Capital*, vol. 1, 399.

43 Ibid., 404.

44 Ibid., 416. Emphasis added.

45 Ibid., 456.

46 For a précis of Marx's theory of historical materialism, see Karl Marx, "Preface: A Contribution to the Critique of Political Economy," in *The Marx-Engels Reader*, ed. Tucker, 4-5; for more on this particular aspect of the theory, see Marx, *Capital*, vol. 3, trans. Samuel Moore and Edward Aveling (Moscow: Progress, 1986), 883-4, and Marx and Engels, *Manifesto of the Communist Party*, 39.

47 For a discussion of Marx's commitment to science, see Edward Andrew, "The Unity of Theory and Practice: The Science of Marx and Nietzsche," in *Political Theory and Praxis: New Perspectives*, ed. Terence Ball (Minneapolis: University of Minnesota Press, 1977), 118-34.

48 Marx and Engels, *The German Ideology*, 54. For a critical discussion of Marx's enthusiasm for technologically induced plenty, see G.A. Cohen, "Self-Ownership, Communism and Equality: Against the Marxist Technological Fix," in *Technology in the Western Political Tradition*, ed. Arthur M. Melzer, Jerry Weinberger, and M. Richard Zinman (Ithaca, NY: Cornell University Press, 1993), 131-61.

49 Marx and Engels, *Manifesto of the Communist Party*, 53.

50 Martin Heidegger, "Introduction: The Exposition of the Question of the Meaning of Being," trans. Joan Stambaugh, in *Basic Writings*, ed. David Farrell Krell (New York: Harper and Row, 1977), 56.

51 Martin Heidegger, *Being and Time*, trans. John Macquarrie and Edward Robinson (New York: Harper and Row, 1962). Originally published as *Sein und Zeit* (1927).

52 Martin Heidegger, "The Question Concerning Technology," trans. William Lovitt, in *Basic Writings*, ed. Krell, 287-317.

53 Ibid., 287.

54 Martin Heidegger, "Only a God Can Save Us Now: An Interview with Martin Heidegger," trans. David Schendler, *Graduate Faculty Philosophy Journal* 6, 1 (1977): 17.

55 Heidegger, "The Question Concerning Technology," 289-91.

56 Ibid., 294.

57 Ibid., 301.

58 Martin Heidegger, *Discourse on Thinking*, trans. John M. Anderson and E. Hans Freund (New York: Harper and Row, 1959), 49, 53.

59 Heidegger, "Only a God Can Save Us Now," 18.

60 Heidegger, "The Question Concerning Technology," 296.

61 Ibid., 298.

62 Heidegger, *Discourse on Thinking*, 50.

63 Ibid., 47.

64 Ibid., 45-8.

65 Heidegger, "The Question Concerning Technology," 309.

66 Heidegger, *Discourse on Thinking*, 56.

67 Heidegger, "Only a God Can Save Us Now," 17.

68 Heidegger, *Discourse on Thinking*, 56.

69 Heidegger, "Only a God Can Save Us Now," 16-7.

70 Heidegger, *Discourse on Thinking*, 51. Heidegger presents a similar account elsewhere, and notes "the revealing never simply comes to an end"; see Heidegger, "The Question Concerning Technology," 298.

71 Heidegger, "Only a God Can Save Us Now," 18.

72 Heidegger, *Discourse on Thinking*, 53. For Heidegger's comments on the Rhine, see "The Question Concerning Technology," 297.

73 Heidegger, "The Question Concerning Technology," 299.

74 Heidegger, *Discourse on Thinking*, 54.

75 Heidegger, "The Question Concerning Technology," 310.

76 Ibid.

77 Martin Heidegger, "The Turning," in *The Question Concerning Technology and Other Essays*, trans. William Lovitt (New York: Harper and Row, 1977), 48.

78 Hubert L. Dreyfus, "Heidegger on Gaining a Free Relation to Technology," in *Technology and the Politics of Knowledge*, ed. Andrew Feenberg and Alastair Hannay (Bloomington: University of Indiana Press, 1995), 99.

79 Heidegger, "The Question Concerning Technology," 314.

80 Heidegger, "Only a God Can Save Us Now," 22.

81 Heidegger, *Discourse on Thinking*, 54.

82 See ibid., translator's note 4.

83 Ibid.

84 Ibid., 55.

85 Ibid.

86 Ibid.

87 Heidegger, "The Question Concerning Technology," 316.

88 See William Christian, *George Grant: A Biography* (Toronto: University of Toronto Press, 1993), 364.

89 George Grant, *Philosophy in the Mass Age* (Toronto: Copp Clark, 1959), vii.

90 Ibid., III.

91 Ibid., 22.

92 Ibid., 44-8.

93 Ibid., 52.

94 David Cayley, ed., *George Grant: In Conversation* (Toronto: Anansi, 1995), 133. See also George Grant, *Technology and Justice* (Toronto: Anansi, 1986), 11-4.

95 Grant, *Technology and Justice*, 13.

96 Cayley, ed., *George Grant: In Conversation*, 133. Emphasis added.

97 George Grant (with Gad Horowitz), "A Conversation on Technology and Man," *Journal of Canadian Studies* 4, 3 (August 1969): 3.

98 Grant, *Technology and Justice*, 36.

99 George Grant, *Technology and Empire: Perspectives on North America* (Toronto: Anansi, 1969), 15.

100 Grant (with Horowitz), "A Conversation on Technology and Man," 3.

101 Grant, *Technology and Empire*, 114, note 3.

102 George Grant, *Lament for a Nation: The Defeat of Canadian Nationalism* (Toronto, McClelland and Stewart, 1965), 56.

103 For a thorough critique of claims about liberal neutrality, see Ronald Beiner, *What's the Matter with Liberalism?* (Berkeley: University of California Press, 1992).

104 Grant, *Lament for a Nation*, 85.

105 Grant, *Technology and Empire*, 26. See also *Lament for a Nation*, 57.

106 Ibid., 138.

107 Ibid., 66.

108 Grant, *Technology and Justice*, 16, 32-4.

109 Grant, *Technology and Empire*, 33.

110 Ibid., 128.

111 Ibid.

112 Ibid., 58.

113 Ibid., 67.

114 Grant, *Lament for a Nation*, 53-4.

115 Grant maintains that a recognition of the virtue of diverse qualities must always be subordinate to the equal regard for persons predicated on "the absolute worth of all men": see George Grant, "An Ethic for Community," in *Social Purpose for Canada*, ed. Michael Oliver (Toronto: University of Toronto Press, 1961), 23-4. On the matter of loving our own, see Grant, *Technology and Empire*, 77-8.

116 See, for example, Grant, *Philosophy in the Mass Age*, 10-1; and *Technology and Empire*, 137. Edward Andrew has referred to this as Grant's residual faith in the "irreducible good of the modern technological project." Edward Andrew, "George Grant on Technological Imperatives," in *Democratic Theory and Technological Society*, ed. R. Day, J. Masciulli, and R. Beiner (New York: M.E. Sharpe, 1988), 301.

117 Cayley, ed., *George Grant: In Conversation*, 142.

118 Langdon Winner, "Citizen Virtues in a Technological Order," in *Technology and the Politics of Knowledge*, ed. Andrew Feenberg and Alastair Hannay (Bloomington: University of Indiana Press, 1995), 67.

Chapter 3: Networks

1 United States, Federal Communications Commission; American Association for the Advancement of Science. Figures cited in "Harper's Index," *Harper's* 294, 1764 (May 1997), 15.

2 For current statistics on Internet penetration, see Matrix Information and Directory Services, "The State of the Internet," *Matrix Maps Quarterly* 401 (January 1997): 1-6 (www.mids.org), and the Internet Society (www.info.isoc.org).

3 See Peter Golding, "World Wide Wedge: Division and Contradiction in the Global Information Infrastructure," *Monthly Review* 48, 3 (July-August 1996): 70-85. Golding (83) reports that, in 1994, there were 0.002 Internet users per 1,000 inhabitants in India (compared with 48.9 in Sweden, for example).

4 For a good account of the numerous inventions that contributed to the development of the modern computer, see Stan Augarten, *BIT by BIT: An Illustrated History of Computers* (New York: Ticknor and Fields, 1984).

5 See Stewart Brand, *The Media Lab: Inventing the Future at MIT* (New York: Penguin, 1988), 152. See also Tom Forester, *High Tech Society* (Cambridge, MA: MIT Press, 1988), 17. See also Alan Turing, "On Computable Numbers; with an Application to the *Entsheidungsproblem*," *Proceedings of the London Mathematical Society*, 2nd ser., 42 (1936): 230-65, and Alan Turing, "Computer Machinery and Intelligence," *Mind* 59 (1950): 433-60. For more on Turing, see J. David Bolter, *Turing's Man: Western Culture in the Computer Age* (Chapel Hill: University of North Carolina Press, 1984).

6 Augarten, *BIT by BIT*, 146.

7   Ibid., 124-5. For the probably apocryphal tale of Philadelphia's unenlightenment, see Forester, *High Tech Society*, 17.

8   Augarten, *BIT by BIT*, 128.

9   John von Neumann, "First Draft of a Report on the EDVAC," in *From ENIAC to UNIVAC: An Appraisal of the Eckert-Mauchly Computers*, ed. Nancy Stern (Bedford, MA: Digital Press, 1981), 177-246.

10   Augarten, *BIT by BIT*, 148-50.

11   Ibid., 155.

12   Ibid., 164.

13   John Halton, "The Anatomy of Computing," in *The Information Technology Revolution*, ed. Tom Forester (Cambridge, MA: MIT Press, 1986), 3-26.

14   Ibid., 5-6.

15   Ibid., 9.

16   Augarten, *BIT by BIT*, 134-5.

17   For the complete hexadecimal system of binary notation, see Halton, "The Anatomy of Computing," 6.

18   Ibid., 10.

19   Augarten, *BIT by BIT*, 140.

20   Ibid., 89-92.

21   Ibid., 90.

22   Ibid., 162.

23   Gene Bylinsky, "Here Comes the Second Computer Revolution," in *The Microelectronics Revolution*, ed. Tom Forester (Oxford: Basil Blackwell, 1980), 3-15. For a helpful illustration of the planar process for creating chips, see Augarten, *BIT by BIT*, 244.

24   Clive Thompson, "Beyond the Microchip," *Report on Business Magazine*, June 1997, 78-86.

25   Most notebook computers sold at the time of writing included a standard sixteen megabytes of RAM and a one-gigabyte hard disk: a byte contains eight bits; a megabyte contains 1 million bytes; a gigabyte contains 1 billion bytes.

26   Mary Gooderham, "Microchip Turning into a Cache Cow," *Globe and Mail*, 18 November 1996, A8, and "Microchip at 25: More Power to It," *Globe and Mail*, 15 November 1996, A10.

27   See "From Toasters to Cars, Chips Part of Daily Life," *Globe and Mail*, 15 November 1996, A10.

28   For an account of the beginnings of the PC, see Paul Freiberger and Michael Swaine, *Fire in the Valley: The Making of the Personal Computer* (Berkeley, CA: McGraw-Hill, 1984).

29   There are many accounts of the history of computer-network technology. The summary that follows here draws on the two I have found most helpful and thorough: Katie Hafner and Matthew Lyon, *Where Wizards Stay*

Up Late: The Origins of the Internet (New York: Simon and Schuster, 1996), and Peter H. Salus, *Casting the Net: From ARPANET to INTERNET and Beyond* (Reading, MA: Addison-Wesley, 1995). Others include Nicholas Baran, *Inside the Information Superhighway Revolution* (Scottsdale, AZ: Coriolis, 1995); Gerard J. Holzmann and Bjørn Pehrson, *The Early History of Data Networks* (Los Alamitos, CA: IEEE Press, 1995); and Irwin Lebow, *Information Highways and Byways: From the Telegraph to the 21st Century* (New York: IEEE Press, 1995).

30 See "Utah Blasts Cut 3 Phone Relays," *New York Times*, 29 May 1961, 1; "F.B.I. Questions 2 in Tower Blasts," *New York Times*, 20 June 1961, 10; and "Jailed in Tower Blasts," *New York Times*, 3 November 1961, 24.

31 Baran authored a number of studies in this area, the most accessible of which is "On Distributed Communications Networks," *IEEE Transactions on Communications Systems* CS-12, 1 (1 March 1964): 1-9. Baran's broader work on distributed networks is collected at RAND's Web site under the title "On Distributed Communications," Memorandum RM3420PR. See www.rand.org/publications/RM/RM3420.

32 Hafner and Lyon, *Where Wizards Stay Up Late*, 66.

33 Figure 3.1 is adapted from a similar diagram in Paul Baran, "On Distributed Communications – Chapter 1: Introduction," Memorandum RM3420PR.

34 For a concise discussion of switching techniques, see Salus, *Casting the Net*, 6-10.

35 Ibid., 41-59; see also ibid., 19-34.

36 For a brief history of IMP design, see ibid., 35-8.

37 Hafner and Lyon, *Where Wizards Stay Up Late*, 145.

38 See Salus, *Casting the Net*, 51-67.

39 For a discussion of the development of the TIP, see Hafner and Lyon, *Where Wizards Stay Up Late*, 171-3. See also Salus, *Casting the Net*, 74.

40 For a concise history of electronic mail, see Salus, *Casting the Net*, 95-8, and Hafner and Lyon, *Where Wizards Stay Up Late*, 187-218.

41 Hafner and Lyon, *Where Wizards Stay Up Late*, 194.

42 As quoted, ibid., 214.

43 For a synopsis of early networking activity in Europe and Japan, see Salus, *Casting the Net*, 85-91.

44 Hafner and Lyon, *Where Wizards Stay Up Late*, 227. For more on the development of the TCP/IP protocol, see Salus *Casting the Net*, 110-14.

45 See Jim Carroll and Rick Broadhead, *The Canadian Internet Handbook*, 1995 edition (Scarborough, ON: Prentice-Hall, 1995), 43-8, 273-5.

46 Statistics Canada, *Household Facilities and Equipment 1996* (64-202-xpd), 51.

47 Diane Marleau, "Stop the Information Revolution, They Want to Get On," *Globe and Mail*, 19 June 1997, A19.

48 Network Wizards, "Internet Domain Survey, January 1999," available online at www.nw.com.

49 Matrix Information and Directory Services, "The State of the Internet," *Matrix Maps Quarterly* 401 (January 1997): 1-6 (www.mids.org).

50 Ibid.

51 Lists of BITNET lists are available via e-mail from listserv@bitnic.cren.net. Lists of Internet lists can be acquired from mail-server@rtfm.mit.edu. These, of course, account only for *public* lists on *two* networks.

52 For a more detailed discussion of the technicalities of mailing lists, see Carroll and Broadhead, *The Canadian Internet Handbook*, 1995 edition, 119-28.

53 See Katie Hafner, "The Epic Saga of the Well," *WIRED* 5, 5 (May 1997): 98-142.

54 See Carroll and Broadhead, *The Canadian Internet Handbook*, 1995 edition, 131-3.

55 Ibid., 132.

56 The Internet Society (www.info.isoc.org).

57 The site was located at cbc.sympatico.ca/index.html.

58 See Theodor Nelson, *Literary Machines* (South Bend, IN: Distributor's, 1987), and *Dream Machines* (Self-published, 1974). For a discussion of Nelson's quixotic vision of an infinite, universal, hypertext library called "Xanadu," see Gary Wolf, "The Curse of Xanadu," *WIRED* 3, 6 (June 1995): 137-52.

59 Joseph Masciulli, "Rousseau versus Instant Government: Democratic Participation in the Age of Telepolitics," in *Democratic Theory and Technological Society*, ed. R.B. Day, Ronald Beiner, and J. Masciulli (London: M.E. Sharpe, 1988), 150.

60 Kevin Kelly, *Out of Control: The Rise of Neo-Biological Civilization* (Reading, MA: Addison Wesley, 1994).

61 See Norbert Wiener, *Cybernetics: Or Control and Communications in the Animal and Machine* (Cambridge, MA: MIT Press, 1948); and Norbert Wiener, *The Human Use of Human Beings: Cybernetics and Society* (Boston: Houghton Mifflin, 1950).

62 James R. Beniger, *The Control Revolution: Technological and Economic Origins of the Information Society* (Cambridge, MA: Harvard University Press, 1986), 7.

63 Ibid., 8.

64 Ibid., 7.

65 Ibid., 8-9.

66 Albert Gore, speaking at the International Telecommunications Union Meeting, Buenos Aires, Argentina, 21 March 1994. As quoted in Ching-Chih Chen, *Planning Global Information Infrastructure* (Norwood, NJ: Ablex, 1995), vii.

67 "WARNING! Community at Play on the Infoway," Stonehaven West press release, 21 September 1995.

68 Northern Telecom, "Vista 350: It's More than a Phone," advertising brochure, September 1995.

69 Ibid.

Chapter 4: The Political Economy of Network Technology 1

1 Derrick de Kerckhove on Marx is quoted in Kevin Kelly, "What Would McLuhan Say?" *WIRED* 4, 10 (October 1996): 148. I doubt very much this is what McLuhan would say.

2 Ken Hirschkop, "Democracy and the New Technologies," *Monthly Review* 48, 3 (July-August 1996): 93.

3 See, for example, Katie Hafner and Matthew Lyon, *Where Wizards Stay Up Late: The Origins of the Internet* (New York: Simon and Schuster, 1996); or Stewart Brand, *The Media Lab: Inventing the Future at MIT* (New York: Penguin, 1988).

4 See Howard Rheingold, *The Virtual Community: Homesteading on the Electronic Frontier* (Reading, MA: Addison-Wesley, 1993).

5 Robert W. McChesney, "Public Broadcasting in the Age of Communication Revolution," *Monthly Review* 47, 7 (December 1995): 19.

6 Daniel Burstein and David Kline, *Road Warriors: Dreams and Nightmares along the Information Superhighway* (New York: Dutton, 1995), 262. In 1995, John Malone's Tele-Communications International (TCI) owned 1,200 cable delivery systems in the United States. Ray Smith was CEO of Bell Atlantic.

7 Ibid., 275-80.

8 See William Leiss, "The Myth of the Information Society," in *Cultural Politics in Contemporary America,* ed. Ian Angus and Sut Jhally (New York: Routledge, 1989), 282-98; David Lyon, *The Information Society: Issues and Illusions* (Cambridge, MA: Polity, 1988); Kevin Robins and Frank Webster, "Cybernetic Capitalism: Information, Technology, Everyday Life," in *The Political Economy of Information,* ed. Vincent Mosco and Janet Wasko (Madison: University of Wisconsin Press, 1988), 44-75; Herbert Schiller, "Information for What Kind of Society?" in *The Information Society: Economic, Social and Structural Issues,* ed. Jerry L. Salvaggio (Hillsdale, NJ: Lawrence Erlbaum, 1989); and Michael Traber, *The Myth of the Information Revolution: Social and Ethical Implications of Communication Technology* (London: Sage, 1986).

9 Burstein and Kline, *Road Warriors,* 32.

10 For concurring remarks, see Gaëtan Tremblay, "The Information Society: From Fordism to Gatesism," *Canadian Journal of Communication* 20, 4 (1995): 473.

11 Ellen Meiksins Wood, "Modernity, Postmodernity or Capitalism?" *Monthly Review* 48, 3 (July-August 1996): 35.

12 See Peter Golding, "World Wide Wedge: Division and Contradiction in the Global Information Infrastructure," *Monthly Review* 48, 3 (July-August 1996): 81.

13 Wood, "Modernity, Postmodernity or Capitalism?" 38.

14 See Karl Marx and Friedrich Engels, *Manifesto of the Communist Party*, trans. Samuel Moore (Moscow: Progress, 1986), 35, note "a."

15 See, for example, Mark Poster, *Foucault, Marxism and History: Mode of Production versus Mode of Information* (Cambridge, MA: Polity, 1984).

16 Executive Office of the President of the United States, *The National Information Infrastructure: Agenda for Action* (Washington, DC: Executive Office of the President, 1993), reprinted in *The Information Revolution*, ed. Donald Altschiller (New York: H.W. Wilson, 1995), 14. This document is also available at nii.nist.gov.

17 David Johnston, Deborah Johnston, and Sunny Handa, *Getting Canada Online: Understanding the Information Highway* (Toronto: Stoddart, 1995), 240-1. For more on this initiative, see Leslie Regan Shade, "Computer Networking in Canada: From CA*net to CANARIE," *Canadian Journal of Communication* 19, 1 (Winter 1994): 53-9.

18 John Macdonald, as quoted in "Manley Announces Plan to Build World's First National Optical Internet High Speed Network to Revolutionize Electronic Data Information Exchange," Industry Canada press release, Ottawa, 25 August 1998. The 1998 Budget is available on-line at www.fin.gc.ca.

19 Heather Menzies, *Whose Brave New World? The Information Highway and the New Economy* (Toronto: Between the Lines, 1996), 52-3.

20 Robert W. McChesney, "The Global Struggle for Democratic Communication," *Monthly Review* 48, 3 (July-August 1996): 5.

21 Conner Middlemann, "Lines of Investors Buzz with Avalanche of Issues," *Financial Times*, 12 June 1995.

22 For a concise summary of the privatization of the Internet, see Nicholas Baran, "Privatization of Telecommunications," *Monthly Review* 48, 3 (July-August 1996): 59-69.

23 Ibid., 62.

24 See McChesney, "The Global Struggle for Democratic Communication," 11; and Herbert Schiller, *Culture Inc.: The Corporate Takeover of Public Expression* (New York: Oxford, 1989).

25 Robert W. McChesney, "The Internet and U.S. Communication Policy-Making in Historical and Critical Perspective," *Journal of Communication* 46, 1 (Winter 1996): 103-4.

26 McChesney, "The Global Struggle for Democratic Communication," 12.

27 McChesney, "The Internet and U.S. Communication Policy-Making in Historical and Critical Perspective," 105.

28 Menzies, *Whose Brave New World?* 54.

29 Dwayne Winseck, "Power Shift? Towards a Political Economy of Canadian Telecommunications and Regulation," *Canadian Journal of Communication* 20, 1 (Spring 1995): 95.

30 Information Highway Advisory Council, *Connection, Community, Content: The Challenge of the Information Highway*, Final Report (Ottawa: Minister of Supply and Services, 1995), 93, recs. 1.2 and 1.3.

31 Ibid., 96, rec. 2.1.

32 Canada, *Building the Information Society: Moving Canada into the 21st Century* (Ottawa: Minister of Supply and Services, 1996), 5-6.

33 Information Highway Advisory Council, *Preparing Canada for a Digital World*, Final Report (Ottawa: Minister of Supply and Services, 1997), 10-1.

34 Winseck, "Power Shift?" 95. Winseck's figures are drawn from Statistics Canada's 1990 report *Telephone Statistics*.

35 Mary Vipond, *The Mass Media in Canada* (Toronto: James Lorimer, 1989), 70. See also Dallas W. Smythe, *Dependency Road: Communications, Capitalism, Consciousness and Canada* (Norwood, NJ: Ablex, 1981). For documentation and analysis of the substantial concentration of media ownership in the United States, see Ben Bagdikian, *The Media Monopoly*, 5th ed. (Boston: Beacon, 1997).

36 Golding, "World Wide Wedge," 74.

37 Ibid. For Canadian evidence of this dynamic, see Winseck, "Power Shift?" 95.

38 Quoted in Doug Halonen, "Malone: Few Will Rule Superhighway," *Electronic Media*, 31 January 1994, 31.

39 See Stephanie Mehta, "AT&T Faces Hurdles in Integrating TCI," *Globe and Mail*, 25 June 1998, B13.

40 There are a number of published accounts of the rise of Microsoft. A selective list includes: James Wallace, *Overdrive: Bill Gates and the Race to Control Cyberspace* (New York: John Wiley, 1997); Randall E. Stross, *The Microsoft Way: The Real Story of How the Company Outsmarts Its Competition* (Reading, MA: Addison-Wesley, 1996); Stephen Manes and Paul Andrews, *Gates: How Microsoft's Mogul Reinvented an Industry – And Made Himself the Richest Man in America* (New York: Doubleday, 1993); and James Wallace and Jim Erickson, *Hard Drive: Bill Gates and the Making of the Microsoft Empire* (New York: John Wiley, 1992). Except where otherwise noted, the information on Microsoft that follows is drawn from an excellent report commissioned by the network advocacy group NetAction: Nathan Newman, "From Microsoft Word to Microsoft World: How Microsoft Is Building a Global Monopoly," NetAction White Paper, 1997. Available on-line at www.netaction.org/msoft/world.

41 Newman, "From Microsoft Word to Microsoft World," 1.

42 Ibid., 2.

43 In May 1998, the US Department of Justice and the Attorneys-General of twenty US states filed formal antitrust charges against Microsoft. In November 1999, the court's findings-of-fact in the trial asserted that Microsoft

indeed used its monopoly position to harm competitors and thwart innovation. A final ruling in the case is expected in 2000.

44 See John R. Wilke and Bryan Gruley, "In Merger Blitz, Regulators' Profiles Rise," *Globe and Mail*, 11 June 1998, B14.

45 McChesney, "The Internet and U.S. Communication Policy-Making in Historical and Critical Perspective," 103.

46 Jack Kapica, "Choosing Less Government or Less Gates," *Globe and Mail*, 23 May 1997, A7.

47 Bill Gates, *The Road Ahead* (New York: Viking, 1995), 157.

48 Ibid., 183.

49 Group of Lisbon, *Limits to Competition* (Cambridge, MA: MIT Press, 1995), 23, 32.

50 Ibid., 37.

51 Ibid., 38.

52 Ibid., 39-44. These authors cite figures showing that unemployment in OECD countries nearly doubled between 1973 and 1991, reaching nearly 7 percent, or more than 30 million people out of work, with roughly half of these deemed "long-term unemployed." These authors suggest that, with the exception of Southeast Asia, the situation in the developing world is even worse.

53 Gates, *The Road Ahead*, 157-8.

54 Wood, "Modernity, Postmodernity or Capitalism?" 37.

55 Ibid.

56 Walter B. Wriston, *The Twilight of Sovereignty: How the Information Revolution Is Transforming Our World* (New York: Macmillan, 1992), 9. For a profile on Wriston, a pioneer in the field of automated banking, see Thomas A. Bass, "The Future of Money," *WIRED* 4, 10 (October 1996): 140-205.

57 Mark Potter and Marc Lee, "The Economic Impacts of the Information Highway: An Overview," discussion paper prepared for the Task Force on Growth, Employment and Competitiveness, Information Highway Advisory Council, Industry Canada, July 1995, 9.

58 Automation is a well-studied phenomenon. For a good start on this literature, see Harry Braverman, *Labour and Monopoly Capital: The Degradation of Work in the Twentieth Century* (New York: Monthly Review, 1974), and David F. Noble, *Forces of Production: A Social History of Industrial Automation* (New York: Knopf, 1984).

59 Tremblay, "The Information Society."

60 See B. Joseph Pine II, *Mass Customization: The New Frontier in Business Competition* (Cambridge, MA: Harvard Business School, 1993), 34.

61 Michael J. Piore and Charles F. Sabel, *The Second Industrial Divide: Possibilities for Prosperity* (New York: Basic Books, 1984), 17.

62 William H. Davidow and Michael S. Malone, *The Virtual Corporation: Structuring and Revitalizing the Corporation for the 21st Century* (New York: HarperCollins, 1993), 136-7.

63 Steven L. Goldman, Roger N. Nagel, and Kenneth Preiss, *Agile Competitors and Virtual Organizations: Strategies for Enriching the Customer* (New York: Van Nostrand Reinhold, 1995), 97.

64 Ibid., 330.

65 See Yasuhiro Monden, *Toyota Production System* (Cambridge, MA: Industrial Engineering and Management Press, 1983).

66 See James Womack, Daniel T. Jones, and Daniel Roos, *The Machine That Changed the World: The Story of Lean Production* (New York: Macmillan, 1990).

67 See Davidow and Malone, *The Virtual Corporation*, 114-9.

68 Goldman, Nagel, and Preiss, *Agile Competitors and Virtual Organizations*, 4. For an analysis of agile manufacturing in the Canadian auto industry, see Bruce Roberts, "From Lean Production to Agile Manufacturing: A New Round of Quicker, Cheaper and Better," in *Re-shaping Work: Union Responses to Technological Change*, ed. Christopher Schenk and John Anderson (Don Mills, ON: Ontario Federation of Labour, 1995), 197-215.

69 Goldman, Nagel, and Preiss, *Agile Competitors and Virtual Organizations*, 7, 27.

70 See Pine, *Mass Customization*, 34-44, for a survey of ten industry categories, ranging from automobile production to personal care to foodstuffs, in which the shift to this production model is occurring.

71 Goldman, Ngel, and Preiss, *Agile Competitors and Virtual Organizations*, 133-81.

72 Ibid., 145.

73 Ibid., 135-6.

74 Ibid., 147.

75 Ibid., 169.

76 Ibid., 149.

77 Ibid., 28.

78 Ibid., 27.

79 Strategis is Canada's largest business Web site, and contains 500,000 pages of searchable text relating to markets, trade, investment opportunities, industrial and business services, and micro-economic research. See strategis.ic.gc.ca. TYMNET is the world's largest international producers network; SWIFT is the network of the Society for Worldwide Interbank Financial Telecommunications; SITA is the Société Internationale des Telecommunications Aéronautique, which serves 300 airlines in 170 countries.

80 John Baldwin and David Sabourin, *Technology Adoption in Canadian Manufacturing* (Ottawa: Minister of Industry, Science and Technology, 1995), Statistics Canada Catalogue No. 88-512. All figures in this paragraph are taken from this report. See 7-8 for highlights.

81 Ibid., 61.

82 Goldman, Nagel, and Preiss, *Agile Competitors and Virtual Organizations*, 183.

83 Ibid., 95.

Chapter 5: The Political Economy of Network Technology 2

1 Karl Marx and Friedrich Engels, *Manifesto of the Communist Party*, trans. Samuel Moore (Moscow: Progress, 1986), 37.

2 Karl Marx, *Capital*, vol. 1, trans. Samuel Moore and Edward Aveling (Moscow: Progress, 1978), 455.

3 David F. Noble, *Progress without People: New Technology, Unemployment, and the Message of Resistance* (Toronto: Between the Lines, 1995), xi.

4 Bill Gates, *The Road Ahead* (New York: Viking, 1995), 254.

5 Marx and Engels, *Manifesto of the Communist Party*, 37.

6 Stanley Aronowitz and William DiFazio, *The Jobless Future: Sci-tech and the Dogma of Work* (Minneapolis: University of Minnesota Press, 1994), 50.

7 Ibid., 6, 3. Emphasis added.

8 Jean-Claude Parrot, "Minority Report," *Connection, Community, Content: The Challenge of the Information Highway*, Final Report of the Information Highway Advisory Council (Ottawa: Minister of Supply and Services Canada, 1995), 218.

9 Mark Potter and Marc Lee, "The Economic Impacts of the Information Highway: An Overview," discussion paper prepared for the Task Force on Growth, Employment and Competitiveness, Information Highway Advisory Council, Industry Canada, July 1995, 15.

10 Ibid.

11 For popular treatments at both of these poles, see Jeremy Rifkin, *The End of Work: The Decline of the Global Labour Force and the Dawn of the Post-Market Era* (New York: Putnam, 1995), and William Bridges, *JobShift: How to Prosper in a Workplace without Jobs* (Reading, MA: Addison-Wesley, 1994).

12 See Noble, *Progress without People*, xii. See also Heather Menzies, *Whose Brave New World? The Information Highway and the New Economy* (Toronto: Between the Lines, 1996), 5.

13 Potter and Lee, "The Economic Impacts of the Information Highway," 15.

14 For a survey of examples in these areas, see Menzies, *Whose Brave New World?* 89-102. See also the case studies in Christopher Schenk and John Anderson, eds., *Re-shaping Work: Union Responses to Technological Change* (Don Mills, ON: Ontario Federation of Labour, 1995).

15 See David Robertson and Jeff Wareham, *Technological Change in the Canadian Auto Industry* (Toronto: Canadian Auto Workers, 1988), cited in Menzies, *Whose Brave New World?* 97.

16 John Baldwin and David Sabourin, *Technology Adoption in Canadian Manufacturing* (Ottawa: Minister of Industry, Science and Technology, 1995), Statistics Canada Catalogue No. 88-512, 8 and Table 16.

17 Bruce Roberts, "From Lean Production to Agile Manufacturing: A New Round of Quicker, Cheaper and Better," in *Re-shaping Work*, ed. Schenk and Anderson, 204.

18 David Sobel, "From Grunt Work to No Work: The Impact of Technological Change on the Building Trades," in *Re-shaping Work*, ed. Schenk and Anderson, 63-4.

19 Ibid., 66.

20 See Manuel Castells and Yuko Aoyama, "Paths towards the Informational Society: Employment Structure in G-7 Countries, 1920-90," *International Labour Review* 133, 1 (1994): 5-33, and Barry Cooper, *Sleepers Awake: Technology and the Future of Work* (Melbourne: Oxford University Press Australia, 1995), 74.

21 Ernest Akyeampong and Jennifer Winters, "International Employment Trends by Industry – a Note," *Perspectives on Labour and Income*, Statistics Canada, Summer 1993, 33-7.

22 Potter and Lee, "The Economic Impacts of the Information Highway," 18.

23 Menzies, *Whose Brave New World?* 6.

24 See John Anderson and Christopher Schenk, "Technology on Trial – Lessons from the Labour Frontlines," in *Re-shaping Work*, ed. Schenk and Anderson, 12.

25 See Group of Lisbon, *Limits to Competition* (Cambridge, MA: MIT Press, 1995), 42.

26 For an excellent discussion of the service economy in Canada, see John Myles, "Post-Industrialism and the Service Economy," in *The New Era of Global Competition: State Policy and Market Power*, ed. Daniel Drache and Meric S. Gertler (Montreal and Kingston: McGill-Queen's University Press, 1991), 350-66.

27 See Andrew Tausz, "Store Tests Do-It-Yourself Checkout," *Globe and Mail*, 9 December 1997, C4.

28 Group of Lisbon, *Limits to Competition*, 42.

29 See Heather Menzies, "Telework, Shadow Work: The Privatization of Work in the New Digital Economy," *Studies in Political Economy* 53 (Summer 1997): 109.

30 Aronowitz and DiFazio, *The Jobless Future*, 5.

31 Potter and Lee, "The Economic Impacts of the Information Highway," 18. Emphasis added.

32 Aronowitz and DiFazio, *The Jobless Future*, 9.

33 Andrew Bibby, "Trade Unions and Telework," report prepared for the International Trade Union Federation, Autumn 1996. The examples that follow are all taken from this report.

34 Ibid.

35 Ibid.

36 See Menzies, "Telework, Shadow Work," 111-2.

37 See Victoria Cross, "Off Our Backs: Frank McKenna's High-Tech Revolution in New Brunswick," *Our Times* 14, 2 (May-June 1995): 20-3. See also Menzies, *Whose Brave New World?* 114.

38 Statistics Canada reports that, in 1995, 48 percent of Canadian workers used computers. The percentage exceeds 50 in Canada's three most economically prosperous provinces: Ontario, Alberta, and British Columbia. See Alanna Mitchell, "Computers Taking Root in the Workplace," *Globe and Mail*, 7 June 1995, A8.

39 Menzies, *Whose Brave New World?* 34. See also Peter Meiksins, "Work, New Technology and Capitalism," *Monthly Review* 48, 3 (July-August 1996): 100.

40 Grant Schellenberg, *The Changing Nature of Part-Time Work* (Ottawa: Canadian Council on Social Development, 1997). Statistical backgrounder available at www.ccsd.ca/bge-pt.htm.

41 Ibid. This compares to two-thirds of all full-time employees in Canada who are covered by such plans.

42 Ibid.

43 Ibid.

44 Canadian Union of Public Employees, *Computer Related Change in the Workplace* (Ottawa: CUPE, 1985).

45 Aronowitz and DiFazio, *The Jobless Future*, 16.

46 Bibby, "Trade Unions and Telework."

47 Statistics Canada, *General Social Survey and General Household Equipment Survey* (Ottawa: Statistics Canada, 1995).

48 Cited in Catherine Mulroney, "Canadians Tiptoe into Telework Era," *Globe and Mail*, 10 June 1997, C1.

49 Ibid.

50 See Parrot, "Minority Report," 221-7.

51 See Menzies, *Whose Brave New World?* 77, 116.

52 Public Service Alliance of Canada, "Telework," PSAC Policy no. 33, 1. See also Theresa Johnson, *Go Home ... and Stay There? A PSAC Response to Telework in the Federal Public Sector* (Ottawa: Public Service Alliance of Canada, 1993).

53 Bibby, "Trade Unions and Telework."

54 See Public Service Alliance of Canada, "Telework," 2, and Parrot, "Minority Report," 221-2.

55 Jan Borowy and Theresa Johnson, "Unions Confront Work Reorganization and the Rise of Precarious Employment: Home-Based Work in the Garment Industry and the Federal Public Service," in *Re-shaping Work*, ed. Schenk and Anderson, 31.

56 See ibid., 32, for examples in the garment industry and public service. See also Menzies, "Telework, Shadow Work," 115.

57 Menzies, "Telework, Shadow Work," 116.

58 See Margaret Oldfield, "Heaven or Hell? Telework and Self-Employment," *Our Times* 14, 2 (May-June 1995): 16-9.

59 See Borowy and Johnson, "Unions Confront Work Reorganization and the Rise of Precarious Employment," 41-3.

60 Carol van Helvoort, as quoted in Oldfield, "Heaven or Hell?" 16.

61 Mary Gooderham, "Technology-Race Casualties Find They've Nowhere to Go," *Globe and Mail,* 11 October 1995, A10. In this article, the owner of a rival operation brags that his company plans to eliminate the labour involved in processing orders altogether by allowing customers to order pizzas via their home computers connected to the Internet.

62 Linda Duxbury, *Study of Public Sector Workers in Ottawa Region,* cited in Borowy and Johnson, "Unions Confront Work Reorganization and the Rise of Precarious Employment," 36.

63 Public Service Alliance of Canada, "Telework," 1.

64 Borowy and Johnson, "Unions Confront Work Reorganization and the Rise of Precarious Employment," 39.

65 Public Service Alliance of Canada, "Telework," 1.

66 Gene I. Rochlin, *Trapped in the Net: The Unanticipated Consequences of Computerization* (Princeton, NJ: Princeton University Press, 1997), 8.

67 Davidow and Malone, *The Virtual Corporation,* 167.

68 Rochlin, *Trapped in the Net,* 9.

69 Aronowitz and DiFazio, *The Jobless Future,* 26-34. The works of Braverman and Thompson are classics in this field: Harry Braverman, *Labor and Monopoly Capital* (New York: Monthly Review, 1974), and E.P. Thompson, *The Making of the English Working Class* (New York: Knopf, 1963).

70 See Frederick Winslow Taylor, *Principles of Scientific Management* (New York: Norton, 1967).

71 Edward Andrew, *Closing the Iron Cage: The Scientific Management of Work and Leisure* (Montreal: Black Rose, 1981), 90.

72 Aronowitz and DiFazio, *The Jobless Future,* 27.

73 See Michael J. Piore and Charles F. Sabel, *The Second Industrial Divide: Possibilities for Prosperity* (New York: Basic Books, 1984); Paul Adler, "Automation, Skill and the Future of Capitalism," *Berkeley Journal of Sociology: A Critical Review* 33 (1988): 1-36; and Paul Adler, "Technology and Us," *Socialist Review* 85 (January-February 1986): 67-96.

74 Adler, "Technology and Us," 77.

75 Menzies, *Whose Brave New World?* 35.

76 Myles, "Post-Industrialism and the Service Economy," 357.

77 Ibid. For a more detailed statistical expression of this proposition, see John Myles, "The Expanding Middle: Some Canadian Evidence on the Deskilling Debate," *Canadian Review of Sociology and Anthropology* 25, 3 (August 1988): 335-64.

78 Aronowitz and DiFazio, *The Jobless Future,* 81-103.

79 Karl Marx and Friedrich Engels, *The German Ideology* (Moscow: Progress, 1976), 53.

80 Rochlin, *Trapped in the Net,* 68.

81 See "Big Brother in the Bathroom," *Harper's* 295, 1771 (December 1997), 28. The system sells for US$1,500.

82 See Shoshana Zuboff, *In the Age of the Smart Machine: The Future of Work and Power* (New York: Basic Books, 1988), 316, and Susan Bryant, "Electronic Surveillance in the Workplace," *Canadian Journal of Communication* 20 (1995): 509.

83 Bryant, "Electronic Surveillance in the Workplace," 507. Emphasis in original.

84 For analyses employing the discourse of panopticism, see Kevin Robins and Frank Webster, "Cybernetic Capitalism: Information, Technology, Everyday Life," in *The Political Economy of Information*, ed. Vincent Mosco and Janet Wasko (Madison: University of Wisconsin Press, 1988), 44-75; Zuboff, *In the Age of the Smart Machine*, 345; Mark Poster, *The Mode of Information: Poststructuralism and Social Context* (Chicago: University of Chicago Press, 1990), especially Chapter 3, "Foucault and Databases"; O.H. Gandy, *The Panoptic Sort: A Political Economy of Personal Information* (Boulder, CO: Westview, 1993); and David Lyon, "An Electronic Panopticon? A Sociological Critique of Surveillance Theory," *Sociological Review* 41, 4 (1993): 653-78.

85 Bryant, "Electronic Surveillance in the Workplace," 510.

86 International Labour Organization, "Monitoring and Surveillance in the Workplace," *Conditions of Work Digest* 12, 1 (1993): 25. This estimate is conservative because it does not include the potentially massive number of workers whose telephone activities are monitored.

87 Cited in Valerie Lawton, "Big Brother at Work," *Toronto Star*, 10 November 1997, C4.

88 Andrew Clement, "Electronic Workplace Surveillance: Sweatshops and Fishbowls," *Canadian Journal of Information Science* 17, 4 (December 1992): 22.

89 Ibid., 20.

90 See Zuboff, *In the Age of the Smart Machine*.

91 Clement, "Electronic Workplace Surveillance," 25.

92 See Geoffrey Rowan, "Surfers Beware: Bosses Can Now Watch You On-Line," *Globe and Mail*, 10 July 1996, B1.

93 For a description of systems such as these, see Gordon Arnaut, "Electronic Big Brother Is on the Job," *Globe and Mail*, 22 October 1996, C1.

94 For the ILO's position on this issue, see International Labour Organization, "Monitoring and Surveillance in the Workplace." The International Trade Union Federation expresses similar concerns: see Bibby, "Trade Unions and Telework." For a review of attempts by Canadian unions to include protection against undue surveillance in their collective agreements, see Bryant, "Electronic Surveillance in the Workplace," 513-5. It should be noted that these agreements, of course, do not protect non-unionized workers.

95 Information Highway Advisory Council, *Connection, Community, Content: The Challenge of the Information Highway* (Ottawa: Minister of Supply and

Services, 1995), 50. As the report points out, an exception in this regard
is the province of Quebec, which in 1994 enacted legislation protecting
personal information in both the public and the private sector, to conform to the specific right to privacy contained in the province's Civil Code.

96 Ibid., 141, recommendations 10.1 and 10.2.

97 Ibid., 179, recommendation 13.28.

98 Canada, *Personal Information Protection and Electronic Documents Act* (Bill
C-6), 2nd Session, 36th Parliament, 48 Elizabeth II, 1999.

99 Ibid., s. 4(1)(b).

100 Ibid., s. 7(1)(a).

101 Ibid., Schedule 1, 4.3.6.

102 Arnaut, "Electronic Big Brother Is on the Job," C12.

103 Clement, "Electronic Workplace Surveillance," 29.

104 For a review of studies of stress and strain conditions in electronically
monitored jobs, see ibid., 30-1. See also Menzies, *Whose Brave New World?*
129-30; Barbara Garson, *The Electronic Sweatshop: How Computers Are
Transforming the Office of the Future into the Factory of the Past* (New York:
Penguin, 1988), 113, 180; and Robert Howard, "Strung Out at the Phone
Company: How AT&T's Workers Are Drugged, Bugged and Coming
Unplugged," *Mother Jones,* August 1981, 39-59.

105 See Robert Howard, *Brave New Workplace* (New York, Viking, 1985), especially Chapter 7, "Labor's Muted Voice," 171-97.

106 Clement, "Electronic Workplace Surveillance," 31. See also Menzies,
Whose Brave New World? 119-22.

107 Clement, "Electronic Workplace Surveillance," 29.

108 Menzies, *Whose Brave New World?* 36.

109 Clement, "Electronic Workplace Surveillance," 32.

110 Nicholas Negroponte, *Being Digital* (New York: Knopf, 1995), 70.

111 Tremblay, "The Information Society," 469.

112 Ibid., 469-70.

113 Burstein and Kline, *Road Warriors,* 265.

114 Derrick de Kerckhove, *The Skin of Culture: Investigating the New Electronic
Reality* (Toronto: Somerville House, 1997), 94.

115 Ibid.

116 See Michael Krantz, "Click Till You Drop," *Time,* 20 July 1998, 14-9, and
Patrick Brethour and Mark Evans, "Builders of the Electronic Mall," *Globe
and Mail,* 11 July 1998, B1.

117 Network Wizards, "Internet Domain Survey, January 1999," available
on-line at www.nw.com.

118 A notable exception to this requirement is traditional catalogue shopping, which, in many ways, prefigured network shopping.

119 For a thorough review of the many complex technicalities of various broadband satellite systems, see John Montgomery, "Fiber in the Sky: The Orbiting Internet," *BYTE* 22, 11 (November 1997): 58-70. For more detailed

discussion of transmission issues and options generally, see David Johnston, Deborah Johnston, and Sunny Handa, *Getting Canada Online: Understanding the Information Highway* (Toronto: Stoddart, 1995).

120 Montgomery, "Fiber in the Sky," 64.

121 Information Highway Advisory Council, *Connection, Community, Content*, 12-3.

122 Ibid., 14.

123 Industry Canada, "Manley Announces Plan to Build World's First National Optical Internet High Speed Network to Revolutionize Electronic Data Information Exchange," press release, Ottawa, 25 August 1998.

124 See Chapter 3, note 55.

125 See David Kline, "The Embedded Internet," *WIRED* 4, 10 (October 1996): 98-106. Kline places the figure of non-PC microprocessors at 90 percent, but cites no source for this estimate.

126 Northern Telecom, "Vista 350 – It's More than a Phone," promotional literature.

127 See Lawrence Surtees, "Holy Grail of Convergence Still Ahead," *Globe and Mail*, 6 May 1997, C1, and Gerry Blackwell, "Great Convergence," *Toronto Star*, 13 March 1997, J1.

128 Mark Evans, "Behind the Screens," *Globe and Mail*, 4 February 1999, D1.

129 See Geoffrey Rowan, "Microsoft On-Line Service to Mimic TV," *Globe and Mail*, 14 October 1996, B1, and Geoffrey Rowan "PCTV," *Report on Business Magazine*, June 1997, 86-92.

130 Mark Evans, "Interactive Television to Launch Next Year," *Globe and Mail*, 13 July 1999, A1.

131 Gates, *The Road Ahead*, 165-6.

132 Robert H. Anderson, Tora K. Bikson, Sally Ann Law, Bridger M. Mitchell, Constantijn W.A. Panis, Padmanabhan Srinagesh, Brent Keltner, Christopher Kedzie, and Joel Pliskin, eds., *Universal Access to E-Mail: Feasibility and Societal Implications*, report prepared for the RAND Corporation (Santa Monica, CA, 1995), no. MR650. Available on-line at www.rand.org/publications/MR/MR650/index.html. Hereinafter cited as RAND report. In what follows I summarize the report generally, and refer to specific authors and chapters in the case of direct quotation only.

133 Robert H. Anderson, Tora K. Bikson, Sally Ann Law, and Bridger M. Mitchell, "Chapter One – Introduction," in RAND report.

134 Ibid.

135 Ibid.

136 Tora K. Bikson and Constantijn W.A. Panis, "Chapter Two – Computers and Connectivity: Current Trends," in RAND report.

137 Bridger M. Mitchell and Padmanabhan Srinagesh, "Chapter Four – Economic Issues," in RAND report; Sally Ann Law and Brent Keltner, "Chapter Five – Civic Networks: The Benefits of On-line Communities," in

ibid.; Robert H. Anderson, Tora K. Bikson, Sally Ann Law, and Bridger M. Mitchell, "Chapter Seven – Conclusions and Recommendations," in ibid.

138 Mitchell and Srinagesh, "Chapter Four – Economic Issues," in ibid.

139 Anderson, Bikson, Law, Mitchell, "Chapter Seven – Conclusions and Recommendations," in ibid.

140 Information Highway Advisory Council, *Connection, Community, Content*, and Executive Office of the President of the United States, *The National Information Infrastructure: Agenda for Action* (Washington, DC: Executive Office of the President, 1993). The US document is also available at nii.nist.gov.

141 Canada, *Building the Information Society: Moving Canada into the 21st Century* (Ottawa: Minister of Supply and Services, 1996), 23-4.

142 Christopher Kedzie, "Chapter Six – International Implications for Global Democratization," in RAND report.

143 Executive Office of the President of the United States, *The National Information Infrastructure.* Emphasis added.

144 Law and Keltner, "Chapter Five – Civic Networks: The Benefits of Online Communities," in RAND report.

145 See Ronald J. Deibert, *Parchment, Printing, and Hypermedia: Communication in World Order Transformation* (New York: Columbia University Press, 1997), 147-55. See also Rochlin, *Trapped in the Net*, 74-107.

146 Joel Kurtzman, *The Death of Money: How the Electronic Economy Has Destabilized the World's Markets and Created Financial Chaos* (New York: Simon and Schuster, 1993), 17.

147 See David Pescovitz, "The Future of Money," *WIRED* 4, 8 (August 1996): 68. In the same issue, see Tom Steinert-Threlkeld, "The Buck Starts Here," 133-5, 194.

148 For an overview of the issues in this area, see Udo Flohr, "Electric Money," *BYTE* 21, 6 (June 1996): 74-84.

149 Mondex is owned by MasterCard International (51 percent) and a number of banks from several countries. Canadian banks with an interest in Mondex include Scotiabank, Bank of Montreal, Toronto Dominion, Canadian Imperial Bank of Commerce, Royal Bank, National Bank of Canada, Canada Trust, and Le Mouvement des caisses desjardins. The technology for the Mondex card was developed by Hitachi. In 1999, there were three Mondex trials under way in Canada, and more than a dozen worldwide. On Mondex see www.mondex.com and www.hitachi-canada.com/mondex. For more on the Canadian trial, see Ann Kerr, "Mondex Trial Getting Mixed Results," *Globe and Mail*, 9 December 1997, C20.

150 www.mondex.com.

151 Mondex, "Mondex First with Electronic Cash Euro," press release, 14 January 1999.

152 See Ann Kerr, "Smart Uses for Smart Cards," *Globe and Mail*, 9 December 1997, C20.

153 For a compendium of the various on-line services offered by Canada's major banks, see "Banking from Home," *Globe and Mail*, 7 October 1997, C2.

154 See Kevin Marron, "How to Make E-banking Pay," *Globe and Mail*, 11 November 1998, C1.

155 See Randy Ray, "Phone Banking Keeps Growing," *Globe and Mail*, 7 October 1997, C2.

156 For a reasonably good description of push software, see Whit Andrews, "Planning for Push," *Internet World*, May 1997, 44-53. See also Jack Kapica, "When 'Push' Comes to Shovelling," *Globe and Mail*, 24 October 1997, A8.

157 PointCast's Canadian rights are owned by the *Globe and Mail* newspaper. Information on the system is available at www.pointcast.ca. Microsoft's Active Desktop, BackWeb and Netscape's Communicator are other examples of popular "push" applications.

158 www.pioneer.pointcast.com/whatis.html.

159 Internet Advertising Bureau, press release, October 1997. Available on-line at www.iab.net/news/newssource4.html.

160 Ibid.

161 Internet Advertising Bureau, *On-line Advertising Effectiveness Study – Executive Summary* (New York: Internet Advertising Bureau, 1997). Available on-line at www.mbinteractive.com/site/iab/ exec.html.

162 Ibid. The IAB surveyed 16,758 respondents randomly sampled from an audience of 1 million visitors to twelve leading Web sites in the United States over a thirteen-day period. Sites sampled included CNN, Compuserve, ESPN SportsZone, Excite, Geocities, HotWired, Looksmart, Lycos, MacWorld, National Geographic Online, People, and Ziff-Davis.

163 Ibid.

164 Ibid.

165 Ibid. Emphasis added.

166 Ibid.

167 Deibert, *Parchment, Printing, and Hypermedia*, 206.

168 Robert W. McChesney, "The Internet and U.S. Communication Policy-Making in Historical and Critical Perspective," *Journal of Communication* 46, 1 (Winter 1996): 113.

169 Robert W. McChesney, "The Global Struggle for Democratic Communication," *Monthly Review* 48, 3 (July-August 1996): 6.

170 Noble, *Progress without People*, 43.

171 McChesney, "The Internet and U.S. Communication Policy-Making in Historical and Critical Perspective," 98.

172 Peter Golding, "World Wide Wedge: Division and Contradiction in the Global Information Infrastructure," *Monthly Review* 48, 3 (July-August 1996): 75.

Chapter 6: A Standing-Reserve of Bits

1 Douglas Rushkoff, "E-mail with Douglas Rushkoff," *Shift* 4, 4 (April 1996): 12. Rushkoff is the author of a book entitled *Cyberia: Life in the Trenches of Hyperspace* (San Francisco: HarperCollins, 1994).

2 See Jerry Weinberger, *Science, Faith, and Politics: Francis Bacon and the Utopian Roots of the Modern Age* (Ithaca, NY: Cornell University Press, 1985). See also William Leiss, *The Domination of Nature* (New York: George Braziller, 1972), Chap. 3.

3 Fredric Jameson, *Postmodernism, or, The Cultural Logic of Late Capitalism* (Durham, NC: Duke University Press, 1991), 7.

4 Jean-François Lyotard, *The Postmodern Condition: A Report on Knowledge*, trans. Geoff Bennington and Brian Massumi (Minneapolis: University of Minnesota Press, 1984), xxiv.

5 Jameson, *Postmodernism*, 14.

6 See ibid., 16-7.

7 Ronald J. Deibert, *Parchment, Printing, and Hypermedia: Communication in World Order Transformation* (New York: Columbia University Press, 1997), 182, 185.

8 Jameson, *Postmodernism*, 12.

9 Mark Poster, *The Mode of Information: Poststructuralism and Social Context* (Chicago: University of Chicago Press, 1990), 128.

10 Mark Poster, *The Second Media Age* (Cambridge, MA: Polity, 1995), 11, 18.

11 Ibid., 33.

12 Ibid., 4.

13 Sherry Turkle, *Life on Screen: Identity in the Age of the Internet* (New York: Simon and Schuster, 1995).

14 Ibid., 10.

15 Ibid., 18.

16 Ibid., 18-9.

17 In this discussion of simulation, Turkle relies heavily on the work of postmodern pioneer Jean Baudrillard. See, in particular, his *Simulations*, trans. Paul Foss, Paul Patton, and Philip Beitchman (New York: Semiotext(e), 1983).

18 Turkle, *Life on Screen*, 23. Emphasis in original.

19 Ibid., 50-73.

20 See Claude Lévi-Strauss, *The Savage Mind* (Chicago: University of Chicago Press, 1968), 16-33.

21 Turkle, *Life on Screen*, 51.

22 Ibid., 178-80.

23 Ibid., 263-4.

24 Ibid., 35-6.

25 Ibid., 138-43.

26 Ibid., 12, 185.

27   Ibid., 186.

28   Ibid., 258.

29   Ibid., 47.

30   Martin Heidegger, "Letter on Humanism," trans. Frank A. Capuzzi, in *Basic Writings*, ed. David Farrell Krell (New York: Harper and Row, 1977), 193.

31   Ibid., 210.

32   Ibid., 199.

33   Martin Heidegger, "The Question Concerning Technology," trans. William Lovitt, in *Basic Writings*, ed. Krell, 314.

34   Ibid., 289-91.

35   Nicholas Negroponte, *Being Digital* (New York: Knopf, 1995), 228.

36   Kevin Kelly, *Out of Control: The Rise of Neo-Biological Civilization* (Reading, MA: Addison-Wesley, 1994), 1, 55.

37   Douglas Rushkoff, "E-mail with Douglas Rushkoff," *Shift* 4, 4 (April 1996): 12.

38   Martin Heidegger, *Discourse on Thinking*, trans. John M. Anderson and E. Hans Freund (New York: Harper and Row, 1959), 50.

39   Martin Heidegger, "Building, Dwelling, Thinking," trans. William Lovitt, in *Basic Writings*, ed. Krell, 328.

40   Simone Weil, *The Need for Roots* (London: Routledge and Kegan Paul, 1952), 41.

41   See William Gibson, *Neuromancer* (New York: Ace, 1984), 6.

42   Weil, *The Need for Roots*, 41.

43   Ibid., 41.

44   Heidegger, *Discourse on Thinking*, 48.

45   Ibid.

46   On the WELL, see Howard Rheingold, *The Virtual Community: Homesteading on the Electronic Frontier* (Reading, MA: Addison-Wesley, 1993); Katie Hafner, "The Epic Saga of the Well," *WIRED* 5, 5 (May 1997): 98-142; and John Perry Barlow, "Is There a There in Cyberspace?" *Utne Reader*, March-April 1995, 53-6.

47   For a comprehensive review of the social-science literature on virtual communities, see Barry Wellman, Janet Salaff, Dimitrina Dimitrova, Laura Garton, Milenia Gulia, and Caroline Haythornthwaite, "Computer Networks as Social Networks: Collaborative Work, Telework, and Virtual Community," *Annual Review of Sociology* 22 (1996): 213-38.

48   See Ray Oldenburg, *The Great Good Place: Cafes, Coffee Shops, Community Centers, Beauty Parlors, General Stores, Bars, Hangouts and How They Get You Through the Day* (New York: Paragon House, 1991). For direct references to Oldenburg's conception of a "third place" by proponents of on-line communities, see Douglas Schuler, *New Community Networks: Wired for Change* (New York: ACM Press, 1996), 42; John Coate, "Cyberspace Innkeeping: Building On-line Community," in *Reinventing Technology*,

Rediscovering Community: Critical Explorations of Computing as a Social Practice, ed. Philip E. Agre and Douglas Schuler (Greenwich, CT: Ablex, 1997), 166; and Howard Rheingold, "The Virtual Community," 63.

49 Coate, "Cyberspace Innkeeping," 168.

50 Ibid., 173-5.

51 Julian Stallabrass, "Empowering Technology: The Exploration of Cyberspace," *New Left Review* 211 (May-June 1995): 16.

52 Wellman et al., "Computer Networks as Social Networks," 220-1.

53 Barlow, "Is There a There in Cyberspace?" 54. For a similar argument, see Coate, "Cyberspace Innkeeping," 166.

54 Roger Gibbins and Carey Hill, "New Technologies and the Future of Civil Society," paper presented to the annual meeting of the Canadian Communication Association, Ottawa, Ontario, 31 May 1998, 20.

55 Wellman et al., "Computer Networks as Social Networks," 232.

56 Ibid., 224.

57 For a collection of case studies on civic networks, see Roza Tsagarousianou, Damian Tambini, and Cathy Bryan, eds., *Cyberdemocracy: Technology, Cities and Civic Networks* (London: Routledge, 1998).

58 Douglas Schuler, "Community Networks: Building a New Participatory Medium," *Communications of the ACM* 37, 1 (January 1994): 39.

59 See Pierre Bourque and Rosaleen Dickson, *Freenet: Canadian Online Access the Free and Easy Way* (Toronto: Stoddart, 1996).

60 See, for example, accounts of the Blacksburg Electronic Village in Andrew Cohill and Andrea Kavanaugh, eds., *Community Networks: Lessons from Blacksburg, Virginia* (Boston: Artech House, 1997).

61 See Robert D. Putnam, "Bowling Alone? America's Declining Social Capital," *Journal of Democracy* 6, 1 (January 1995): 65-78. Douglas Schuler makes explicit the connection between Putnam's diagnosis and community networks as a potential cure in *New Community Networks: Wired for Change*, 5-6, 36-7.

62 See Schuler, *New Community Networks*, 11-22.

63 Ibid., 43.

64 Ibid., 25.

65 Ibid., 26, 33.

66 See Gibbins and Hill, "New Technologies and the Future of Civil Society," 14-8.

67 Joseph Weizenbaum, *Computer Power and Human Reason: From Judgment to Calculation* (San Francisco: W.H. Freeman, 1976).

68 Ibid., 11.

69 Ibid., 226-7, 258-80.

70 J. David Bolter, *Turing's Man: Western Culture in the Computer Age* (Chapel Hill: University of North Carolina Press, 1984).

71 Ibid., 71, 73.

72 Ibid., 132.

73 Ibid., 220-2.
74 Turkle, *Life on Screen*, 19.
75 Ibid., 137.
76 This perspective on networked computing is expressed most clearly in Kevin Kelly, *Out of Control: The Rise of Neo-Biological Civilization* (Reading, MA: Addison-Wesley, 1994).
77 Ibid., 22-3.
78 Connectivity, or "connectionism," is emphasized by Kelly (186), by Turkle in *Life on Screen* (135), and by Derrick de Kerckhove in *Connected Intelligence: The Arrival of the Web Society* (Toronto: Somerville House, 1997).
79 Kerckhove, *Connected Intelligence*, xxiii.
80 Ibid., xxxi.
81 David Lyon, *The Electronic Eye: The Rise of Surveillance Society* (Minnesota: University of Minnesota Press, 1994), 4. Emphasis added.
82 Ibid., 11.
83 There is a burgeoning literature on these and related topics. Lyon's study is perhaps the most thoughtful, but there are others of note, including: Oscar H. Gandy, *The Panoptic Sort: A Political Economy of Personal Information* (Boulder, CO: Westview, 1993); David Flaherty, *Protecting Privacy in Surveillance Societies* (Chapel Hill: University of North Carolina Press, 1989); Roger Clarke, "Information and Dataveillance," *Communications of the ACM* 31, 5 (1988): 498-512; Kenneth Laudon, *The Dossier Society: Value Choices in the Design of National Information Systems* (New York: Columbia University Press, 1986); David Burnham, *The Rise of the Computer State* (London: Weidenfeld and Nicolson, 1983); and Malcolm Warner and Michael Stone, *The Data Bank Society: Organizations, Computers and Social Freedom* (London: George Allen and Unwin, 1970); Bruce Phillips, *Annual Report: Privacy Commissioner, 1992-1993* (Ottawa: Privacy Commissioner of Canada, 1993).
84 See Lyon, *The Electronic Eye*, 37. See also Christopher Dandeker, *Surveillance Power and Modernity* (Cambridge, MA: Polity, 1990), and James B. Rule, *Private Lives and Public Surveillance* (London: Allen Lane, 1973), especially Chapter 1, "Social Control and Modern Social Structure."
85 Max Weber, "Bureaucracy," in *From Max Weber: Essays in Sociology*, trans. and ed. H.H. Gerth and C. Wright Mills (New York: Oxford University Press, 1958), 196-266. Note especially Weber's emphasis on the importance of "the files" in complex bureaucratic administrations (197); George Orwell, *Nineteen Eighty-Four* (New York: Harcourt, Brace and World, 1949); Michel Foucault, *Discipline and Punish: The Birth of the Prison*, trans. Alan Sheridan (New York: Vintage, 1995).
86 Rule, *Private Lives and Public Surveillance*, 22-3.
87 Gandy, *The Panoptic Sort*, 53.
88 Ibid., 63.
89 Lyon, *The Electronic Eye*, 81.

90 See ibid., 169, 220.

91 For a review in which privacy concerns are central, see Ann Cavoukian and David Tapscott, *Who Knows? Safeguarding Your Privacy in a Networked World* (Toronto: Random House, 1995). Chapter 8, "Medical Privacy," is a particularly useful supplement to Lyon's fourfold classification. For a Foucaultian perspective on networks as "superpanopticons," see Poster, *The Mode of Information,* especially Chapter 3, "Foucault and Databases."

92 Lyon, *The Electronic Eye,* 137.

93 Ibid., 156.

94 Ann Cavoukian, *Data Mining: Staking a Claim on Your Privacy* (Toronto: Information and Privacy Commissioner, Ontario, 1998), 4.

95 Andrew Tausz, "Data Warehouses Store Wealth of Information," *Globe and Mail,* 9 December 1997, C1. A terabyte contains 1 trillion bytes, which means a 24-terabyte database contains 192 trillion bits. "Tera" derives from the Greek *teras,* meaning "monster."

96 For a step-by-step description of the data-mining process, see Joseph P. Bigus, *Data Mining with Neural Networks* (New York: McGraw-Hill, 1996), 9-11.

97 Cavoukian, *Data Mining,* 8.

98 Kimberley Noble, "The Data Game," *Maclean's,* 17 August, 1998, 15.

99 Ibid., 16.

100 Cavoukian, *Data Mining,* 8.

101 Noble, "The Data Game," 16.

102 Tausz, "Data Warehouses Store Wealth of Information," C1.

103 Ibid.

104 Salam Alaton, "Mountains of Data Help Marketers Make Connection," *Globe and Mail,* 4 April, 1995, B28.

105 Noble, "The Data Game," 16.

106 Bill Cross, "Virtual Election Campaigns: Campaign Communication in the Fourth Canadian Party System," paper presented at the annual meeting of the Atlantic Provinces Political Studies Association, Charlottetown, PEI, 25-7 September 1998.

107 Canada, *Personal Information Protection and Electronic Documents Act* (Bill C-6), 2nd Session, 36th Parliament, 48 Elizabeth II, 1999.

108 John Manley, "Third Reading Speaking Notes: Bill C-6, Personal Information Protection and Electronic Documents Act," 22 October 1999. Available on-line at www.ic.gc.ca.

109 Lyon, *The Electronic Eye,* 24, 76.

110 Ibid., 86.

111 Ibid., 100.

112 Cavoukian, *Data Mining,* 17.

113 John Grace, *Annual Report: Canadian Privacy Commissioner* (Ottawa: Ministry of Supply and Services, 1988), 4. Emphasis added.

114 Poster, *The Mode of Information,* 97-8.

115 Heidegger, "The Question Concerning Technology," 309.

116 Ibid., 314.

117 Many of these have already been cited in this and previous chapters. For a collection of scholarly and popular critiques of network technology, see *Resisting the Virtual Life: The Culture and Politics of Information*, ed. James Brook and Iain A. Boal (San Francisco: City Lights, 1995). Other examples of popular, if somewhat nostalgic, reticence surrounding this technology include Clifford Stoll, *Silicon Snake Oil: Second Thoughts on the Information Highway* (New York: Doubleday, 1995), and Kirkpatrick Sale, *Rebels against the Future: The Luddites and Their War on the Industrial Revolution; Lessons for the Computer Age* (Reading, MA: Addison-Wesley, 1995).

118 Information Highway Advisory Council, *Connection, Community, Content*, viii.

119 Kelly, *Out of Control*, 257.

120 Heidegger, "Letter on Humanism," 221.

121 Heidegger, *Discourse on Thinking*, 54.

Chapter 7: Government, Politics, and Democracy

1 As quoted in Madelaine Drohan, "Nations See Internet as Threat to Security," *Globe and Mail*, 3 February 1996, A12.

2 Ibid.

3 Daniel Burstein and David Klein, *Road Warriors: Dreams and Nightmares along the Information Superhighway* (New York: Dutton, 1995), 113.

4 John Perry Barlow, as quoted in Evan Solomon, "Unlikely Messiah," *Shift* 4, 1 (1995): 31.

5 Kevin Kelly, *Out of Control: The Rise of Neo-biological Civilization* (Reading, MA: Addison-Wesley, 1994), 90.

6 See Anne Wells Branscomb, "Jurisdictional Quandaries for Global Networks," in *Global Networks: Computers and International Communication*, ed. Linda Harasim (Cambridge, MA: MIT Press, 1993), 83-103.

7 Langdon Winner, *Autonomous Technology: Technics-out-of-Control as a Theme in Political Thought* (Cambridge, MA: MIT Press, 1977), 34.

8 Langdon Winner, *The Whale and the Reactor: A Search for Limits in an Age of High Technology* (Chicago: University of Chicago Press, 1986), 54.

9 Walter B. Wriston, *The Twilight of Sovereignty: How the Information Revolution Is Transforming Our World* (New York: Macmillan, 1992). Wriston is the former CEO of Citibank/Citicorp.

10 Ibid., 8-9.

11 Ibid., 148.

12 See William Stallings, *Protect Your Privacy: A Guide for PGP Users* (New York: Prentice-Hall, 1995), and Simson Garfinkel, *PGP: Pretty Good Privacy* (New York: O'Reilly and Associates, 1995).

13 Industry Canada, *A Cryptography Policy Framework for Electronic Commerce* (Ottawa: Task Force on Electronic Commerce, February 1998).

14 Phil Zimmerman, as quoted in Jeff Elliot, "Fighting the (Pretty) Good Fight," *Infobahn*, June 1995, 34.

15 Jonathan Wallace and Mark Mangan, *Sex, Laws and Cyberspace: Freedom and Censorship on the Frontiers of the Online Revolution* (New York: Henry Holt, 1997), 48.

16 See Peter Cassidy, "Reluctant Hero," *WIRED* 4, 6 (June 1996): 121.

17 Electronic Frontier Foundation, "EFF and Privacy Organizations Issue Statement at Launch of New Industry-Led Alliance on Encryption Controls," press release, 4 March 1998. Available on-line at www.eff.org.

18 For a good review of the Clipper scheme, see Dorothy Denning, "Resolving the Encryption Dilemma: The Case for Clipper," *Technology Review*, July 1995. Available on-line at www.techreview.com/articles/july95/Denning.html.

19 Brock Meeks, "The Cyber Rights Report Card," *WIRED* 4, 10 (October 1996): 94.

20 See "Enabling Privacy, Commerce, Security and Public Safety in the Global Information Infrastructure," Executive Office of the President, Washington, DC, 20 May 1996. For more on cryptography policy in the United States, including the key-escrow proposals, see *Cryptography's Role in Securing the Information Society*, ed. Kenneth Lam and Herbert Lin (Washington, DC: National Research Council, 1996).

21 Between 1996 and 1998, no fewer than seven separate bills pertaining to digital cryptography were introduced in the US Congress. Included among these are the Safety and Freedom through Encryption Act (HR 695), the Encrypted Communications Privacy Act (S376), the Promotion of Commerce Online in the Digital Era (S377), the Secure Public Networks Act (S909), and the E-Privacy Act (S2067).

22 Senator Leahy, speaking to the 1996 version of the Encrypted Communications Privacy Act (s. 1587). Statement available on-line at www.eff.org/pub/privacy/key_escrow/Clipper_III/crypto_bills_1996/leahy_S1587_95_intro.statement.

23 John Simons, "U.S. to Allow Export of Encryption Technology," *Globe and Mail*, 19 October 1998, B13.

24 See Edmund Andrews, "Europeans Reject US Plan on Electronic Cryptography," *New York Times*, 9 October 1997, D4. For a good comparative review of cryptography policy, see Richard C. Barth and Clint N. Smith, "International Regulation of Encryption: Technology Will Drive Policy," in *Borders in Cyberspace: Information Policy and the Global Information Infrastructure*, ed. Brian Kahin and Charles Nesson (Cambridge, MA: MIT Press, 1997), 283-99.

25 Industry Canada, *A Cryptography Policy Framework for Electronic Commerce*.

26 See Communications Security Establishment, "Government of Canada Public Key Infrastructure: White Paper," (Ottawa: Ministry of Supply and Services, 1997). Available on-line at www.cse-cst.gc.ca/cse/english/gov.html.

27 See Richard Ericson and Kevin Haggerty, *Policing the Risk Society* (Toronto: University of Toronto Press, 1997).

28 See *Janet Reno, Attorney General of the United States, et al.* v. *American Civil Liberties Union,* United States Supreme Court. See also Graham Fraser, "Top U.S. Court Defends Internet Freedom," *Globe and Mail,* 27 June 1997, A1.

29 See Darren McDermott, "Singapore Curbs Flow of Internet Material," *Globe and Mail,* 7 March 1996, A11.

30 See Robert Everett-Green, "Sweeping U.S. Censorship Bill May Defeat Itself," *Globe and Mail,* 12 March 1996, C4.

31 Allan J. MacEachen and Jean Robert Gauthier, *Canada's Foreign Policy: Principles and Priorities for the Future,* Report of the Special Joint Committee of the Senate and House of Commons Reviewing Canada's Foreign Policy (Ottawa: Parliamentary Publications Directorate, 1994), 1. On the fate of the nation-state and the challenges of globalization, see David Elkins, *Beyond Sovereignty: Territoriality and Political Economy in the Twenty-First Century* (Toronto: University of Toronto Press, 1995); Samuel Barkin and Bruce Cronin, "The State and the Nation: Changing Norms and Rules of Sovereignty in International Relations," *International Organization* 48, 1 (Winter 1994): 107-30; Louis W. Pauly, "Capital Mobility, State Autonomy, and Political Legitimacy," *Journal of International Affairs* 48, 2 (Winter 1995): 369-89; John Gerard Ruggie, "Territoriality and Beyond: Problematizing Modernity in International Relations," *International Organization* 47, 1 (Winter 1993): 139-74; and Robert Cox, "The Global Political Economy and Social Choice," in *The New Era of Global Competition: State Policy and Market Power,* ed. Daniel Drache and Meric Gertler (Montreal and Kingston: McGill-Queen's University Press, 1991), 335-49.

32 Klaus Lenk, "The Challenge of Cyberspatial Forms of Human Interaction to Territorial Governance and Policing," in *The Governance of Cyberspace: Politics, Technology and Global Restructuring,* ed. Brian D. Loader (London: Routledge, 1997), 126-35.

33 On Singapore, see McDermott, "Singapore Curbs Flow of Internet Material," A11. On the German legislation, see Jordan Bonfante, "The Internet Trials," *Time,* 14 July 1997, 32-3.

34 For a survey of such opinions, see John Lorinc, "Information Revolution Puts Copyright under Siege," *Globe and Mail,* 29 October 1996, D1, and Andrew Coyne, "The Information Highwayman Comes Riding," *Globe and Mail,* 1 October 1994, B1. See also Anne W. Branscomb, "Who Owns Creativity?" in *Computers in the Human Context: Information Technology, Productivity and People,* ed. Tom Forester (Cambridge, MA: MIT Press, 1989), 407-14.

35 See Kevin Marron, "Electronic Commerce Raises Some Taxing Questions," *Globe and Mail*, 9 December 1997, C3.

36 Heather Scofield, "OECD Pledges Light Touch on Internet," *Globe and Mail*, 15 October 1998. D1.

37 David R. Johnston and David G. Post, "The Rise of Law on the Global Network," in *Borders in Cyberspace: Information Policy and the Global Information Infrastructure*, ed. Brian Kahin and Charles Nesson (Cambridge, MA: MIT Press, 1997), 23.

38 Ibid., 3, 34.

39 Ibid., 28. Emphasis added.

40 Ibid., 36. See also Joel R. Reidenberg, "Governing Networks and Rule-Making in Cyberspace," in *Borders in Cyberspace: Information Policy and the Global Information Infrastructure*, ed. Brian Kahin and Charles Nesson (Cambridge, MA: MIT Press, 1997), 90.

41 George Grant, *Lament for a Nation: The Defeat of Canadian Nationalism* (Toronto: McClelland and Stewart, 1965), 43.

42 George Grant, *Technology and Empire: Perspectives on North America* (Toronto: Anansi, 1969), 114.

43 Grant, *Lament for a Nation*, 64.

44 Ibid., 53.

45 Ibid., 56.

46 Ibid., 54.

47 Ibid.

48 See Greg Richards, "Businesslike Government: The Ultimate Oxymoron?" *Optimum: The Journal of Public Sector Management* 27, 1 (1996): 21-5.

49 Richard Simeon, George Hoberg, and Keith Banting, "Globalization, Fragmentation, and the Social Contract," in *Degrees of Freedom: Canada and the United States in a Changing World*, ed. Keith Banting, George Hoberg, and Richard Simeon (Montreal and Kingston: McGill-Queen's University Press, 1997), 389-416.

50 Ibid., 390.

51 Canada, *Building the Information Society: Moving Canada into the 21st Century* (Ottawa: Minister of Supply and Services, 1996), 1-2.

52 Ibid., 1.

53 Johnston and Post, "The Rise of Law on the Global Network," 8.

54 Wriston, *The Twilight of Sovereignty*, 136.

55 Canada, *Building the Information Society*, 3.

56 Grant, *Lament for a Nation*, 76. Here Grant is referring specifically to Quebec, but it is fair to say the point is general.

57 On Canadian broadcasting and telecommunications policy, see Marc Raboy, *Missed Opportunities: The Story of Canada's Broadcasting Policy* (Montreal and Kingston: McGill-Queen's University Press, 1990), and Robert E. Babe, *Telecommunications in Canada: Technology, Industry and Government* (Toronto: University of Toronto Press, 1990).

58 Canada, Broadcasting Act (1991), s. 3.1(d)(ii).

59 Canada, Telecommunications Act (1993), s. 7.

60 Marc Raboy, "Cultural Sovereignty, Public Participation, and Democratization of the Public Sphere: The Canadian Debate on the New Information Infrastructure," in *National Information Infrastructure Initiatives: Vision and Policy Design*, ed. Brian Kahin and Ernest Wilson (Cambridge, MA: MIT Press, 1997), 200-1.

61 Ibid., 199.

62 Mary Vipond, *The Mass Media in Canada* (Toronto: James Lorimer, 1989), 119. See also Dallas Smythe, *Dependency Road: Communications, Capitalism, Consciousness and Canada* (Norwood, NJ: Ablex, 1981).

63 Johnston and Post, "The Rise of Law on the Global Network," 23.

64 Canada, *Building the Information Society*, 11.

65 Information Highway Advisory Council, *Preparing Canada for a Digital World*, 59.

66 Canada, *Building the Information Society*, 12.

67 Ibid., 15.

68 Information Highway Advisory Council, *Connection, Community, Content*, x.

69 Ibid., 94-100, recommendations 1.5, 2.1, 2.15; and 122-4, recommendations 7.1, 7.5.

70 Canada, *Building the Information Society*, 2.

71 Information Highway Advisory Council, *Preparing Canada for a Digital World*, 13, 38.

72 Ibid., 28, 34.

73 Ibid., 22.

74 Executive Office of the President of the United States, *A Short Summary of the Telecommunications Reform Act of 1996* (Washington, DC, 1996).

75 Office of Science and Technology Policy, "President Clinton Welcomes Plan to Strengthen U.S. Leadership in Information Technology," press release (Washington, DC: 10 August 1998).

76 Ibid.

77 European Commission, *Europe's Way to the Information Society: An Action Plan* (Brussels: European Commission, 1994). Emphasis added.

78 Japan Telecommunications Council, Ministry of Posts and Telecommunications, *Vision 21 for Info-Communications* (Tokyo: Ministry of Posts and Telecommunications, 1997).

79 Wriston, *The Twilight of Sovereignty*, 45.

80 Grant, *Lament for a Nation*, 50.

81 For a number of case studies of network use at the local level, see *Cyberdemocracy: Technology, Cities and Civic Networks*, ed. Roza Tsagarousianou, Damian Tambini, and Cathy Bryan (London: Routledge, 1998). For a Canadian study based on the city of Calgary, Alberta, see Roger Gibbins and

Carey Hill, "New Technologies and the Future of Civil Society," paper presented to the annual meeting of the Canadian Communication Association, Ottawa, Ontario, 31 May 1998.

82 Pippa Norris, "Who Surfs? New Technology, Old Voters and Virtual Democracy in America," paper presented at the third annual meeting of the John F. Kennedy Visions of Governance for the Twenty-First Century, Mount Washington Hotel, Bretton Woods, 19-22 July 1998.

83 Plato, *Gorgias*, trans. W. Hamilton (Harmondsworth: Penguin, 1960), 465d.

84 See Daniel Girard, "Tories Hit the Internet to Sell Deal to Union," *Toronto Star*, 8 February 1996, A8.

85 See Barbara Ehrenreich, "Spinning the Poor into Gold: How Corporations Seek to Profit from Welfare Reform," *Harper's* 295, 1767 (August 1997), 44-51.

86 On Canada's Schoolnet program, see Canada, *Building the Information Society*, 22.

87 Ibid., 23.

88 Information Highway Advisory Council, *Preparing Canada for a Digital World*, 2.

89 Lawrence K. Grossman, *The Electronic Republic: Reshaping Democracy in the Information Age* (New York: Viking, 1995), 153.

90 Ibid., 154.

91 George Grant, *Technology and Justice* (Toronto: Anansi, 1986), 21.

bibliography

Adler, Paul. "Automation, Skill and the Future of Capitalism." *Berkeley Journal of Sociology: A Critical Review* 33 (1988): 1-36.

–. "Technology and Us." *Socialist Review* 85 (January-February 1986): 67-96.

Aeschylus. *Prometheus Bound.* Trans. Herbert W. Smyth. London: Heinemann, 1922.

Akyeampong, Ernest, and Jennifer Winters. "International Employment Trends by Industry – A Note." *Perspectives on Labour and Income,* Statistics Canada, Summer 1993, 33-7.

Alaton, Salam. "Mountains of Data Help Marketers Make Connection." *Globe and Mail,* 4 April 1995, B28.

Anderson, Benedict. *Imagined Communities: Reflections on the Origin and Spread of Nationalism.* London: Verso, 1991.

Anderson, Robert H., Tora K. Bikson, Sally Ann Law, Bridger M. Mitchell, Constantijn W.A. Panis, Padmanabhan Srinagesh, Brent Keltner, Christopher Kedzie, and Joel Pliskin, eds. *Universal Access to E-Mail: Feasibility and Societal Implications.* Report prepared for the RAND Corporation, No. MR650. Santa Monica, CA, 1995.

Andrew, Edward. *Closing the Iron Cage: The Scientific Management of Work and Leisure.* Montreal: Black Rose, 1981.

–. "George Grant on Technological Imperatives." In *Democratic Theory and Technological Society,* ed. R. Day, J. Masciulli, and R. Beiner, 299-315. New York: M.E. Sharpe, 1988.

–. "The Unity of Theory and Practice: The Science of Marx and Nietzsche." In *Political Theory and Praxis: New Perspectives,* ed. Terence Ball, 118-34. Minneapolis: University of Minnesota Press, 1977.

Andrews, Edmund. "Europeans Reject US Plan on Electronic Cryptograpy." *New York Times,* 9 October 1997, D4.

Andrews, Whit. "Planning for Push." *Internet World,* May 1997, 44-53.

Aristotle. *Nichomachean Ethics.* Trans. Martin Ostwald. Indianapolis: Liberal Arts Press, 1962.

Arnaut, Gordon. "Electronic Big Brother Is on the Job." *Globe and Mail,* 22 October 1996, C1.

Aronowitz, Stanley, and William DiFazio. *The Jobless Future: Sci-tech and the Dogma of Work.* Minneapolis: University of Minnesota Press, 1994.

Arterton, F. Christopher. *Teledemocracy: Can Technology Protect Democracy?* Newbury Park, CA: Sage, 1987.

Augarten, Stan. *BIT by BIT: An Illustrated History of Computers.* New York: Ticknor and Fields, 1984.

Babe, Robert E. *Telecommunications in Canada: Technology, Industry and Government.* Toronto: University of Toronto Press, 1990.

Bacon, Francis. "*De Sapientia Veterum* or, Wisdom of the Ancients." In *The Philosophical Works of Francis Bacon.* Trans. James Spedding; ed. John M. Robertson, 821-58. London: George Routledge and Sons, 1905.

Bagdikian, Ben. *The Media Monopoly.* 5th ed. Boston: Beacon, 1997.

Baldwin, John, and David Sabourin. *Technology Adoption in Canadian Manufacturing.* Statistics Canada Catalogue No. 88-512. Ottawa: Minister of Industry, Science and Technology, 1995.

"Banking from Home." *Globe and Mail,* 7 October 1997, C2.

Baran, Nicholas. *Inside the Information Superhighway Revolution.* Scottsdale, AZ: Coriolis, 1995.

–. "Privatization of Telecommunications." *Monthly Review* 48, 3 (July-August 1996): 59-69.

Baran, Paul. "On Distributed Communications Networks." *IEEE Transactions on Communications* CS-12, 1 (1 March 1964): 1-9.

Barkin, Samuel, and Bruce Cronin. "The State and the Nation: Changing Norms and Rules of Sovereignty in International Relations." *International Organization* 48, 1 (Winter 1994): 107-30.

Barlow, John Perry. "Is There a There in Cyberspace?" *Utne Reader,* March-April 1995, 53-6.

Barney, Darin David. "Push-Button Populism: The Reform Party and the Real World of Teledemocracy." *Canadian Journal of Communication* 21, 3 (Summer 1996): 381-413.

Barth, Richard C., and Clint N. Smith. "International Regulation of Encryption: Technology Will Drive Policy." In *Borders in Cyberspace: Information Policy and the Global Information Infrastructure,* ed. Brian Kahin and Charles Nesson, 283-99. Cambridge, MA: MIT Press, 1997.

Bass, Thomas A. "The Future of Money." *WIRED* 4, 10 (October 1996): 140-205.

Baudrillard, Jean. *The Ecstasy of Communication.* Trans. Bernard and Caroline Schultze. Montreal: Semiotext(e), 1988.

–. *Simulations.* Trans. Paul Foss, Paul Patton, and Philip Beitchman. New York: Semiotext(e), 1983.

Beiner, Ronald. "Ethics and Technology: Hans Jonas' Theory of Responsibility." In *Democratic Theory and Technological Society,* ed. Richard B. Day, Ronald Beiner, and Joseph Masciulli, 336-54. Armonk, NY: M.E. Sharpe, 1988.

–. "Introduction." In *Democratic Theory and Technological Society*, ed. Richard B. Day, Ronald Beiner, and Joseph Masciulli, ix-xii. Armonk, NY: M.E. Sharpe, 1988.

–. *Political Judgment*. London: Methuen, 1983.

–. *What's the Matter with Liberalism?* Berkeley: University of California Press, 1992.

Beniger, James R. *The Control Revolution: Technological and Economic Origins of the Information Society*. Cambridge, MA: Harvard University Press, 1986.

Bibby, Andrew. "Trade Unions and Telework." Report prepared for the International Trade Union Federation, Autumn 1996.

"Big Brother in the Bathroom." *Harper's* 295, 1771 (December 1997), 28.

Bigus, Joseph. *Data Mining with Neural Networks*. New York: McGraw-Hill, 1996.

Blackwell, Gerry. "Great Convergence." *Toronto Star*, 13 March 1997, J1.

Bolter, J. David. *Turing's Man: Western Culture in the Computer Age*. Chapel Hill: University of North Carolina Press, 1984.

Bonfante, Jordan. "The Internet Trials." *Time*, 14 July 1997, 32-3.

Bourque, Pierre, and Rosaleen Dickson. *Freenet: Canadian Online Access the Free and Easy Way*. Toronto: Stoddart, 1996.

Brand, Stewart. *The Media Lab: Inventing the Future at MIT*. New York: Penguin, 1988.

Branscomb, Anne Wells. "Jurisdictional Quandaries for Global Networks." In *Global Networks: Computers and International Communication*, ed. Linda Harasim, 83-103. Cambridge, MA: MIT Press, 1993.

–. "Who Owns Creativity?" In *Computers in the Human Context: Information Technology, Productivity and People*, ed. Tom Forester, 407-14. Cambridge, MA: MIT Press, 1989.

Braverman, Harry. *Labour and Monopoly Capital: The Degradation of Work in the Twentieth Century*. New York: Monthly Review, 1974.

Brethour, Patrick, and Mark Evans. "Builders of the Electronic Mall." *Globe and Mail*, 11 July 1998, B1.

Bridges, William. *JobShift: How to Prosper in a Workplace without Jobs*. Reading, MA: Addison-Wesley, 1994.

Brook, James, and Iain A. Boal, eds. *Resisting the Virtual Life: The Culture and Politics of Information*. San Francisco: City Lights, 1995.

Browning, Graeme. *Electronic Democracy: Using the Internet to Influence American Politics*. Wilton, CT: On-line Press, 1996.

Bryant, Susan. "Electronic Surveillance in the Workplace." *Canadian Journal of Communication* 20 (1995): 505-21.

Burnham, David. *The Rise of the Computer State*. London: Weidenfeld and Nicolson, 1983.

Burstein, Daniel, and David Klein. *Road Warriors: Dreams and Nightmares along the Information Superhighway*. New York: Dutton, 1995.

Butler, Pierce. *The Origin of Printing in Europe*. Chicago: University of Chicago Press, 1940.

Bylinsky, Gene. "Here Comes the Second Computer Revolution." In *The Microelectronics Revolution,* ed. Tom Forester, 3-15. Oxford: Basil Blackwell, 1980.

Canada. *Building the Information Society: Moving Canada into the 21st Century.* Ottawa: Minister of Supply and Services, 1996.

–. *Personal Information Protection and Electronic Documents Act* (Bill C-6). 2nd Session, 36th Parliament, 48 Elizabeth II, 1999.

Canadian Union of Public Employees. *Computer Related Change in the Workplace.* Ottawa: CUPE, 1985.

Carroll, Jim, and Rick Broadhead. *The Canadian Internet Handbook.* 1995 edition. Scarborough, ON: Prentice-Hall, 1995.

Cassidy, Peter. "Reluctant Hero." *WIRED* 4, 6 (June 1996): 121.

Castells, Manuel, and Yuko Aoyama. "Paths towards the Informational Society: Employment Structure in G-7 Countries 1920-1990." *International Labour Review* 133, 1 (1994): 5-33.

Cavoukian, Ann. *Data Mining: Staking a Claim on Your Privacy.* Toronto: Information and Privacy Commissioner, Ontario, 1998.

Cavoukian, Ann, and David Tapscott. *Who Knows? Safeguarding Your Privacy in a Networked World.* Toronto: Random House, 1995.

Cayley, David, ed. *George Grant: In Conversation.* Toronto: Anansi, 1995.

Chandrasekaran, Raju, and Elizabeth Corcoran. "Justice Department Says Microsoft in Contempt." *Washington Post,* 18 December 1997, E1.

Chen, Ching-Chih. *Planning Global Information Infrastructure.* Norwood, NJ: Ablex, 1995.

Christian, William. *George Grant: A Biography.* Toronto: University of Toronto Press, 1993.

Clarke, Roger. "Information and Dataveillance." *Communications of the ACM* 31, 5 (1988): 498-512.

Clement, Andrew. "Electronic Workplace Surveillance: Sweatshops and Fishbowls." *Canadian Journal of Information Science* 17, 4 (December 1992): 18-45.

Coate, John. "Cyberspace Innkeeping: Building On-line Community." In *Reinventing Technology, Rediscovering Community: Critical Explorations of Computing as a Social Practice,* ed. Philip E. Agre and Douglas Schuler, 165-90. Greenwich, CT: Ablex, 1997.

Cohen, G.A. "Self-Ownership, Communism and Equality: Against the Marxist Technological Fix." In *Technology in the Western Political Tradition,* ed. Arthur M. Melzer, Jerry Weinberger, and M. Richard Zinman, 131-61. Ithaca, NY: Cornell University Press, 1993.

Cohill, Andrew, and Andrea Kavanaugh, eds. *Community Networks: Lessons from Blacksburg, Virginia.* Boston: Artech House, 1997.

Communications Security Establishment. "Government of Canada Public Key Infrastructure: White Paper." Ottawa: Ministry of Supply and Services, 1997. Available on-line at www.cse-cst.gc.ca/cse/english/gov.html.

Cooper, Barry. *Sleepers Awake: Technology and the Future of Work.* Melbourne: Oxford University Press Australia, 1995.

Cox, Robert. "The Global Political Economy and Social Choice." In *The New Era of Global Competition: State Policy and Market Power,* ed. Daniel Drache and Meric Gertler, 335-49. Montreal and Kingston: McGill-Queen's University Press, 1991.

Coyne, Andrew. "The Information Highwayman Comes Riding." *Globe and Mail,* 1 October 1994, B1.

Cross, Bill. "Virtual Election Campaigns: Campaign Communication in the Fourth Canadian Party System." Paper presented at the annual meeting of the Atlantic Provinces Political Studies Association, Charlottetown, PEI, 25-7 September 1998.

Cross, Victoria. "Off Our Backs: Frank McKenna's High-Tech Revolution in New Brunswick." *Our Times* 14, 2 (May-June 1995): 20-3.

Dandeker, Christopher. *Surveillance Power and Modernity.* Cambridge, MA: Polity, 1990.

Davidow, William H., and Michael S. Malone. *The Virtual Corporation: Structuring and Revitalizing the Corporation for the 21st Century.* New York: HarperCollins, 1993.

Deibert, Ronald J. *Parchment, Printing, and Hypermedia: Communication in World Order Transformation.* New York: Columbia University Press, 1997.

Denning, Dorothy. "Resolving the Encryption Dilemma: The Case for Clipper." *Technology Review* (July 1995). Available on-line at www.techreview.com/articles/july95/Denning.html.

Dreyfus, Hubert L. "Heidegger on Gaining a Free Relation to Technology." In *Technology and the Politics of Knowledge,* ed. Andrew Feenberg and Alastair Hannay, 97-107. Bloomington: University of Indiana Press, 1995.

Drohan, Madelaine. "Nations See Internet as Threat to Security." *Globe and Mail,* 3 February 1996, A12.

Duxbury, Linda. *Study of Public Sector Workers in Ottawa Region.* Ottawa: Carleton University Press, 1995.

Ehrenreich, Barbara. "Spinning the Poor into Gold: How Corporations Seek to Profit from Welfare Reform." *Harper's* 295, 1767 (August 1997): 44-51.

Eisenstein, Elizabeth L. *The Printing Press As an Agent of Change: Communications and Cultural Transformations in Early Modern Europe.* Cambridge: Cambridge University Press, 1979.

Electronic Frontier Foundation. "EFF and Privacy Organizations Issue Statement at Launch of New Industry-led Alliance on Encryption Controls." Press release, 4 March 1998.

Elkins, David. *Beyond Sovereignty: Territoriality and Political Economy in the Twenty-First Century.* Toronto: University of Toronto Press, 1995.

Elliot, Jeff. "Fighting the (Pretty) Good Fight." *Infobahn* (June 1995): 33-9.

Ericson, Richard, and Kevin Haggerty. *Policing the Risk Society.* Toronto: University of Toronto Press, 1997.

European Commission. *Europe's Way to the Information Society: An Action Plan.* Brussels: European Commission, 1994.

Evans, Mark. "Behind the Screens." *Globe and Mail*, 4 February 1999, D1.

–."Interactive Television to Launch Next Year." *Globe and Mail*, 13 July 1999, A1.

Everett-Green, Robert. "Sweeping U.S. Censorship Bill May Defeat Itself." *Globe and Mail*, 12 March 1996, C4.

Executive Office of the President of the United States. *Enabling Privacy, Commerce, Security and Public Safety in the Global Information Infrastructure.* Washington, DC: Executive Office of the President, 20 May 1996.

–. *The National Information Infrastructure: Agenda for Action.* Washington, DC: Executive Office of the President, 1993.

–. *A Short Summary of the Telecommunications Reform Act of 1996.* Washington, DC: Executive Office of the President, 1996.

"F.B.I. Questions 2 in Tower Blasts." *New York Times*, 20 June 1961, 10.

Flaherty, David. *Protecting Privacy in Surveillance Societies.* Chapel Hill: University of North Carolina Press, 1989.

Flohr, Udo. "Electric Money." *BYTE* 21, 6 (June 1996): 74-84.

Forbes, H.D. "Dahl, Democracy and Technology." In *Democratic Theory and Technological Society*, ed. Richard B. Day, Ronald Beiner, and Joseph Masciulli. Armonk, NY: M.E. Sharpe, 1988.

Forester, Tom. *High Tech Society.* Cambridge, MA: MIT Press, 1988.

Foucault, Michel. *Discipline and Punish: The Birth of the Prison.* Trans. Alan Sheridan. New York: Vintage, 1995.

Fraser, Graham. "Top U.S. Court Defends Internet Freedom." *Globe and Mail*, 27 June 1997, A1.

Freiberger, Paul, and Michael Swaine. *Fire in the Valley: The Making of the Personal Computer.* Berkeley: McGraw-Hill, 1984.

Gandy, O.H. *The Panoptic Sort: A Political Economy of Personal Information.* Boulder, CO: Westview, 1993.

Garfinkel, Simson. *PGP: Pretty Good Privacy.* New York: O'Reilly and Associates, 1995.

Garson, Barbara. *The Electronic Sweatshop: How Computers Are Transforming the Office of the Future into the Factory of the Past.* New York: Penguin, 1988.

Gates, Bill. *The Road Ahead.* New York: Viking, 1995.

Gibbins, Roger, and Carey Hill. "New Technologies and the Future of Civil Society." Paper presented to the annual meeting of the Canadian Communication Association, Ottawa, Ontario, 31 May 1998.

Gibson, William. *Neuromancer.* New York: Ace, 1984.

Gilder, George. *Life after Television: The Coming Transformation of Media and American Life.* New York: Norton, 1994.

Girard, Daniel. "Tories Hit the Internet to Sell Deal to Union." *Toronto Star*, 8 February 1996, A8.

Golding, Peter. "World Wide Wedge: Division and Contradiction in the Global Information Infrastructure." *Monthly Review* 48, 3 (July-August 1996): 70-85.

Goldman, Steven L., Roger N. Nagel, and Kenneth Preiss. *Agile Competitors and Virtual Organizations: Strategies for Enriching the Customer.* New York: Van Nostrand Reinhold, 1995.

Gooderham, Mary. "Concerns Ease over Internet Security." *Globe and Mail,* 7 October 1997, C1.

–. "Microchip at 25: More Power to It." *Globe and Mail,* 15 November 1996, A10.

–. "Microchip Turning into a Cache Cow." *Globe and Mail,* 18 November 1996, A8.

–. "Technology-Race Casualties Find They've Nowhere to Go." *Globe and Mail,* 11 October 1995, A10.

Grace, John. *Annual Report: Canadian Privacy Commissioner.* Ottawa: Ministry of Supply and Services, 1988.

Grant, George. "An Ethic for Community." In *Social Purpose for Canada,* ed. Michael Oliver, 3-26. Toronto: University of Toronto Press, 1961.

–. *Lament for a Nation: The Defeat of Canadian Nationalism.* Toronto: McClelland and Stewart, 1965.

–. *Philosophy in the Mass Age.* Toronto: Copp Clark, 1959.

–. *Technology and Empire: Perspectives on North America.* Toronto: Anansi, 1969.

–. *Technology and Justice.* Toronto: Anansi, 1986.

Grant, George (with Gad Horowitz). "A Conversation on Technology and Man." *Journal of Canadian Studies* 4, 3 (August 1969): 3-6.

Gray, Chris Hables. *Postmodern War: The New Politics of Conflict.* London: Guilford, 1997.

Grossman, Lawrence K. *The Electronic Republic: Reshaping Democracy in the Information Age.* New York: Viking, 1995.

Group of Lisbon. *Limits to Competition.* Cambridge, MA: MIT Press, 1995.

Hafner, Katie. "The Epic Saga of the Well." *WIRED* 5, 5 (May 1997): 98-142.

Hafner, Katie, and Matthew Lyon. *Where Wizards Stay Up Late: The Origins of the Internet.* New York: Simon and Schuster, 1996.

Halonen, Doug. "Malone: Few Will Rule Superhighway." *Electronic Media,* 31 January 1994, 31.

Halton, John. "The Anatomy of Computing." In *The Information Technology Revolution,* ed. Tom Forester, 3-26. Cambridge, MA: MIT Press, 1986.

Hansen, Alvin H. "The Technological Interpretation of History." *Quarterly Journal of Economics* 36 (November 1921): 72-83.

Heidegger, Martin. *Being and Time.* Trans. John Macquarrie and Edward Robinson. New York: Harper and Row, 1962.

–. "Building, Dwelling, Thinking." Trans. William Lovitt. In *Basic Writings,* ed. David Farrell Krell, 319-40. New York: Harper and Row, 1977.

–. *Discourse on Thinking.* Trans. John M. Anderson and E. Hans Freund. New York: Harper and Row, 1959.

–. "Introduction: The Exposition of the Question of the Meaning of Being." Trans. Joan Stambaugh. In *Basic Writings,* ed. David Farrell Krell, 41-89. New York: Harper and Row, 1977.

–. *An Introduction to Metaphysics.* Trans. Ralph Manheim. New Haven, CT: Yale University Press, 1959.

–. "Letter on Humanism." Trans. Frank A. Capuzzi. In *Basic Writings,* ed. David Farrell Krell, 189-242. New York: Harper and Row, 1977.

–. "Only a God Can Save Us Now: An Interview with Martin Heidegger." Trans. David Schendler. New School of Social Research, *Graduate Faculty Philosophy Journal* 6, 1 (1977): 5-27.

–. "The Question Concerning Technology." Trans. William Lovitt. In *Basic Writings,* ed. David Farrell Krell, 283-318. New York: Harper and Row, 1977.

–. "The Turning." In *The Question Concerning Technology and Other Essays,* trans. William Lovitt, 36-49. New York: Harper and Row, 1977.

Heilbroner, Robert. "Do Machines Make History?" *Technology and Culture* 8, 3 (July 1967): 335-45.

Held, David. *Models of Democracy.* 2nd ed. Cambridge, MA: Polity, 1996.

Herman, Edward S., and Noam Chomsky. *Manufacturing Consent: The Political Economy of the Mass Media.* New York: Pantheon, 1988.

Hesiod. "Works and Days." In *Hesiod: The Poems and Fragments,* trans. A.W. Mair, 1-30. Oxford: Clarendon Press, 1908.

Hirschkop, Ken. "Democracy and the New Technologies." *Monthly Review* 48, 3 (July-August 1996): 86-98.

Holzmann, Gerard J., and Bjørn Pehrson. *The Early History of Data Networks.* Los Alamitos, CA: IEEE Press, 1995.

Hood, Webster F. "The Aristotelian versus the Heideggerian Approach to the Problem of Technology." In *Philosophy and Technology: Readings in the Philosophical Problem of Technology,* ed. Carl Mitcham and Robert Mackey, 347-63. New York: Free Press, 1972.

Howard, Robert. *Brave New Workplace.* New York: Viking, 1985.

–. "Strung Out at the Phone Company: How AT&T's Workers Are Drugged, Bugged and Coming Unplugged." *Mother Jones,* August 1981, 39-59.

Industry Canada. *A Cryptography Policy Framework for Electronic Commerce.* Ottawa: Task Force on Electronic Commerce, February 1998.

Information Highway Advisory Council. *Connection, Community, Content: The Challenge of the Information Highway.* Ottawa: Minister of Supply and Services, 1995.

–. *Preparing Canada for a Digital World.* Final Report. Ottawa: Minister of Supply and Services, September 1997.

Innis, Harold A. *The Bias of Communication.* Toronto: University of Toronto Press, 1991.

International Labour Organization. "Monitoring and Surveillance in the Workplace." *Conditions of Work Digest* 12, 1 (1993).

Internet Advertising Bureau. *On-line Advertising Effectiveness Study – Executive Summary.* New York: Internet Advertising Bureau, 1997.

"Jailed in Tower Blasts." *New York Times,* 3 November 1961, 24.

Jameson, Fredric. *Postmodernism, or, The Cultural Logic of Late Capitalism.* Durham, NC: Duke University Press, 1991.

Japan Telecommunications Council. Ministry of Posts and Telecommunications. *Vision 21 for Info-Communications.* Tokyo: Ministry of Posts and Telecommunications, 1997.

Johnson, Theresa. *Go Home ... and Stay There? A PSAC Response to Telework in the Federal Public Sector.* Ottawa: Public Service Alliance of Canada, 1993.

Johnston, David, Deborah Johnston, and Sunny Handa. *Getting Canada Online: Understanding the Information Highway.* Toronto: Stoddart, 1995.

Johnston, David R., and David G. Post. "The Rise of Law on the Global Network." In *Borders in Cyberspace: Information Policy and the Global Information Infrastructure,* ed. Brian Kahin and Charles Nesson, 3-47. Cambridge, MA: MIT Press, 1997.

Kapica, Jack. "Choosing Less Government or Less Gates." *Globe and Mail,* 23 May 1997, A7.

–. "When 'Push' Comes to Shovelling." *Globe and Mail,* 24 October 1997, A8.

Kelly, Kevin. *Out of Control: The Rise of Neo-Biological Civilization.* Reading, MA: Addison-Wesley, 1994.

–. "What Would McLuhan Say?" *WIRED* 4, 10 (October 1996): 148.

Kerckhove, Derrick de. *Connected Intelligence: The Arrival of the Web Society.* Toronto: Somerville House, 1997.

–. *The Skin of Culture: Investigating the New Electronic Reality.* Toronto: Somerville House, 1995.

Kerr, Ann. "Mondex Trial Getting Mixed Results." *Globe and Mail,* 9 December 1997, C20.

–. "Smart Uses for Smart Cards." *Globe and Mail,* 9 December 1997, C20.

Kline, David. "The Embedded Internet." *WIRED* 4, 10 (October 1996): 98-106.

Krantz, Michael. "Click Till You Drop." *Time,* 20 July 1998, 14-9.

Kurtzman, Joel. *The Death of Money: How the Electronic Economy Has Destabilized the World's Markets and Created Financial Chaos.* New York: Simon and Schuster, 1993.

Lam, Kenneth, and Herbert Lin, eds. *Cryptography's Role in Securing the Information Society.* Washington, DC: National Research Council, 1996.

Landes, David S. *Revolution in Time: Clocks and the Making of the Modern World.* Cambridge, MA: Belknap, 1983.

–. *The Unbound Prometheus: Technological Change and Industrial Development in Western Europe from 1750 to the Present.* Cambridge: Cambridge University Press, 1969.

Landow, George. *Hypertext: The Convergence of Contemporary Critical Theory and Technology.* Baltimore, MD: Johns Hopkins University Press, 1992.

Laudon, Kenneth. *The Dossier Society: Value Choices in the Design of National Information Systems.* New York: Columbia University Press, 1986.

Lawton, Valerie. "Big Brother at Work." *Toronto Star,* 10 November 1997, C4.

Lebow, Irwin. *Information Highways and Byways: From the Telegraph to the 21st Century.* New York: IEEE Press, 1995.

Leiss, William. *The Domination of Nature.* New York: George Braziller, 1972.

–. "The Myth of the Information Society." In *Cultural Politics in Contemporary America,* ed. Ian Angus and Sut Jhally, 282-98. New York: Routledge, 1989.

Leiss, William, Sut Jhally, and Stephen Kline. *Social Communication in Advertising: Persons, Products and Images of Well-Being.* Scarborough, ON: Nelson, 1990.

Lenk, Klaus. "The Challenge of Cyberspatial Forms of Human Interaction to Territorial Governance and Policing." In *The Governance of Cyberspace: Politics, Technology and Global Restructuring,* ed. Brian D. Loader, 126-35. London: Routledge, 1997.

Lévi-Strauss, Claude. *The Savage Mind.* Chicago: University of Chicago Press, 1968.

Lopez, Barry. "Before the Temple of Fire." *Harper's* 296, 1771 (January 1998).

Lorinc, John. "Information Revolution Puts Copyright under Siege." *Globe and Mail,* 29 October 1996, D1.

Luke, Timothy W. *Screens of Power: Ideology, Domination and Resistance in Informational Society.* Chicago: University of Illinois Press, 1989.

Lyon, David. *The Electronic Eye: The Rise of Surveillance Society.* Minneapolis: University of Minnesota Press, 1994.

–. "An Electronic Panopticon? A Sociological Critique of Surveillance Theory." *Sociological Review* 41, 4 (1993): 653-78.

–. *The Information Society: Issues and Illusions.* Cambridge, MA: Polity, 1988.

Lyotard, Jean-François. *The Postmodern Condition: A Report on Knowledge.* Trans. Geoff Bennington and Brian Massumi. Minneapolis: University of Minnesota Press, 1984.

McChesney, Robert W. "The Global Struggle for Democratic Communication." *Monthly Review* 48, 3 (July-August 1996): 1-20.

–. "The Internet and U.S. Communication Policy-Making in Historical and Critical Perspective." *Journal of Communication* 46, 1 (Winter 1996): 98-124.

–. "Public Broadcasting in the Age of Communication Revolution." *Monthly Review* 47, 7 (December 1995): 1-19.

McDermott, Darren. "Singapore Curbs Flow of Internet Material." *Globe and Mail,* 7 March 1996, A11.

MacEachen, Allan J., and Jean Robert Gauthier. *Canada's Foreign Policy: Principles and Priorities for the Future.* Report of the Special Joint Committee of the Senate and House of Commons Reviewing Canada's Foreign Policy. Ottawa: Parliamentary Publications Directorate, 1994.

McKenna, Barrie. "Microsoft Avoids Contempt Charges in Browser Deal." *Globe and Mail,* 23 January 1998, B6.

Mackenzie, Donald. "Marx and the Machine." *Technology and Culture* 25, 3 (July 1984): 473-502.

McLuhan, Marshall. *The Gutenberg Galaxy: The Making of Typographical Man.* Toronto: University of Toronto Press, 1962.

Manes, Stephen, and Paul Andrews. *Gates: How Microsoft's Mogul Reinvented an Industry – And Made Himself the Richest Man in America.* New York: Doubleday, 1993.

Manley, John. "Third Reading Speaking Notes: Bill C-6, Personal Information Protection and Electronic Documents Act." 22 October 1999. Available online at www.ic.gc.ca.

Marleau, Diane. "Stop the Information Revolution, They Want to Get On." *Globe and Mail,* 19 June 1997, A19.

Marron, Kevin. "Electronic Commerce Raises Some Taxing Questions." *Globe and Mail,* 9 December 1997, C3.

–. "How to Make E-banking Pay." *Globe and Mail,* 11 November 1998, C1.

Marx, Karl. *Capital.* Vol. 1. Trans. Samuel Moore and Edward Aveling. Moscow: Progress, 1978.

–. *Capital.* Vol. 3. Trans. Samuel Moore and Edward Aveling. Moscow: Progress, 1986.

–. "The Difference between Democritean and Epicurean Philosophies of Nature." In *The Marx-Engels Reader,* 2nd ed., ed. Robert C. Tucker, 9-11. New York: Norton, 1978.

–. "Economic and Philosophic Manuscripts of 1844." In *The Marx-Engels Reader,* 2nd ed., ed. Robert C. Tucker, 66-125. New York: Norton, 1978.

–. "The *Grundrisse.*" In *The Marx-Engels Reader,* 2nd ed., ed. Robert C. Tucker, 221-93. New York: Norton, 1978.

–. *The Poverty of Philosophy.* Trans. Harry Quelch. Moscow: Progress, 1984.

–. "Preface: A Contribution to the Critique of Political Economy." In *The Marx-Engels Reader,* 2nd ed., ed. Robert C. Tucker, 3-6. New York: Norton, 1978.

Marx, Karl, and Friedrich Engels. *The German Ideology.* Moscow: Progress, 1976.

–. *Manifesto of the Communist Party.* Trans. Samuel Moore. Moscow: Progress, 1986.

Masciulli, Joseph. "Rousseau versus Instant Government: Democratic Participation in the Age of Telepolitics." In *Democratic Theory and Technological Society,* ed. R.B. Day, Ronald Beiner, and J. Masciulli, 150-64. London: M.E. Sharpe, 1988.

Matrix Information and Directory Services. "The State of the Internet." *Matrix Maps Quarterly,* 401 (January 1997): 1-6.

Meeks, Brock. "The Cyber Rights Report Card." *WIRED* 4, 10 (October 1996): 94.

Mehta, Stephanie. "AT&T Faces Hurdles in Integrating TCI." *Globe and Mail,* 25 June 1998, B13.

Meiksins, Peter. "Work, New Technology and Capitalism." *Monthly Review* 48, 3 (July-August 1996): 99-114.

Menzies, Heather. "Telework, Shadow Work: The Privatization of Work in the New Digital Economy." *Studies in Political Economy* 53 (Summer 1997): 103-23.

–. *Whose Brave New World? The Information Highway and the New Economy.* Toronto: Between the Lines, 1996.

Middlemann, Conner. "Lines of Investors Buzz with Avalanche of Issues." *Financial Times,* 12 June 1995.

Milner, Brian. "Regulators Attack Microsoft." *Globe and Mail,* 19 May 1998, B1.

Mitchell, Alanna. "Computers Taking Root in the Workplace." *Globe and Mail,* 7 June 1995, A8.

Monden, Yasuhiro. *Toyota Production System.* Cambridge, MA: Industrial Engineering and Management Press, 1983.

Monroe, James A. *The Democratic Wish: Popular Participation and the Limits of American Government.* New York: Basic Books, 1992.

Montgomery, John. "Fiber in the Sky: The Orbiting Internet." *BYTE* 22, 11 (November 1997): 58-70.

Mulroney, Catherine. "Canadians Tiptoe into Telework Era." *Globe and Mail,* 10 June 1997, C1.

Mumford, Lewis. *Technics and Civilization.* New York: Harcourt, Brace and World, 1963.

Myles, John. "The Expanding Middle: Some Canadian Evidence on the Deskilling Debate." *Canadian Review of Sociology and Anthropology* 25, 3 (August 1988): 335-64.

–. "Post-Industrialism and the Service Economy." In *The New Era of Global Competition: State Policy and Market Power,* ed. Daniel Drache and Meric S. Gertler, 350-66. Montreal and Kingston: McGill-Queen's University Press, 1991.

Negroponte, Nicholas. *Being Digital.* New York: Knopf, 1995.

Nelson, Theodor. *Dream Machines.* Self-published, 1974.

–. *Literary Machines.* South Bend, IN: Distributor's, 1987.

Network Wizards. "Internet Domain Survey, January 1999." www.nw.com.

Neumann, John von. "First Draft of a Report on the EDVAC." In *From ENIAC to UNIVAC: An Appraisal of the Eckert-Mauchly Computers,* ed. Nancy Stern, 177-246. Bedford, MA: Digital Press, 1981.

Newman, Nathan. "From Microsoft Word to Microsoft World: How Microsoft Is Building a Global Monopoly." NetAction White Paper, NetAction/The Tides Center, San Francisco, 1997. Available on-line at www.netaction.org/msoft/world.

Nietzsche, Friedrich. *The Gay Science.* Trans. Walter Kaufmann. New York: Random House, 1974.

Noble, David F. *Forces of Production: A Social History of Industrial Automation.* New York: Knopf, 1984.

–. *Progress without People: New Technology, Unemployment, and the Message of Resistance.* Toronto: Between the Lines, 1995.

Noble, Kimberley. "The Data Game." *Maclean's,* 17 August 1998, 14-19.

Norris, Pippa. "Who Surfs? New Technology, Old Voters and Virtual Democracy in America." Paper presented at the third annual meeting of the John F.

Kennedy Visions of Governance for the Twenty-First Century, Mount Washington Hotel, Bretton Woods, 19-22 July 1998.

Office of Science and Technology Policy. "President Clinton Welcomes Plan to Strengthen U.S. Leadership in Information Technology." Press release, 10 August 1998, Washington, DC.

Oldenburg, Ray. *The Great Good Place: Cafes, Coffee Shops, Community Centers, Beauty Parlors, General Stores, Bars, Hangouts and How They Get You Through the Day.* New York: Paragon House, 1991.

Oldfield, Margaret. "Heaven or Hell? Telework and Self-Employment." *Our Times* 14, 2 (May-June 1995): 16-19.

Orwell, George. *Nineteen Eighty-Four.* New York: Harcourt, Brace and World, 1949.

Parenti, Michael. *Inventing Reality: The Politics of the Mass Media.* New York: St. Martin's Press, 1986.

Parrot, Jean-Claude. "Minority Report." In *Connection, Community, Content: The Challenge of the Information Highway.* Final Report of the Information Highway Advisory Council, 215-27. Ottawa: Minister of Supply and Services Canada, 1995.

Pauly, Louis W. "Capital Mobility, State Autonomy, and Political Legitimacy." *Journal of International Affairs* 48, 2 (Winter 1995): 369-89.

Pescovitz, David. "The Future of Money." *WIRED* 4, 8 (August 1996): 68.

Phillips, Bruce. *Annual Report: Privacy Commissioner, 1992-1993.* Ottawa: Privacy Commissioner of Canada, 1993.

Pine, B. Joseph II. *Mass Customization: The New Frontier in Business Competition.* Cambridge, MA: Harvard Business School, 1993.

Piore, Michael J., and Charles F. Sabel. *The Second Industrial Divide: Possibilities for Prosperity.* New York: Basic Books, 1984.

Plato. *Gorgias.* Trans. W. Hamilton. Harmondsworth: Penguin, 1960.

Poster, Mark. *Foucault, Marxism and History: Mode of Production versus Mode of Information.* Cambridge, MA: Polity, 1984.

–. *The Mode of Information: Poststructuralism and Social Context.* Chicago: University of Chicago Press, 1990.

–. *The Second Media Age.* Cambridge, MA: Polity, 1995.

Potter, Mark, and Marc Lee. "The Economic Impacts of the Information Highway: An Overview." Discussion paper prepared for the Task Force on Growth, Employment and Competitiveness, Information Highway Advisory Council, Industry Canada, July 1995.

Public Service Alliance of Canada. "Telework." PSAC Policy no. 33. Internal policy document.

Putnam, Robert D. "Bowling Alone? America's Declining Social Capital." *Journal of Democracy* 6, 1 (January 1995): 65-78.

Raboy, Marc. "Cultural Sovereignty, Public Participation, and Democratization of the Public Sphere: The Canadian Debate on the New Information Infrastructure." In *National Information Infrastructure Initiatives: Vision and*

Policy Design, ed. Brian Kahin and Ernest Wilson, 190-216. Cambridge, MA: MIT Press, 1997.

–. *Missed Opportunities: The Story of Canada's Broadcasting Policy*. Montreal and Kingston: McGill-Queen's University Press, 1990.

Raggio, Olga. "The Myth of Prometheus: Its Survival and Metamorphoses up to the Eighteenth Century." *Journal of the Warbourg and Courtauld Institutes* 21, 1 (1958): 44-62.

Ray, Randy. "Phone Banking Keeps Growing." *Globe and Mail*, 7 October 1997, C2.

Reidenberg, Joel R. "Governing Networks and Rule-Making in Cyberspace." In *Borders in Cyberspace: Information Policy and the Global Information Infrastructure*, ed. Brian Kahin and Charles Nesson, 84-105. Cambridge, MA: MIT Press, 1997.

Rheingold, Howard. "The Virtual Community." *Utne Reader*, March-April 1995, 63.

–. *The Virtual Community: Homesteading on the Electronic Frontier*. Reading, MA: Addison-Wesley, 1993.

Richards, Greg. "Businesslike Government: The Ultimate Oxymoron?" *Optimum: The Journal of Public Sector Management* 27, 1 (1996): 21-5.

Rifkin, Jeremy. *The End of Work: The Decline of the Global Labour Force and the Dawn of the Post-Market Era*. New York: Putnam, 1995.

Roberts, Bruce. "From Lean Production to Agile Manufacturing: A New Round of Quicker, Cheaper and Better." In *Re-shaping Work: Union Responses to Technological Change*, ed. Christopher Schenk and John Anderson, 197-215. Don Mills, ON: Ontario Federation of Labour, 1995.

Robertson, David, and Jeff Wareham. *Technological Change in the Canadian Auto Industry*. Toronto: Canadian Auto Workers, 1988.

Robins, Kevin, and Frank Webster. "Cybernetic Capitalism: Information, Technology, Everyday Life." In *The Political Economy of Information*, ed. Vincent Mosco and Janet Wasko, 44-75. Madison: University of Wisconsin Press, 1988.

Rochlin, Gene I. *Trapped in the Net: The Unanticipated Consequences of Computerization*. Princeton, NJ: Princeton University Press, 1997.

Rosenberg, Nathan. "Marx as a Student of Technology." *Monthly Review* 28, 3 (Summer 1976): 56-77.

Rousseau, Jean-Jacques. "A Discourse on the Moral Effects of the Arts and Sciences." In *The Social Contract and Discourses*, trans. G.D.H. Cole, 1-29. London: Everyman, 1988.

Rowan, Geoffrey. "Microsoft On-Line Service to Mimic TV." *Globe and Mail*, 14 October 1996, B1.

–. "PCTV." *Report on Business Magazine*, June 1997, 86-92.

–. "Surfers Beware: Bosses Can Now Watch You On-Line." *Globe and Mail*, 10 July 1996, B1.

Ruggie, John Gerard. "Territoriality and Beyond: Problematizing Modernity in International Relations." *International Organization* 47, 1 (Winter 1993): 139-74.

Rule, James B. *Private Lives and Public Surveillance.* London: Allen Lane, 1973.

Rushkoff, Douglas. *Cyberia: Life in the Trenches of Hyperspace.* San Francisco: HarperCollins, 1994.

–. "E-mail with Douglas Rushkoff." *Shift* 4, 4 (April 1996): 12.

Sale, Kirkpatrick. *Rebels against the Future: The Luddites and Their War on the Industrial Revolution; Lessons for the Computer Age.* Reading, MA: Addison-Wesley, 1995.

Sallot, Jeff. "Ottawa Weaving Tight Web Privacy Law." *Globe and Mail,* 27 January 1998, A1.

Salus, Peter H. *Casting the Net: From ARPANET to INTERNET and Beyond.* Reading, MA: Addison-Wesley, 1995.

Schellenberg, Grant. *The Changing Nature of Part-time Work.* Ottawa: Canadian Council on Social Development, 1997.

Schenk, Christopher, and John Anderson, eds. *Re-shaping Work: Union Responses to Technological Change.* Don Mills, ON: Ontario Federation of Labour, 1995.

Schiller, Herbert. *Culture Inc.: The Corporate Takeover of Public Expression.* New York: Oxford, 1989.

–. "Information for What Kind of Society?" In *The Information Society: Economic, Social and Structural Issues,* ed. Jerry L. Salvaggio. Hillsdale, NJ: Lawrence Erlbaum, 1989.

–. "Information Superhighway: Paving over the Public." *Z Magazine,* March 1994, 48.

Schuler, Douglas. "Community Networks: Building a New Participatory Medium." *Communications of the ACM* 37, 1 (January 1994): 39-51.

–. *New Community Networks: Wired for Change.* New York: ACM Press, 1996.

Scofield, Heather. "OECD Pledges Light Touch on Internet." *Globe and Mail,* 15 October 1998, D1.

Shade, Leslie Regan. "Computer Networking in Canada: From CA*net to CANARIE." *Canadian Journal of Communication* 19, 1 (Winter 1994): 53-69.

Shaw, William M. "'The Handmill Gives You the Feudal Lord': Marx's Technological Determinism." *History and Theory* 18 (1979): 155-76.

Shelley, Mary. *Frankenstein, or, The Modern Prometheus.* London: Oxford University Press, 1969.

Shenk, David. *Data Smog: Surviving the Information Glut.* New York: HarperCollins, 1998.

Simeon, Richard, George Hoberg, and Keith Banting. "Globalization, Fragmentation, and the Social Contract." In *Degrees of Freedom: Canada and the United States in a Changing World,* ed. Keith Banting, George Hoberg, and Richard Simeon, 389-416. Montreal and Kingston: McGill-Queen's University Press, 1997.

Simons, John. "U.S. to Allow Export of Encryption Technology." *Globe and Mail*, 19 October 1998, B13.

Slaton, Christa Daryl. *Televote: Expanding Citizen Participation in the Quantum Age*. New York: Praeger, 1992.

Smythe, Dallas W. *Dependency Road: Communications, Capitalism, Consciousness and Canada*. Norwood, NJ: Ablex, 1981.

Solomon, Evan. "Unlikely Messiah." *Shift* 4, 1 (1995): 31.

Sombart, Werner. *Der Moderne Kapitalismus*. 4 vols. Munich: Duncker and Humbolt, 1928.

Stallabrass, Julian. "Empowering Technology: The Exploration of Cyberspace." *New Left Review* 211 (May-June 1995): 3-32.

Stallings, William. *Protect Your Privacy: A Guide for PGP Users*. New York: Prentice-Hall, 1995.

Statistics Canada. *General Social Survey and General Household Equipment Survey* (11-612-mpe). Ottawa: Statistics Canada, 1995.

–. *Household Facilities and Equipment 1996* (64-202-xpd). Ottawa: Statistics Canada, 1996.

Stoll, Clifford. *Silicon Snake Oil: Second Thoughts on the Information Highway*. New York: Doubleday, 1995.

Stross, Randall E. *The Microsoft Way: The Real Story of How the Company Outsmarts Its Competition*. Reading, MA: Addison-Wesley, 1996.

Surtees, Lawrence. "Holy Grail of Convergence Still Ahead." *Globe and Mail*, 6 May 1997, C1.

Tausz, Andrew. "Data Warehouses Store Wealth of Information." *Globe and Mail*, 9 December 1997, C1.

–. "Store Tests Do-It-Yourself Checkout." *Globe and Mail*, 9 December 1997, C4.

Taylor, Frederick Winslow. *Principles of Scientific Management*. New York: Norton, 1967.

Thompson, Clive. "Beyond the Microchip." *Report on Business Magazine*, June 1997, 78-86.

Thompson, E.P. *The Making of the English Working Class*. New York: Knopf, 1963.

Toffler, Alvin, and Heidi Toffler. *Creating a New Civilization: The Politics of the Third Wave*. Atlanta, GA: Tower, 1995.

Traber, Michael. *The Myth of the Information Revolution: Social and Ethical Implications of Communication Technology*. London: Sage, 1986.

Tremblay, Gaëtan. "The Information Society: From Fordism to Gatesism." *Canadian Journal of Communication* 20, 4 (1995): 461-82.

Tsagarousianou, Roza, Damian Tambini, and Cathy Bryan, eds. *Cyberdemocracy: Technology, Cities and Civic Networks*. London: Routledge, 1998.

Turing, Alan. "Computer Machinery and Intelligence." *Mind* 59 (1950): 433-60.

–. "On Computable Numbers, with an Application to the *Entscheidungsproblem*." *Proceedings of the London Mathematical Society*, 2nd ser., 42 (1936): 230-65.

Turkle, Sherry. *Life on Screen: Identity in the Age of the Internet*. New York: Simon and Schuster, 1995.

"Utah Blasts Cut 3 Phone Relays." *New York Times*, 29 May 1961, 1.

Vipond, Mary. *The Mass Media in Canada*. Toronto: James Lorimer, 1989.

Wallace, James. *Overdrive: Bill Gates and the Race to Control Cyberspace*. New York: John Wiley, 1997.

Wallace, James, and Jim Erickson. *Hard Drive: Bill Gates and the Making of the Microsoft Empire*. New York: John Wiley, 1992.

Wallace, Jonathan, and Mark Mangan. *Sex, Laws and Cyberspace: Freedom and Censorship on the Frontiers of the Online Revolution*. New York: Henry Holt, 1997.

Warner, Malcolm, and Michael Stone. *The Data Bank Society: Organizations, Computers and Social Freedom*. London: George Allen and Unwin, 1970.

Weber, Max. "Bureaucracy." In *From Max Weber: Essays in Sociology*, trans. and ed. H.H. Gerth and C. Wright Mills, 196-266. New York: Oxford University Press, 1958.

Weil, Simone. *The Need for Roots*. London: Routledge and Kegan Paul, 1952.

Weinberger, Jerry. *Science, Faith, and Politics: Francis Bacon and the Utopian Roots of the Modern Age*. Ithaca, NY: Cornell University Press, 1985.

Weizenbaum, Joseph. *Computer Power and Human Reason: From Judgment to Calculation*. San Francisco, CA: W.H. Freeman, 1976.

Wellman, Barry, Janet Salaff, Dimitrina Dimitrova, Laura Garton, Milenia Gulia, and Caroline Haythornthwaite. "Computer Networks as Social Networks: Collaborative Work, Telework, and Virtual Community." *Annual Review of Sociology* 22 (1996): 213-38.

White, Ted, MP, Reform Party of Canada. Personal correspondence, 22 November 1995.

Wiener, Norbert. *Cybernetics: Or Control and Communications in the Animal and Machine*. Cambridge, MA: MIT Press, 1948.

–. *The Human Use of Human Beings: Cybernetics and Society*. Boston: Houghton Mifflin, 1950.

Wilke, John R., and Bryan Gruley. "In Merger Blitz, Regulators' Profiles Rise." *Globe and Mail*, 11 June 1998, B14.

Wilke, John R., and David Banke. "Appeals Court Rules for Microsoft." *Globe and Mail*, 24 June 1998, B12.

Wilson, David L. "Microsoft vs. US Goes to Court Friday." *San Jose Mercury News*, 1 December 1997.

Winner, Langdon. *Autonomous Technology: Technics-out-of-Control as a Theme in Political Thought*. Cambridge, MA: MIT Press, 1977.

–. "Citizen Virtues in a Technological Order." In *Technology and the Politics of Knowledge*, ed. Andrew Feenberg and Alastair Hannay, 65-84. Bloomington: University of Indiana Press, 1995.

–. *The Whale and the Reactor: A Search for Limits in an Age of High Technology*. Chicago: University of Chicago Press, 1986.

Winseck, Dwayne. "Power Shift? Towards a Political Economy of Canadian Telecommunications and Regulation." *Canadian Journal of Communication* 20, 1 (1995): 81-106.

Wolf, Gary. "The Curse of Xanadu." *WIRED* 3, 6 (June 1995): 137-52.

Womack, James, Daniel T. Jones, and Daniel Roos. *The Machine That Changed the World: The Story of Lean Production.* New York: Macmillan, 1990.

Wood, Ellen Meiksins. "Modernity, Postmodernity or Capitalism?" *Monthly Review* 48, 3 (July-August 1996): 21-39.

Wriston, Walter B. *The Twilight of Sovereignty: How the Information Revolution Is Transforming Our World.* New York: Macmillan, 1992.

Zuboff, Shoshana. *In the Age of the Smart Machine: The Future of Work and Power.* New York: Basic Books, 1988.

index

Accessibility, of computer networks: access devices, 169-70; and "information highway," 167

Adler, Paul, 152

Advanced Research Projects Agency (ARPA): communications research, 68; establishment by US Defense Department, 68

Advertising: on computer networks, 181-7; on television, 181-2, 184

Agenda for Action (NII), 113, 173

Agile production methods, 125, 127, 128-9

Air Miles program, data mining by, 227-8

Alienation: replaced by fragmentation, in postmodern life, 197; and virtual communities, 212

Altair 8800 (personal computer), 67

Aluminum, 9

America Online (AOL): as commercial computer network, 110; news and discussion groups provider, 91; purchase of NSFNet, 112

American Management Association, on workplace electronic surveillance, 157

American Republican Army (anti-technology group), 68

American Telephone and Telegraph Company. *See* AT&T

Americans for Computer Privacy, 242

AND (in Boolean algebra), 64, 65

Andrew, Edward, 27-8, 152

AOL. *See* America Online (AOL)

Apple Computers: as new elite corporation, 106; as pioneer in personal computers, 67

Applications, for computer networks: electronic mail, 76-8; FTP (file transfer protocol), 76; telnet, 76

Aristotle: on democracy, 23; on *logos*, 28, 30, 55; on politics and technology, 54-5; on science of politics, 237; on *technē*, 28, 30, 32-4, 34, 55

Aronowitz, Stanley, 134, 139-40, 144

ARPANET: initial host computers, 73; origins of, 73-6; replacement by NSFNET, 79

Art: and digital media, 192-5; postmodern, as assemblage, 198, 203

Artificial intelligence, enabled by networks, 201

Asynchronous communication, of computer networks, 86

Atomic bomb, 11

AT&T: Global Information Systems Architecture, 128; purchase of TCI conglomerate, 116; sabotage of microwave relay stations, 68

Audio recording, as transcendence of space and time, 12
Auto industry, downsizing due to computerized technology, 136
Automobiles, as consumer goods, 9

Bacon, Francis: on benefits of science, 196; cryogenic experiments, 10; and myth of Prometheus, 6
Bakelite plastic, 9
Bandwidth: increase, financing, 169; transmission limitations, 168
Banking: banking machines, as computer network input terminals, 83; cost savings of on-line transactions, 181; debit cards, and data trail, 176; electronic workplace monitoring, 161; employment decrease in, due to network technology, 138-9; on-line, and transaction security, 179-80; smart cards, 178
Banks, Russell, 192
Banner advertising, on computer networks, 185-7
Baran, Nicholas, 112
Baran, Paul, 68
Barlow, John Perry, 3, 4, 238
BBS (Bulletin Board Service), 81. *See also* Discussion groups, on-line; Mailing lists, on-line
Because It's Time Network (BITNET), 79
Beiner, Ronald, 16, 25-6
Being Digital (Negroponte), 208
Bell Canada, electronic workplace surveillance, 158
Benefits, lacking in fringe jobs, 143
Beniger, James, 98, 99-100
Bentham, Jeremy, 156
Berners-Lee, Tim, 92
Bernstein, Daniel, 241-2

Bible, as instrument of socialization, 12
Bill C-6 (Personal Information Protection and Electronic Documents Act) (Canada), 160-1, 229
Binary digital notation, 62-3, 72-3
Biology, and new technologies, 10
BITNET (Because It's Time Network), 79
Boeing, and Teledesic program, 168
Bolter, David, 220
Books, historical impact of, 12
Boole, George, 64
Boolean algebra, 64-5
Braverman, Harry, 151
Broadcast technologies: audiences as product, 13; convergence with telecommunications, 257; indifference to content, 85; and socialization, 13; and stimulation of consumption, 13. *See also* Push technology; Television
Broadcasting Act (Canada), 255, 258
Bryant, Susan, 156
Bulletin Board Service (BBS), 81. *See also* Discussion groups, on-line; Mailing lists, on-line

Calculation by computers, as replacement of human judgment, 219-20
Call centres: electronic workplace monitoring, 157-8; location in low-wage zones, 141-2
Cameras, as computer network input terminals, 83
Canada: CA*net, 78, 79, 112; cultural protectionism, 255, 256, 258-9; cultural values, and pressure of network homogeneity, 252-3; de-institutionalization of labour, 143; employment patterns, by sector, 137; and

globalization, 253; government stand on encryption, 243-4; government support of technological development, 111, 254-60; NSFNET connection at University of Toronto, 79; privatization of computer networks, 112-14; social policy, and influence of globalization, 253; telecommunications policy, 255-7; unemployment rate, 135. *See also* Canadian Network for the Advancement of Research in Industry and Education (CANARIE); Information Highway Advisory Council (IHAC)

Canada Employment Centres, replacement by kiosks, 139

Canadian Auto Workers, study of downsizing in auto industry, 136

Canadian Broadcasting Corporation web site, 94-6

Canadian Council on Social Development, on declining full-time employment, 143-4

Canadian Labour Code, silent on electronic workplace surveillance, 159

Canadian Network for the Advancement of Research in Industry and Education (CANARIE), 111

Canadian Radio-Television and Telecommunications Commission (CRTC), 255, 258

Canadian Standards Association (CSA), "National standard for the Protection of Personal Information," 160

Canadian Union of Public Employees (CUPE), on decrease in full-time positions, 144

CANARIE (Canadian Network for the Advancement of Research in Industry and Education), 111

CA*net, 78, 79, 112

Capital: mobility, and deterritorialization of work, 142-3; movable, as opposite of rootedness, 209

Capitalism: and flexible manufacturing, 125-7; "friction-free," according to Bill Gates, 121-4; and liberalism, 251-3; and machines, according to Marx, 36-40; national boundaries as political formality, 251; and network mode of production, 104-31; and network technology, 107-8, 123-4, 187-8

"Capstone" (Clipper chip encryption), 242

Cards, and digital financial transactions, 176

Caribbean Islands, and data-processing industry, 141

Carpal tunnel syndrome, 161

Cash registers, as computer network input terminals, 83

CBS television network, owned by Westinghouse, 116

CD-ROM (Compact Disc-Read Only Memory), 80-1

Central processing unit (CPU), of computer, 64-5

Chemical warfare, 11

Church, influence replacement by broadcast technologies, 12

Circuit switching, 70-1

Clement, Andrew, 158

Clinton, Bill, 261

Clipper chip encryption, 242-3

Coal, and industrialization, 9

"Colossus" (antecedent of computers), 59

Command and Control Research group (US), 68

CommerceNet, 129

Communication technologies: and consciousness, 11-14; definition of communication, 29-30; and democratic aspirations, 20-1. *See also* Technological change, historical

Communications Act (US), 113

Communications Decency Act (US), 244, 245

Communications industry, employment decrease in, 138

Communism: as egalitarian distribution of abundance, 40. *See also* Marx, Karl

Communities, on-line: attachment, and virtual communities, 212-16; compared with real communities, 211-12, 216; and democratic character of computer networks, 21; marginal, communitarian cultures, 105-6; as representative of localized communities, 216-18

Compact Disc-Read Only Memory (CD-ROM), 80-1

Compression, and transmission speed, 168

Computer networks: advertising techniques, 181-7; applications, 79-80; audience creation challenge, for marketers, 181-2, 184-5; and calculation, 219-31; and capitalist mode of production, 104-31; as commercial technology, 163-87; as communications utilities, 85-92, 102; concentrated ownership of, 114-15; as control utilities, 97-101, 103, 120-4, 222; and converging digital technologies, 170-1, 257; data mining by, 224-31; decentralization, appearance of, 149-51; and democracy, 17-21, 24-6, 189-91, 239, 245-6, 262, 264, 266-7; economic effects, 105-6, 189-90; electronic surveillance by, 222-31; encryption, and law enforcement, 240-4; and globalization, 246-51; government of, 236-46; and homogeneous state, 251-64; horizontal integration, 116; as information utilities, 80-4, 102; input terminals, 83-4; integration of production, consumption, and exchange, 164; international networks, 78-9; and knowledge, 3; and marketability of goods, 181-7; as mode of transaction, 175-81; most prevalent in North America and Europe, 58; ownership of, 109-10, 119-20; portrayal as challenge to existing political hierarchy, 19-21; portrayal as ungovernable, 238-40, 242, 245; privatization, 112-14; and production practices, 124-31; proprietary, 110, 179; as public space, 110, 218; and social considerations, 231-5; speed advantage over other communications media, 87; as stand-in for human activities and experience, 264-8; support of inequalities created by capitalism, 189-90; TCP/IP (Transmission Control Protocol/Internet Protocol) standard, 78; technical indifference to content, 85; transaction facilitation, 165-7; vertical inte-gration, 115-19. *See also* Communities, on-line; Computer networks, history; Computers; Computers, history; Network technology; Surveillance, electronic

Computer networks, history: alternative links as "redundancy," 68-70; application software, 76-9; binary digital notation, 72-3; defence research origins, 67-78, 111; design, 67-78, 101-3; distributed network design, 68-70;

packet switching, 71-2. *See also* Computer networks; Computers; Computers, history; Network technology

Computers: binary digital notation, 62-4; characteristics, 61-70; "Colossus," 59; compared with calculating machines, 61; CPU (central processing unit), 64-5; definition, 61, 65; floppy disk, 62; hard disk, 62; input devices, 64; memory, 61-2

Computers, history: ENIAC, 59-60; Manchester Mark I, 60; miniaturization of, 66-7; UNIVAC (Universal Automatic Computer), 60-1, 66

Consciousness, and technology, 11-14

Consent, manufacture of, 13

Construction industry, downsizing, 136-7

Consumption: and data mining, 224-31; as freedom, under capitalist liberalism, 263; manufacture of needs, 13; and socialization, 225

"Control Revolution," 98-9

CPU (central processing unit), of computer, 64-5

Craftsmanship, disruption of, 151-2

Credit cards, 176

Criminal Code of Canada, on employee surveillance, 159

CRTC (Canadian Radio-Television and Telecommunications Commission), 255, 258

Cryogenics, 10

Cryptology, digital. *See* Encryption, digital

CSNET (National Science Foundation), 78

Cybernetics: Greek origins of word, 98, 236; as mechanical and unconscious control, 236

Cyberpunk science fiction, 210

Cyberspace: government of, 236-46; Greek origins of word, 98

Data mining, 224-31. *See also* Privacy

Databases: as collections of information, 81-2; cross-industry, 129; networked, 81-4; searching, 84; view of world as, by network technology, 209

Data-processing industry: effect of digital scanning on employment rates, 141; migration to lower wage countries, 140-1

Davies, Donald, 68

Da Vinci. *See* Leonardo

Debit cards, 176

Deibert, Ronald, 15, 175, 187, 198-9

Democracy: according to Aristotle, 23; and capitalist philosophy, 189; characteristics, 23; and citizen participation, 22-3; and computer networks, 17-21, 24-6, 189-91, 239, 245-6, 262, 264, 266-7; definition, 22, 189; as different from liberalism, 23, 263-4; and equality of opportunity, 22; and interconnectivity, 173-5. *See also* Liberty

Deregulation, and globalization, 122

Derrida, Jacques, 14

Deterritorialization, of work, 140-3

DiFazio, William, 134, 139-40, 144

Digital Revolution: and social change, 18; subject to traditional political and economic control, 107

Digitization: of human activities, 82-4; of information, by computers, 63-4; scanning, effect on data-processing industry, 141; special effects, and reality, 192-5; of voice, 87

Discussion groups, on-line, 90. *See also* Communities, on-line

Disks, of computer: floppy, 62; hard, 62

Disney Corporation, vertical integration of, 115-16

Distributed networks, 70

DOS, compared with "point and click" interfaces, 201

Downloading, from networks, 81

Dreamworks Interactive, 117

Dreamworks SKG, 117

Eckert, Presper, 59

Eckert-Mauchly Computer Corporation, 60

Economics: global, and waning sovereignty of nation-states, 246-51; and public good, 51, 52; universal capitalism and computer networks, 123-4

EDI (electronic data interchange), 166

Egoyan, Atom, 192

Eisenhower, Dwight D., 68

Electric power, 9

Electronic bulletin boards, 105

Electronic Communications Privacy Act (US), on employee surveillance, 159

Electronic data interchange (EDI), 166

Electronic door locks, as computer network input terminals, 83

Electronic Frontier Foundation (EFF), 3, 242

Electronic mail: ephemerality of content, 87-8; mailing lists, 89; origins, 76, 78; popularity of, 76, 78; as synchronous and asynchronous, 86-7; universal access feasibility, 171-5; usage rates, 89

Electronic networks. *See* Computer networks; Internet; Network technology

The Electronic Republic (Grossman), 20, 266

Elevators, as computer network input terminals, 83

Ellul, Jacques, 48

E-mail. *See* Electronic mail

Empeiria, in Greek philosophy, 30-2

Employment: benefits, decrease in, 143; contingent work (part-time, temporary, term, self-employment), 143-4. *See also* Telework

Encryption, digital: Clipper chip, 242-3; Data Encryption Standard, 240; and law enforcement, on computer networks, 240-4; PGP encryption software case, 240-1; "Snuffle" program, 241-2; unregulated, supported by civil libertarian organizations, 242

Encyclopedia Britannica (on CD-ROM), 80

ENIAC (Electronic Numerical Integrator and Calculator), 59-60

Enterprise Integration Network, 129

Entertainment: industries, and economy, 106; as socialization, 13

European Community: Euro currency and Mondex smart card, 178; and issues of encryption and data privacy, 243; support for network technology development, 261

European Nuclear Research Centre, 92

Factory, development of, 37

Ferrant (company), 60

Fibre-optic lines, and bandwidth increase, 168

FidoNet, 79

File transfer protocol (FTP), 76

Finance industry, employment decrease in, 138. *See also* Banking

Fire, and myth of Prometheus, 4-5
Firearms, 11
Flexible manufacturing: mobility of capital, 125-6; specialization and customization, 126-7
Floppy disk, of computer, 62
Ford Motor Company, agile production methods, 128
Foucault, Michel, 14, 97, 156
Frankenstein (Shelley), 3, 10
Freedom. *See* Liberty
Freenets, 217
Freeware, 81
FTP (file transfer protocol), 76
Fuels, and technology, 8-9

Gandy, Oscar, 223
Gates, Bill: and Canadian Schoolnet initiative, 260; on "friction-free" capitalism, 121-4; "Gatesist" production practices, 125; as member of new elite, 106; and Teledesic program, 168; on transactive attributes of network technology, 171. *See also* Microsoft Corporation
General Electric, ownership of NBC television network, 116
GII (Global Information Infrastructure), 100
Gilder, George, 18
Gingrich, Newt, 21
Glass technologies, 6-7
Global Information Infrastructure (GII), 100
Globalization: and computer networks, 187-8, 246-51; and dismantling of national control of finance practices, 121-4; liberalization, deregulation and privatization as keys to, 122
Golding, Peter, 115, 190-1
Gore, Al, 100

Government: absence, as good for capitalists, 245; and Clipper chip encryption, 242-3; computer networks as stand-in for, 100, 266; and control of computer networks, 236-46; employment decrease in, 138; financial support of network infrastructure, 169; homogeneous state, and network technology, 251-64; law enforcement role, and opposition to non-standard encryption, 240-2; surveillance techniques, and social programs, 230; use of network technology, 21, 139; waning national sovereignty, in global political economy, 246-51
Grant, George: on democracy, 262-3; on goodness, undermined by technology, 51, 52; on judgments and right action, 48-9; on liberalism as corollary of technology, 51-2; on modern humanity's relation to Nature, 49, 50; on network technology and homogeneity, 51-2, 53, 251-3; on neutrality of computer technology, 267-8; on technology and modernity, 48-54; on technology, liberalism, and politics, 56
Gray, Chris, 17
Greek terms: *alētheia*, 42; *apophainesthai*, 42; *empeiria*, 264-5; *epistēmē*, 33; *legein*, 42; *logos*, 27, 28, 42, 55, 58; *nous*, 33; *phronēsis*, 33; *poiēsis*, 42, 47; *politikē*, 33, 34; *sophia*, 33; *technai*, 264-5; *technē*, 27, 28, 30, 31, 33, 34, 48, 55, 58
Grossman, Lawrence, 20, 266
Group of Lisbon, on effects of globalization, 122
Grupo Azteca, agile production methods, 128

Handicrafts, replacement by machines, 36

Hansard (on CD-ROM), 80

Hard disk, of computer, 62

Heidegger, Martin: on Aristotle and causality, 42-3; on essence of technology, 27, 230, 231, 333-5; on man as meditative being, 44; on man's relationship with technology, 44-8; on ontological significance of technology, 204-7; philosophical influence on Grant, 48; reduction of experience by technology, 195; on saving power of technology, 231-2; on technology and being, 40-8; on technology as enframing, 43, 56; on technology's denial of rootedness, 209; on technology's replacement of meditation by calculation, 219

High Performance Computing and Communications Act (US), 260

High technology industry, employment increase in, 137

Historical materialism, 35-6

Hobbes, Thomas, 195, 196

Hölderlin, Friedrich, 45

Hope: in myth of Prometheus, 4-6; and technology, 5-6, 25-6

Household appliances, as network access devices, 169-70

HTML (Hypertext Markup Language), 94

Human Resources Development kiosks (Canada), 139

Hydro-electricity, 9

Hygiene Guard program, 155-6

Hypermedia, 98

Hypertext: Greek origins of word, 95; and interactivity, 164; links, among Web documents, 94-5; and postmodern critical theory, 16

Hypertext Markup Language (HTML), 94

IAB (Internet Advertising Bureau), 184

IBM Corporation, 67

Identity: "additional" self, composed of binary bits, 231; construction, on Internet, 199; fit between postmodernist philosophy and network technology, 198-204; fragmentation, in postmodernist philosophy, 197-8; mutability, as possibility in virtual communities, 215; as represented by a Web site, 203

IHAC. *See* Information Highway Advisory Council (IHAC) (Canada)

IMPs (Interface Message Processors), 73-4

In Home Network, 100-1

Individualism, and broadcast technologies, 13

Industrial Revolution: and consumer goods, 12; and mechanization of production, 7, 8

Information: gathering, on World Wide Web (WWW), 211; Latin origins of word, 29. *See also* Databases

"Information highway," 100, 167. *See also* Computer networks; Network technology

Information Highway Advisory Council (IHAC) (Canada): on effects of network technology, 18; on network technology as "enabler," 124; on privatizing network development, 113, 114, 169, 259; pro-competition stand, 116; on technological advances and employment, 134, 140, 145; on universal access to electronic

mail, 173; on workplace surveillance and privacy, 159-60
Information Processing Techniques Office (IPTO) (US), 68, 73
Information society: as class society in capitalist system, 108; as not revolutionary, 19, 107
Input devices, 64, 83-4
Institute for Advanced Study (US), 60
Insurance industry: electronic workplace monitoring, 161; employment decrease in, 138
Intel: 4004 microprocessor, 66; input devices for digital transaction cards, 179
Intellectual property, uncertain status of, 248
Interactivity, of computer networks, 164-5
Interface Message Processors (IMPs), 73-4
Internal combustion engines, 9
International Federation of Trade Unions: on migration of work to lower wage countries, 140-1; on teleworkers, 145
International Labour Organization, on teleworkers, 145
International Traffic in Arms Regulations (ITAR) (US), 241
Internet: content, 58; distributed ownership, 110-11; growth, 80; and identity construction, 199-204; origins, 78-9; as publicly developed resource, 112; and universal capitalism, 123. *See also* World Wide Web (WWW)
Internet Advertising Bureau (IAB), 184
Internet Explorer Web browser, 118-19
Internet Relay Chat (IRC), 87
IPTO (Information Processing Techniques Office) (US), 68, 73

IRC (Internet Relay Chat), 87
Ireland, as "call centre" of Europe, 141

Jameson, Fredric, 17, 197
Jobs, Steve, 106
Johnson, David, 250
Journey's End Hotels, agile service provision, 128
Just-in-time: delivery, 125-7; manufacturing, 127

Kelly, Kevin, 234, 239
Kerckhove, Derrick de, 104, 221
King, William Lyon Mackenzie, 263
Knowledge, and printing technology, 12
Knowledge industries, and economy, 106

Labour: and machines, according to Marx, 36-40; non-specialized, according to Marx, 154. *See also* Employment; Telework
Landow, George, 14
Language: and Being, according to Heidegger, 205-6; and identity, in postmodernist philosophy, 196
LANs (Local Area Networks), 79
Lean production model, 125, 127
LEO system, 168
Leonardo da Vinci, 11
Letter on Humanism (Heidegger), 27
Lévi-Strauss, Claude, 200
Liberalism: and capitalism, 51-2, 251-3, 263-4; definition, 23, 196; as different from democracy, 23, 263-4
Libertarianism, and computer networks, 242
Liberty: of acquisition and consumption, under capitalist liberalism, 263; equated with openness of computer networks,

239; and rootedness, 210. *See also* Democracy
Life on the Screen: Identity in the Age of the Internet (Turkle), 199
Local Area Networks (LANs), 79
Lyon, David, 222, 230
Lyotard, Jean-François, 14, 197

McChesney, Robert, 113, 188
Machiavelli, Niccolo, 195
Machines: and capitalism, according to Marx, 36-40; and labour, according to Marx, 36-40
McKenna, Frank, 141
McLuhan, Marshall, 56
Mail, electronic. *See* Electronic mail
Mailing lists, on-line, 89-90
Malone, John, 116
Manchester Mark I (computer), 60
Manhattan Project, 60
Manley, John, 229
Manning, Preston, 21
Manufacturing industries: agile production methods, 125, 127, 128-9; automation, 136; employment decrease in, 137; "Gatesist" production practices, 125; mechanization, 135-6; networked computerization, 136
Marx, Karl: doubtful of relationship between capitalism and democracy, 189; on effect of mode of production on society, 104, 132; and historical materialism, 35-6; on machines, labour, and capitalism, 36-40; and myth of Prometheus, 6; on non-neutrality of mechanization, 56; not a technological determinist, 34-5; on ownership of means of production, 108; on technologies of capitalism, 34-40; on technology and nature, 36; utopian view of non-specialized labour, 154;

view of human nature, compared with postmodernism, 196
Mass pluralism, and broadcast technologies, 13
MasterCard International, data mining by, 226-7
Mathematics of infinity, 7, 8
Mauchly, John, 59
Mechanization, of industrial production, 7, 8, 56, 135-6
Media Lab (MIT), 238
Medicine, and technology, 10
Meisel, John, 257
Memory, of computer: RAM (Random Access Memory), 62; ROM (Read Only Memory), 61-2
Microchips, 66
Microprocessors, 66-7
Micro-sites, and advertising on computer networks, 185
Microsoft Corporation: acquisitions and mergers, 117; attempt at Web browser market dominance, 118-19; input devices for digital transaction cards, 179; investment in Rogers Communication, 171; investment in satellite technology, 117; monopolist policies, 116-19; as new elite corporation, 106. *See also* Gates, Bill
Microsoft Network (MSN), 91, 110, 171
Mode of production. *See* Production, mode of
Modem (modulator/demodulator): function, 74; and IMPs, 74
The Modern Prometheus (Shelley), 6
Mondex smart cards, 177-9
Money exchange, through computer networks: and consumer transactions, 175-81; proprietary networks, 179; verification challenges in open networks, 178-80

MSN (Microsoft Network), 91
MSNBC on-line news service, 117
MUDs (Multi-User Domains),
202-4
Multicasting, by computer net-
works, 88
Multimedia, and interactivity, 164
Multi-User Domains (MUDs),
202-4
Murrow, Edward R., 61
Myles, John, 152-3

National Information Infrastructure
(NII) (US), 111; pro-competition
stand, 116; on universal access to
electronic mail, 173
National Science Foundation (US),
78, 79
"National Standard for the Protec-
tion of Personal Information"
(CSA), 160
Nature: and modern humanity,
according to Grant, 49, 50;
relationship to, under auspices of
technology, 43; as resource for
exploitation, 207; as standing-
reserve of binary bits, 209-11, 219
NBC television network, owned by
General Electric, 116
Negroponte, Nicholas, 208, 238
Nelson, Ted, 95
"The Net." See Internet
Netscape, 110, 179
Net/Tech, 155-6
Network technology: accessibility,
167-75; bandwidth, 79-80, 168-9;
and capitalist mode of production, 104-31; commercial deployment of, 167-87; and democracy,
17-21; distribution obstacles, 171;
essence, as binary bits, 204-9;
essential for virtual corporations,
130-1; fibre optics, 79; and
homogeneous state, 251-64; as

imitative of human activities and
experience, 206, 207, 208;
interconnectivity and inter-
operability standards, 119-20;
marketability, 181-7; mode of
transaction, 175-81; multicasting,
88; opposed to rootedness, 209-
18; as postmodern technology, 14-
17; prevalence of, 106; reinforcing
market links, 118; reliability,
compared with other media, 86;
synchronous and asynchronous,
86-7; and work, effect on, 132-63.
See also Computer networks
Networks. See Computer networks
Neumann, John von, 60
Newman, Nathan, 117, 118
News, as political instruction, 13
Newsgroups, 90
Next Generation Internet Research
Act (US, 1998), 260
Nichomachean Ethics (Aristotle), 32,
33
Nietzsche, Friedrich: and myth of
Prometheus, 6; philosophical
influence on Grant, 48
Noble, David, 133, 189
Northern Telecom: call centres in
New Brunswick, 141-2; VISTA
350 Interactive Telephone, 169-
70
NOT (in Boolean algebra), 64, 65
NSFNet (National Science Founda-
tion), 79, 112
Nuclear power: atomic bomb, 11; as
fuel source, 8, 9

On-line discussion groups, 21
Optical character recognition, effect
on data-processing industry, 141
OR (in Boolean algebra), 64, 65
Organization of Economic Cooperation, on taxation of virtual
revenue, 249

Oxford English Dictionary (on CD-ROM), 80

Packet switching, 71-2
Pandora, and myth of Prometheus, 4
Parrot, Jean-Claude, 134
PCs. *See* Personal computers
Personal computers: appearance of, 67; ownership, and means of production, 108-9
Personal Information Protection and Electronic Documents Act (Canada, Bill C-6), 160-1, 229
PGP (Pretty Good Privacy) encryption software, 240-1
Philippines, and data-processing industry, 140-1
Photography, as transcendence of space and time, 12
Pizza Pizza, 148
Plastics: as human-made matter, 10; invention of, 9
Plato: on distinction between *technai* and *empeiriai*, 264-5; philosophical influence on Grant, 48; on *technē* and *logos*, 28, 30, 31-2, 34
Pluralism: appearance of, according to Grant, 51-2; genuine versus mass, 13
Plutonium, 9
PointCast servers, and advertising on the Web, 183-4
Political instruction, as news, 13
Politics: of computer networks, 16; election predictions by computer, 60-1; and technology, 55-7
Pornography: distribution, argued as free speech issue, 241, 244; and issue of encryption, 242; and law enforcement, 239; and transnational boundaries of computer networks, 247-8
Post, David, 250

Poster, Mark, 14, 16, 17, 199, 231
Postmodernism: art, as assemblage of fragmented parts, 198, 203; attitude to traditional knowledge, 14; and computer network technology, 14-15, 97; and identity, 196-204; lack of critical distance from information technology, 17; philosophical tenets, 197
Pretty Good Privacy (PGP) encryption software, 240-1
Printing, impact of, 12
Privacy: concerns, and data mining, 229; difficulty of monitoring degree of employee surveillance, 159-60; and electronic workplace surveillance, 159
Privatization: of computer networks, 112-14; and globalization, 122
Production, mode of: capitalist, and network technology, 104-31; effect on social relations, 35-6; practices, and computer networks, 124-31
Programs, of computer, 61
Prometheus, myth of, 4-6
Public Service Alliance of Canada, on teleworkers, 148
Purolator couriers, call centres in New Brunswick, 142
Push technology, and advertising on computer networks, 182-4

"The **Q**uestion of Technology" (Heidegger), 41-8

Raboy, Mark, 256
RAM (Random Access Memory), of computer, 61-2
RAND Corporation: and ARPANET, 73; on feasibility of universal access to electronic mail, 171, 173, 174; research on reliable communications infrastructure, 68

Reality, representation by digital media, 192-5
"Redundancy," of computer networks, 68-70
Reform Party of Canada, 21
Repetitive-strain injuries, 161
Retail industry, employment decrease in, 138
Rheingold, Howard, 21
Robotics, 8
Rogers Communication, investment by Microsoft, 171
ROM (Read Only Memory), of computer, 61-2
Room sensors, as computer network input terminals, 83
Rootedness: compared with attachment in virtual communities, 212-16; effect of network technology, 209-18; and liberty, 210
Rousseau, Jean-Jacques, 12, 196
Rule, James, 223

Satellite technology: and bandwidth increases, 168; investment by Microsoft Corporation, 117
Scanning, of text, 82
Schoolnet, 259
Schuler, Douglas, 217
Science fiction, cyberpunk, 210
Searching: of networked computer databases, 84; search engines, on World Wide Web (WWW), 93. See also Data mining
Service industries: and economy, 106; employment increase in, 137-8; gains offset by unemployment in traditional service areas, 138-40
Shakespeare, William, 80
Shareware, 81
Shelley, Mary, 6
Singapore, prohibition against pornography on computer networks, 244, 248

SITA network, 129
Skills, in networked environment, 154-5
Smart cards: and consumer transactions, 176-9; and electronic surveillance of employees, 158-9
Social relations: effect of mode of production on, 35-6; on-line communities, 212-16
Socialization: and consumption, 225; through "entertainment," 13
Socrates, 31-2
Software, dominance of Microsoft platform, 117
Soviet Union, 67
Space, transcendence: through communication technologies, 12; through printing, 12; through transportation advances, 9
Sputnik satellite, 67, 168
Standards, for network technology, 119-20
Stanford Research Institute, as ARPANET "host," 73
Statistics Canada: survey of Canadian manufacturing technology, 129-30; on teleworkers, 145
Stonehaven West model "wired" community, 169-70
Strategis network, 129
Strauss, Leo, 48
Supermarkets: electronic monitoring of cashiers, 157; use of network technology, 138-9
Surveillance, electronic: consumer marketplace, 224-31; as crucial administrative and disciplinary function, 223; as essence of computer networks, 222-31; personal information in data files, 223-4; and view of world as standing-reserve of bits, 230; workplace, 155-8, 159-61, 162-3

Sweden, technologically advanced surveillance in, 230
The Sweet Hereafter, 192-4
SWIFT network, 129
Switching: circuit, 70-1; packet, 71-2
Synchronous communication, of computer networks, 86

Task Force on Financial Services (Canada), 180
Taxation issues, and virtual revenue, 249
Taylorist mass production, 151, 152
TCI, purchase by AT&T Corporation, 116
TCP/IP (Transmission Control Protocol/Internet Protocol), 78
Technique, definition, 32
Technological change, historical: biological technologies, 10; consciousness and knowledge, 11-14; fuels, 8-9; glass, 6-7; healing arts, 10; industrial production, 7; materials, synthetic, 9-10; mathematics of infinity, 7-8; mechanization, 8; time, regularization of, 7; transportation, 7; urbanization, 7
Technological determinism, 34-5
Technology: and Being, according to Heidegger, 40-8; and creation of homogeneity, according to Grant, 51-2, 53; and desire for transcendence, 6-8; as element in forces of production, according to Marx, 35-6; Greek origins of word, 27-8; as not neutral, according to Heidegger, 41-2; and politics, according to Aristotle, 34; and politics, according to Grant, 55-7. *See also* Computer networks; Network technology; Technological change, historical

Telecommunications: convergence with broadcasting, 257; electronic workplace monitoring, 161; industry mergers, 119
Telecommunications Act (Canada), 113, 114, 256
Telecommunications Act (US, 1996), 113, 260
Teledesic system, of Low Earth Orbit digital transmission satellites, 117
Telegraphy, and transcendence of space and time, 12
Telephone: companies, use of data mining, 228; telephones, as computer network input terminals, 83; and transcendence of space and time, 12
Television: advertising techniques, 187; and computer input devices (PCTVs or WebTVs), 170-1; creation of audience with consumer needs, 163, 184; digital signals, 170; success of, 13-14; transactions limited to marketing, 181
Telework: abuse of, 146-8; and de-institutionalization, 144-8; as undemocratic aspect of computer networks, 189-90
Teleworkers: executives and professionals, 145, 146-7; as female ghetto, 145, 147; lack of benefits, 146; lack of control over own work, 150-1; low-level administration, clerical and service, 145, 147; overhead costs borne by, 146; and trade unions, 145, 146, 147, 148-9
Telnet application, for computer networks, 76
Tendinitis, 161
Terminal Interface Processors (TIPs), 74-6

Text, digitization and scanning of, 82

"Thomas" on-line document system, 21

Thompson, E.P., 151

Thread, in on-line discussions, 90

Time: regularization of, 7, 8; transcendence through printing technology, 12; transcendence through transportation advances, 9

TIPs (Terminal Interface Processors), 74-6

Toffler, Alvin and Heidi, 18

Total quality management (TQM), 125, 149-51

Trade unions: on decrease in full-time positions, 144; on migration of work to lower wage countries, 140-1; objection to electronic workplace surveillance, 159; on teleworkers, 145, 146, 147, 148-9

Traffic Operator Position System (Bell Canada), 158

Transmission Control Protocol/Internet Protocol (TCP/IP), 78

Transportation technologies, 7, 8

Tremblay, Gaëtan, 164

"Turing's Man," 220

Turkle, Sherry: on body as "meat," as contrast with binary bits, 210; on intelligence of computer networks, 220; on personal identity and computer networks, 199-204

TYMNET network, 129

The **U**nbound Prometheus (Nietzsche), 6

Unemployment, as result of technology, 134-5

Uniform Resource Locator (URL), 92, 93

United States: development of Internet by, 68-79; employment patterns, by sector, 137; encryption, and law enforcement on computer networks, 240-4; global influence of, through computer networks, 252-3; investment in network infrastructure, 111; legislation in favour of network development and deregulation, 260; privatization of computer networks, 112-14; "Thomas" on-line document system, 21. *See also* National Information Infrastructure (NII) (US)

UNIVAC computer, 66

University of California, at Santa Barbara, as ARPANET "host," 73

University of California, Los Angeles (UCLA), as ARPANET "host," 73

University of Utah, as ARPANET "host," 73

Uranium, 9

Urbanization, 7

URL (Uniform Resource Locator), 92, 93

USENET (on-line discussion groups), 79, 91-2

Video-on-demand, 168, 171

Vipond, Mary, 114, 257

Virtual communities. *See* Communities, on-line

Virtual corporations, 125, 127, 128, 209-10

Visual Basic programming language, 117

Wal-Mart, data mining by, 228

Warfare, technology of, 10-11

Wassenaar Arrangement, 243

"The Web." *See* World Wide Web (WWW)

Web site, and presentation of identity, 203

Weber, Max, 97
Weil, Simone, 48, 210, 211-12
Weizenbaum, Joseph, 219-20
Westinghouse, ownership of CBS
    television network, 116
Whole Earth 'Lectronic Link
    (WELL): and democratic character
    of computer networks, 21;
    discussion groups, 90-1
Wiener, Norbert, 98
Winner, Langdon, 55, 239
"Wired homes," 100-1
Women, as teleworkers, 145, 147
Wood, Ellen Meiksins, 123
Wood, replacement by metals, 9
Word-processing industry, elec-
    tronic monitoring in, 157
Word-processing software, domi-
    nance of Microsoft products, 117
Work: and alienation, under
    capitalism, 132-3; craftsmanship,
    devaluation of, 151; de-institution-
    alization of, 143-8; de-skilling, by
    mechanization, automation, and
    computerization, 151-5; disappear-
    ance, as result of technology, 133-
    43; migration (deterritorialization)
    of, 140-3; performance monitor-
    ing, 157-63; technology and labour
    practices, 132; telework, and de-
    institutionalization, 144-8; total
    quality management (TQM),
    critique of, 149-51. See also
    Workplace electronic surveillance
Workers, enslavement to machines,
    according to Marx, 37, 133

Workplace electronic surveillance:
    enhanced by computer networks,
    155-8; and right to privacy, 159-61;
    social effects, 162-3
World Economic Forum, 1996, 238
World Wide Web (WWW): as
    asynchronous medium, 93-4;
    collapse of distinction between
    information and communication,
    92; compared with television, 93-
    4; as disembodied, unrooted
    information base, 211; download-
    ing contents, 92; dynamic
    interface with users, 94; as
    expanding database, 92; history,
    92; HTML (Hypertext Markup
    Language), 94; hypertext links
    among documents, 94, 96;
    interactive nature of, 96-7; as
    medium of transaction, 97;
    merging of information produc-
    tion, consumption, and commu-
    nication, 96-7, 103; multicasting,
    94; multimedia Web pages, 93;
    search engines, 93; URL (Uni-
    form Resource Locator), 92, 93
Wriston, Walter, 18, 20, 239, 240,
    262
Writing, postmodern, 210
WWW. See World Wide Web
    (WWW)

Yahoo!, 110

Zeus, and myth of Prometheus, 4
Zimmerman, Paul, 240-1

Set in Meta and Scala by Artegraphica Design Co.

Printed and bound in Canada by Friesens

Copy editor: Beverley Endersby

Designer: Richard Bingham, Flex Media Design and Communications

Indexer: Annette Lorek

Proofreader: Darlene Money